Blockchain Technologies

Series Editors

Dhananjay Singh⬥, Department of Electronics Engineering, Hankuk University of Foreign Studies, Yongin-si, Korea (Republic of)

Jong-Hoon Kim, Kent State University, Kent, USA

Madhusudan Singh⬥, Endicott College of International Studies, Woosong University, Daejeon, Korea (Republic of)

This book series provides details of blockchain implementation in technology and interdisciplinary fields such as Medical Science, Applied Mathematics, Environmental Science, Business Management, and Computer Science. It covers an in-depth knowledge of blockchain technology for advance and emerging future technologies. It focuses on the Magnitude: scope, scale & frequency, Risk: security, reliability trust, and accuracy, Time: latency & timelines, utilization and implementation details of blockchain technologies. While Bitcoin and cryptocurrency might have been the first widely known uses of blockchain technology, but today, it has far many applications. In fact, blockchain is revolutionizing almost every industry. Blockchain has emerged as a disruptive technology, which has not only laid the foundation for all crypto-currencies, but also provides beneficial solutions in other fields of technologies. The features of blockchain technology include decentralized and distributed secure ledgers, recording transactions across a peer-to-peer network, creating the potential to remove unintended errors by providing transparency as well as accountability. This could affect not only the finance technology (crypto-currencies) sector, but also other fields such as:

Crypto-economics Blockchain
Enterprise Blockchain
Blockchain Travel Industry
Embedded Privacy Blockchain
Blockchain Industry 4.0
Blockchain Smart Cities,
Blockchain Future technologies,
Blockchain Fake news Detection,
Blockchain Technology and It's Future Applications
Implications of Blockchain technology
Blockchain Privacy
Blockchain Mining and Use cases
Blockchain Network Applications
Blockchain Smart Contract
Blockchain Architecture
Blockchain Business Models
Blockchain Consensus
Bitcoin and Crypto currencies, and related fields

The initiatives in which the technology is used to distribute and trace the communication start point, provide and manage privacy, and create trustworthy environment, are just a few examples of the utility of blockchain technology, which also highlight the risks, such as privacy protection. Opinion on the utility of blockchain technology has a mixed conception. Some are enthusiastic; others believe that it is merely hyped. Blockchain has also entered the sphere of humanitarian and development aids e.g. supply chain management, digital identity, smart contracts and many more. This book series provides clear concepts and applications of Blockchain technology and invites experts from research centers, academia, industry and government to contribute to it.

If you are interested in contributing to this series, please contact madhusudan.singh@oit.edu **OR** loyola.dsilva@springer.com

Ron van der Meyden · Michael J. Maher

Simple Agreements for Future Equity (SAFE)

Smart Contracts for Venture Finance

 Springer

Ron van der Meyden
School of Computer Science
and Engineering
UNSW Sydney
Sydney, NSW, Australia

Michael J. Maher
Reasoning Research Institute
Canberra, ACT, Australia

ISSN 2661-8338   ISSN 2661-8346 (electronic)
Blockchain Technologies
ISBN 978-981-96-3919-9   ISBN 978-981-96-3920-5 (eBook)
https://doi.org/10.1007/978-981-96-3920-5

This Springer imprint is published by the registered company Springer Nature Singapore Pte Ltd.
The registered company address is: 152 Beach Road, #21-01/04 Gateway East, Singapore 189721,
Singapore

If disposing of this product, please recycle the paper.

Preface

Distributed Ledger Technology (DLT), also known as "blockchain", provides a novel form of computational infrastructure for securing ownership records of a range of assets. The emergence of this class of systems was driven by the success of Bitcoin [1] as a decentralized digital currency platform that supports enforcement of simple rules concerning the transfer of value. Subsequent DLT platforms such as Ethereum have incorporated more expressive languages for representing such rules, enabling a realization of Szabo's vision of "smart contracts" [2]: commercially relevant multi-party agreements enforced by code running on computer networks. This development was soon applied to develop "Decentralized Autonomous Organisations" and "Initial Coin Offering" smart contracts, which were used in crowdfundings raising substantial amounts of cryptocurrency value for projects developing blockchain technology and its applications.

These crowdfundings were often of legally questionable status. Following a significant amount of fraud, regulators have asserted their powers over blockchain-based fundraising, and have started to develop a regulatory stance to this new technology. The focus of the application of DLT is therefore increasingly moving towards regulated forms of assets, for which multi-party trust has historically been managed using legal contracts.

Benefits of the digital representation of equity rights on distributed ledger platforms are perceived to include:

- Improved efficiencies from the establishment of a consensus "single source of truth" concerning ownership of rights. Current processes in financial markets frequently involve expensive manual reconciliation of inconsistent records dispersed across multiple organizations, delaying trades and allowing central parties to extract rents from collateral held in trust during the delay.
- Enabling the rights holder or their proxy to uniquely control ownership and transfer of the asset via possession of a private cryptographic key.[1]

---

[1] Since keys are easily distributed, it may be controversial to regulators that direct control by the rights holder is a benefit. Instead, they may require not just the possibility of unique ownership but also unique custody of the asset by a *known* entity. Although it is less within the ethos of the

- Elimination of counter-party risks by ensuring atomicity of exchange transactions, for example, a swap of an asset for digital currency.
- Improved security and availability of the rights ownership data, through use of cryptographic mechanisms and avoidance of single points of failure.
- Automating the enforcement of rules constraining the issuance and transfer of rights, such as vesting schedules and lockups. Exchanges may require, for protection of their customers, constraints on management's power to create and issue new shares without a shareholder vote. By placing these events under the control of a smart contract whose state is accepted to be a legally valid representation of the ownership of the company, investors receive a stronger guarantee of compliance with such rules.
- Increasing global accessibility of investment opportunities and the creation of a 24/7 market.

This raises the question of how regulated assets can be supported using DLT technology.

Some simple forms of commercial agreement can be completely enforced using smart contracts: a swap of a digital asset represented on a blockchain for that chain's cryptocurrency is one example [3]. There is an emerging view, however, that the typical representation of a legal instrument on a DLT platform will require natural language text as well as smart contract code, a hybrid form of representation for which the term *smart legal contract* has come into use [4]. On this view, claims of immutability notwithstanding, the legal system will have powers to overrule actions performed on the blockchain. This position was foreshadowed already in Ian Grigg's notion of "Ricardian Contracts" [5].

The methodology of smart (legal) contract development is still nascent, and raises many questions. To what extent can the contracts defining regulated assets be represented in smart contract code? Which of the provisions of commercial agreements can be automatically enforced using code on DLT systems, and which require representation in natural language? What issues, more generally, need to be addressed in developing a smart legal contract for a regulated instrument?

This book describes the outcomes of a project in which we have studied these questions by means of a case study of a legal contract that has become widespread in the financing of early stage ventures: Y Combinator's Simple Agreement for Future Equity (SAFE) [6].

Y Combinator is a Silicon Valley investment fund that has been called "the world's pre-eminent startup accelerator" [7], and has provided early stage funding to many well known companies, among them AirBnB, Coinbase, Dropbox, Gitlab, Reddit, and Stripe. Developed in order to simplify negotiations between seed investors and startup founders, SAFE contracts are like convertible bonds but omit their provisions for payment of interest, simply offering either a return of the investment or a conversion of the contract to shares at the time of a priced equity round.

---

cryptocurrency community, custody of the key by a proxy trusted by the regulator may be a way to address these concerns.

Due to the formality of the corporate finance domain, and the relative brevity and simplicity of SAFEs, they are, *prima facie*, a good candidate for a study intended to elucidate the nature of smart contract representation of legal contracts. The detailed development of a smart contract representation of legal contracts, can, nevertheless, be non-trivial. The precision and formality of code requires that financial operation of a contract first needs to be very well understood. This requires the careful consideration of a significant number of issues for the representation to adequately capture the legal dynamics of the source contracts. Indeed, our work of attempting to formalize SAFE contracts using smart contracts raised many issues.

The outcomes of our effort to understand these issues are described in the present work. In addition to the original motivation of a case study in smart legal contract development, we believe that the understanding developed in the course of our work clarifies legal and financial issues that may be known to experts working frequently with these contracts, but which are not readily apparent in publicly available works discussing SAFE contracts.

We expect that readers will come to this work from a variety of perspectives and disciplinary interests, including Finance, Law, Computer Science, and Entrepreneurship. The work, accordingly, has been structured into parts, to help the reader to focus on the parts of most interest to them.

Part I deals with the background and motivation for the development of SAFE contracts, and introduces the legal text of Y Combinator's SAFEs. This part will be of interest to all readers.

Parts II and III are concerned with an analysis of SAFE contracts from a financial perspective. We expect that readers whose main interest is Finance, Accounting or Financial Contracts will wish to read these parts. We discuss some subtleties that do not appear to be explicit in available explanations of SAFEs directed at startup founders, so entrepreneurs intending to use SAFEs to fund their ventures may also find these sections helpful. There are multiple versions of SAFE contracts, depending on which of two parameters, the Cap and the Discount, are included. We primarily use the SAFE version with Cap but no Discount as the main focus throughout the book, though we do consider other versions at points in the discussion.

Part II analyzes situations in which a company has issued a single SAFE contract and then, later, raises a priced equity round in which the SAFE contract converts to equity. In order to understand the way that SAFE contracts and other forms of convertible instruments create some complexities in the calculations involved in an equity round, we first describe, in Chap. 3 the equations governing an equity round in which the company has not issued convertible instruments, and how these equations need to be modified once convertible instruments need to be taken into account.

As we describe in Chap. 4, the terms of a Pre-Money SAFE raise unanticipated questions about the meaning of the financial notion of "pre-money valuation" of a company in a priced equity round, which plays a key role in the terms relating to conversion of the SAFE to shares. There are multiple ways in which these questions are answered in practice, raising the question of which should be supported in a smart contract code representation. An understanding of the conversion of a single Post-Money SAFE is developed in Chap. 5.

The existence of several alternative ways to calculate the conversion of a SAFE means that a choice must be made at the time of the equity round. There are multiple stakeholders in this choice: the founders (and other owners of common stock), the SAFE investor, and the new equity round investors. SAFEs are not explicit with respect to which method should be used, so the situation has a game-theoretic nature. Chapter 6 discusses the nature of this game, and its outcomes under the possible distributions of power or cooperation of the stakeholders. One of the outcomes of this analysis is the conclusion that there is, effectively, one canonical conversion method that should be used by rational players. Departures from this method can lead to complex game-theoretical situations. We also consider a way that the company and a new investor can "game" the construction of an equity round so as to disadvantage a SAFE investor. If the equity round is structured into two separate rounds at different prices, the SAFE investor can be left in a situation where their post-round stake in the company is significantly smaller than it would have been in an honestly constructed single round. This constitutes an "attack" on the SAFE that is legally questionable, but significant in the context of a smart contract implementation.

Part III continues the financial analysis of SAFEs by a consideration of scenarios in which multiple SAFEs have been issued. Chapter 7 extends the analysis of the conversion of SAFEs during an equity financing, using the canonical rational method, to scenarios in which the company has issued multiple SAFE contracts.

The existence of multiple SAFEs also requires consideration of game-theoretic issues in the case of Liquidity Events (e.g., an acquisition of the startup). In Liquidity Events, a SAFE investor is offered two options concerning the determination of their payout. One expects that an investor will always choose the option that maximizes their payout. However, it turns out that the payout may depend also on choices made by other investors, giving Liquidity Events an inherently game-theoretic nature. We present our analysis of this game in Chap. 8. Dissolution events (winding up the affairs of the company) are somewhat simpler in this regard. Chapter 9 covers this topic.

Our analysis of multiple SAFE scenarios uncovers some situations that, we argue, are not properly handled in Y Combinator's SAFE contracts, potentially requiring ad hoc resolutions in some cases. This leads us to propose alternative SAFE terms in Chap. 10.

We summarize the key points from the financial analysis of Parts II and III in Chap. 11, particularly with respect to the issues that they raise for smart contract representation of SAFEs. Readers who are interested primarily in issues around smart contract representations of legal contracts might prefer to skip to this chapter after reading the material in Part I.

In Part IV, we turn to a discussion of the representation of SAFE contracts as smart contracts. We begin with an introduction, in Chap. 12, to the topic of blockchain and smart contracts.

Chapter 13 discusses a number of impediments to straightforward translation of a SAFE contract to code. Related to the issue of precision and formality is that the source legal text contains open textured terms, not straightforwardly formalizable, and potentially subject to shifting interpretations as a result of legal rulings. We describe strategies for dealing with these difficulties in Chap. 14. A key element of our approach is to code the smart contract representation of the SAFE in a form that allows multiple different understandings to be accommodated. In the actual implementation of code components of the smart legal contract, as with any software engineering project, decisions need to be made concerning the architecture of the solution. In Chap. 15, we give a description of our architecture, which has been designed to deal with a number of further issues, including the open structure of the context in which the SAFE operates, and the fact that Post-Money versions of the SAFE state declarative constraints instead of giving explicit computations for certain operations. The architecture has moreover been designed so that certain properties relating to the correctness and security of the code can be established based on local reasoning about specific components in the architecture. A Solidity implementation of the architecture and a number of SAFE contracts is available at https://github.com/ RonVanderMeyden/SAFESmartContracts.git.

We turn to an evaluation of our representation of SAFE contracts in Chap. 16. One aspect of the financial operation of the SAFE concerns the freedom that equity round investors and the company have to negotiate the terms of equity rounds, with the SAFE investor not necessarily a party to these negotiations. In analyzing this aspect, we have found that a simple translation of the SAFE to smart contract code leaves the SAFE investor vulnerable to "gaming" of the SAFE contract by the company and the equity round investor. As described in Sect. 16.3, the "attack" identified in Chap. 6 implies that the SAFE investor cannot rely upon a smart contract alone to protect their interests. The source legal text contains language that gives the SAFE investor legal recourse in such a situation, but its effect inherently cannot be captured in smart contract code. This is one example in which use of a smart legal contract, that retains the original legal terms, is critical.

The use of smart contracts and DLT as part of multi-party arrangements that have legal significance raises many general questions of law that we do not attempt to address in this book (see [4, 8] for extensive treatments). We do, however, in Chap. 17, consider a number of issues specific to SAFE smart contracts. In particular, we lay out the demarcation between work done in code, and work done by natural language components of the smart legal contract we propose. Some further legal issues specific to the use of smart contracts for the equity financing domain are also discussed.

Blockchain platforms come in a variety of forms: public and private (or permissioned), which differ with respect to the entities that are able to participate in the platform as validators or users. Chapter 18 discusses privacy and the appropriateness of different platforms for SAFE implementation.

Our attention has been mainly on the three events that terminate a SAFE. In Chap. 19 we discuss issues arising in the encoding of other clauses of SAFEs.

Finally, Chap. 20 concludes the book as a whole. A key theme emerging from the work of this project has been to illustrate the way that undertaking a smart contract implementation of a legal instrument forces a greater degree of precision than is usual in the design of contracts. We outline considerations that will need to be addressed in carrying this project further, including discussing circumstances in which the effort required for such precision is worthwhile, and giving general remarks on the discipline of Smart Legal Contract Engineering, of which the present work might be considered an example.

Source Material

This work is based on technical reports by the authors: [9–12].
SAFE contracts are (Creative Commons) Copyright Y Combinator.
We thank IEEE for permission to use parts of [13] which appear in Chap. 15.

© IEEE, 2021.

Used with permission of the IEEE under license number 5915700440858.

Extended excerpts from the following Y Combinator SAFEs (and others) also appear in the text.
Y Combinator's original SAFE with Valuation Cap

© Y Combinator, 2013.

The original is available at https://web.archive.org/web/20180831020232//http://www.yco mbinator.com/docs/SAFE_Cap.rtf

Y Combinator's Post-Money SAFE with Valuation Cap

© Y Combinator 2018, 2021, 2023.

Y Combinator owns the copyright to this SAFE. It is presented under the Creative Commons Attribution-No Derivatives 4.0 License [14]. Under this license, Y Combinator disclaims all warranties (see [14]). The original is available at https://www.ycombinator.com/documents

Terminological Conventions

Y Combinator's documents refer to Simple Agreements for Future Equity as "Safes" or "safes". In this book, we prefer to use the fully capitalized "SAFE", to avoid confusing the acronym and the English word. However, we follow the convention of any source texts in the context of quotations.

Y Combinator also calls the 2013 SAFEs, "original Safes". In this book, we will refer to them as "Pre-Money SAFEs", in contrast to the Post-Money SAFEs that Y Combinator introduced later.

Sydney, Australia	Ron van der Meyden
Canberra, Australia	Michael J. Maher
December 2024	

Acknowledgements This work was supported in part by a UNSW Goldstar grant. Franck Cassez and Peter Hoefner contributed to discussions at early stages of this work. Parts of this work are informed by the Honours thesis of William Coulter [15]. The authors thank Alex Cook of Herbert Smith Freehills, and Romain de Spoelberch and Vitaly Spassky of Polymorphic Capital for informative discussions about SAFEs.

Michael Maher thanks his wife Robyn for her support during the writing of this book. In addition to his role at the Reasoning Research Institute, Michael has an adjunct position at Griffith University and an honorary position at UNSW.

References

1. Nakamoto S (2008) Bitcoin: A peer-to-peer electronic cash system. Available at https://bitcoin.org/bitcoin.pdf. Accessed on Dec 2024
2. Szabo N (1997) The idea of smart contracts. https://nakamotoinstitute.org/the-idea-of-smart-contracts/. Accessed Dec 2024
3. Herlihy M (2018) Atomic cross-chain swaps. In Proc. ACM Symp. on Distributed Computing. Version at https://arxiv.org/abs/1801.09515
4. Allen JG, Hunn P (eds) (2022) Smart Legal Contracts: Computable Law in Theory and Practice. Oxford University Press, Oxford
5. Grigg I (2004) The Ricardian contract. In Proceedings of the First IEEE International Workshop on Electronic Contracting, pp 25–31. IEEE
6. Levy C (2018) Safe financing documents. Online, https://www.ycombinator.com/documents/#safe. Accessed Sep 2019
7. Crowther D, Coulman W (2024) The world's most important accelerator is deep into AI. Online, https://sherwood.news/tech/startup-accelerator-y-combinator-is-deep-into-ai/. Accessed May 2024
8. DiMatteo LA, Cannarsa M, and Poncibò C (2019) editors. The Cambridge handbook of smart contracts, Blockchain technology and digital platforms. Cambridge Law Handbooks. Cambridge University Press
9. van der Meyden R (2021) A game theoretic analysis of liquidity events in convertible instruments. arXiv https://arxiv.org/abs/2111.12237
10. van der Meyden R (2023) On conversion of multiple safe contracts (working paper). Technical Report
11. van der Meyden R, Maher MJ (2021) Can SAFE contracts be smart? Technical Report
12. van der Meyden R, Maher MJ (2021) Simple agreements for future equity—not so simple? Technical Report
13. van der Meyden R, Maher MJ (2021) Architecture for smart SAFE contracts. In 3rd Conference on Blockchain Research and Applications for Innovative Networks and Services (BRAINS). IEEE
14. Creative Commons. Attribution-Noncommercial-NoDerivatives 4.0 International License. https://creativecommons.org/licenses/by-nd/4.0/. Accessed Dec 2024
15. Coulter W (2021) Building Smart SAFEs. Honours thesis, UNSW School of Computer Science and Engineering

Competing Interests The authors have no competing interests to declare that are relevant to the content of this manuscript.

Contents

Part I An Introduction to SAFE Contracts

1   The Need for SAFEs 3
    1.1   The Need for Money 3
    1.2   A Little History 5
    1.3   Convertible Notes 7
    1.4   Simple Agreements for Future Equity 9
    1.5   Current Use of SAFEs 11
    1.6   Conclusion 12
    References 13

2   SAFE Contract Terms 15
    2.1   Summary of SAFE Contract Structure 15
    2.2   Pre-Money SAFEs 17
        2.2.1   Equity Financing 17
        2.2.2   Liquidity Events 20
        2.2.3   Dissolution Events 22
    2.3   Post-Money SAFEs 23
        2.3.1   Equity Financing 24
        2.3.2   Liquidity Events 25
        2.3.3   Dissolution Events 27
    2.4   Termination 28
    References 28

Part II Analysis: Single SAFE Scenarios

3   Accounting Views of Convertible Instruments 31
    3.1   Standard Equity Raise 31
    3.2   An Inconsistency 34
    3.3   Two Accounting Views 35
        3.3.1   Liability View 35
        3.3.2   Cap Table View 36

	3.3.3	Cap Table View (Alternate)	37
	3.3.4	Accounting Views: Conclusions	38
	References		40

4   Methods for Converting a Pre-Money SAFE 41

4.1   Formalizing Conversion for the SAFE with Cap Only 42

4.2   Standard Method 43

4.3   Percent-Ownership and Discounted Valuation Methods 47

4.3.1   Percent-Ownership Method 47

4.3.2   Discounted Valuation Method 50

4.3.3   Comparison of Percent-Ownership and Discounted
          Valuation, and Limitations 52

4.4   Dollars Invested Method 53

4.5   Two-Stage (Zero-Money Round) Approach 55

4.5.1   Zero-Money Round (Discounted Valuation
          Method) 56

4.5.2   Zero-Money Round (Standard Method) 57

4.6   Summary 59

References 60

5   Conversion of Post-Money SAFEs 61

5.1   Conversion in Terms of Price 62

5.2   Conversion in Terms of Pre-Money Valuation 64

5.3   Other Conversion Methods 66

5.4   Summary 68

References 68

6   Game-Theoretic Aspects of SAFE Conversion 69

6.1   Introduction 69

6.2   Conservative Methods 70

6.3   Deferred Interpretation 75

6.4   Gaming the Two-Stage Process 80

6.5   Summary 85

References 86

Part III   Analysis: Multiple SAFE Scenarios

7   Equity Financing with Multiple SAFEs 89

7.1   Introduction 89

7.2   The Multiple SAFE Scenario 91

7.2.1   Pre-Money SAFE with Cap Only 91

7.2.2   Post-Money SAFE with Cap Only 93

7.3   Discounted Valuation Method Revisited 94

7.4   Discounted Valuation Method: Pre-Money SAFEs 96

7.5   Discounted Valuation Method: Post-Money SAFEs 99

7.6   Comparison 103

  7.7 Mixed SAFE Types 106
  7.8 Conclusion 107
  Reference 108

8 **Liquidity Events with Multiple SAFEs** 109
  8.1 Liquidity Events 109
  8.2 The Liquidity Event Game 110
  8.3 SAFE Contract Variants 113
       8.3.1 Pre-Money SAFE with Cap 113
       8.3.2 Pre and Post-Money SAFE with Discount Only 114
       8.3.3 Post-Money SAFE with Cap 115
  8.4 Nash Equilibria in the General Game 118
  8.5 Special Case: Pre-Money SAFEs 127
  8.6 Special Case: Post-Money SAFEs with Cap 131
  8.7 Mixed SAFE Types 133
  8.8 Related Work and Open Problems 135
  8.9 Dividends 136
  8.10 Conclusion 139
  References 140

9 **Dissolution Events** 141
  References 143

10 **Towards a Better SAFE** 145
  10.1 Weaknesses in SAFEs 145
  10.2 Resolving Conflicting Entitlements 147
  References 150

11 **Summary of the Analysis** 151
  11.1 Summary 151
  11.2 Challenges in Implementing SAFEs 155

Part IV   SAFE Smart (Legal) Contracts

12 **Blockchain and Smart Contracts** 159
  12.1 Blockchain 159
  12.2 Smart Contracts 163
  12.3 Ethereum 164
  12.4 Ethereum's Programming Environment 166
  12.5 Smart Contract Applications 169
  References 171

13 **Difficulties in Formalizing SAFE Contracts** 173
  13.1 Indeterminacy in Legal Text 173
  13.2 Indeterminacy in SAFEs 175
  13.3 Incompleteness 177
  13.4 Open Structure 178

13.5   Legal Context 179
13.6   Nuncospectivity 180
References 181

14   Strategies for Formalization 183
14.1   Increasing Precision 184
14.2   Limitation of Fact Patterns 184
14.3   Use of a Human Oracle 185
14.4   Propose-and-Verify Implementations 187
       14.4.1   Dealing with Incompletely Stated Relationships 187
       14.4.2   Representation of Declaratively Expressed
                Relationships 188
       14.4.3   Price Computations 189
14.5   Programmable Open Structure 189
14.6   Previous Work 191
14.7   Combination with a Legal Contract 193
References 193

15   Architecture for SAFE Smart Contracts 195
15.1   Design Patterns 195
       15.1.1   Atomic Swap 195
       15.1.2   Controller 196
       15.1.3   Payout 198
15.2   Architecture and Implementation 199
       15.2.1   Company Financial Structure 201
       15.2.2   Safe Contract Representation 204
       15.2.3   SAFE Controller 211
       15.2.4   Atomic Swap Implementation of the Equity Round 214
       15.2.5   Liquidity Events 217
       15.2.6   Dissolution Events 220
15.3   Security Requirements 221
15.4   Limitations of the Implementation 225
15.5   Summary 227
References 227

16   Evaluation 229
16.1   Evaluation Methodologies 229
16.2   Risks Controlled 232
16.3   Risks Not Coverable by Smart Contracts Alone 234
16.4   Summary 235
References 236

17   Interaction with Law 237
17.1   Applicability of Law 237
17.2   Legal Status of Smart Contracts 238
17.3   Legal Activation 240

17.4   Towards a Smart Legal Contract 242
17.5   Adjudication 245
17.6   Enabling Response to Legal Orders and Variations 246
17.7   Conclusion 251
References 251

18   Privacy and Platform Issues 253
18.1   Policy 253
18.2   Public Blockchains 254
18.3   Permissioned Blockchains 256
18.4   Conclusion 257
References 257

19   Issues in Implementing Other SAFE Clauses 259
19.1   Execution of Equity Financing Documents 259
19.2   Pro Rata Rights 261
19.3   Company and Investor Representations 262
19.4   Miscellaneous Provisos 264
19.5   Most Favored Nation SAFEs 266
19.6   Summary 267
References 268

20   Conclusion 269
20.1   Summary 269
20.2   Future Work 272
References 274

Appendix: Working for Comparisons of Proportional Shareholding 275

Index 281

Symbols and Abbreviations

c	Valuation cap of SAFE
c_i	Valuation cap of SAFE i
C	Set of SAFEs choosing to Convert in a Liquidity Event
$\mathbf{C}$	Set of all convertible instruments in a Liquidity Event Game
$\mathsf{cash}(K)$	The total payout to the SAFEs K that take the Cashout option
$\mathsf{conv}(K)$	The total payout to the SAFEs C that take the Convert option
d	Discount rate of SAFE with Discount (10% discount means $d = 0.9$)
D	Dividend for each share of common stock
E	Total entitlements/claims for all SAFEs
E_i	Entitlement/claim for SAFE i
k	Number of SAFEs outstanding
K	Set of SAFEs choosing to Cashout in a Liquidity Event
$\overline{K}$	Set of SAFEs choosing to Convert in a Liquidity Event
m_{safe}	Purchase price of SAFE, principal
m_i	Purchase price of SAFE i
m_{new}	Amount of money raised in an equity financing round
$m(K)$	Sum of all Cashouts of SAFEs in K
o_f	The founders' proportion of shares, post-money
o_i	The proportion of shares held by SAFE investor i, post-money
o_{new}	The new (equity round) investor's proportion of shares, post-money
o_{safe}	The SAFE holder's proportion of shares, post-money
$\mathcal{P}()$	Powerset
p	Price of shares used in conversion of SAFE
p_i	Price of shares used in conversion of SAFE i
p_{fair}	The fair market price of a share of common stock
P_{liq}	The Liquidity Price
p_{new}	Price of new shares in an equity round
p_{safe}	The Safe Price
$\mathbb{R}$	The real numbers

s_{common}	The number of common shares in the company
s_C	Shares for convertible instrument C
$s(C)_i$	Number of shares for SAFE i in liquidity conversion when converting SAFEs C
s_f	Number of the founders' shares
s_i	Number of shares issued in conversion of SAFE i
s_{new}	Number of shares issued to new equity round investor(s)
s_{safe}	Number of shares issued in conversion of SAFE
S_{liq}	The Liquidity Capitalization
S_{post}	Company capitalization, post-money
S_{pre}	Company capitalization, pre-money
$U_i(K)$	Payout for SAFE i in a Liquidity Event with SAFEs K choosing Cashout
v_A	Value of company assets
v_C	Value of convertible instrument C
v_i	Valuation of SAFE i
v_I	Inherent valuation of company (excluding SAFEs)
v_L	Value of company liabilities
v_{post}	Post-money valuation of company
v_{pre}	Pre-money valuation of company
v_{pre}^-	Discounted pre-money valuation of company
v_{safe}	Valuation of SAFE
V	Value available for distribution in a Liquidity or Dissolution Event
w_i	The minimum of the pre-money valuation of the company and the valuation cap of SAFE i
$\alpha_i, \beta, \gamma_i$	Parameters in the Liquidity Event game
ξ	$\sum_{i=1}^{k} \frac{m_i}{c_i}$, minimum proportion of company committed to post-money SAFEs with Cap
Γ	The minimum of the γ_i parameters in the Liquidity Event game
2R	Two equity rounds: m_{safe} at valuation c then m_{new} at v_I
DLT	Distributed Ledger Technology
$DV(C)$	Discounted Valuation method with Cap Table view
DI	Dollars Invested method
IPO	Initial Public Offering
MFN	Most Favored Nation
SAFE	Simple Agreement for Future Equity
$Std(L)$	Standard method with Liability view
VC	Venture Capital
$ZMR(DV, C)$	Zero-Money Round method with Discounted Valuation method and Cap Table view
$ZMR(Std, L)$	Zero-Money Round method with Standard method and Liability view

Part I
An Introduction to SAFE Contracts

Chapter 1
The Need for SAFEs

1.1 The Need for Money

It has been an enduring adage, since ancient times, that you need to spend money to make money.[1] This has never been more true for an entrepreneur with a startup enterprise. The calls for money can come from all quarters. Among others:

Intellectual Property At the start of a business, intellectual property might be the main asset. It needs to be protected. This involves establishing intellectual property rights through patents, copyrights, trademarks, and other, more specialized, rights that have been established in law. It also includes identifying, and maneuvering around, others' rights, including licensing those rights, where necessary. Even where the intellectual property is to be protected as a trade secret, it requires the development of procedures and legal instruments, both to protect the secret and to be seen protecting the secret should the issue come to court.

Research, Design, and Development An initial idea or prototype can require substantial further research and development before it becomes marketable. It also needs design work to elevate the user experience. Further, if it is a manufactured product, the design and development of the manufacturing process may be needed. Even if manufacture is outsourced, finding a reliable manufacturer, specifying the requirements, and negotiating a deal can be expensive. And then there is the cost of manufacture itself.

Marketing, Sales, and Distribution It is necessary to identify and evaluate markets for the startup's products, and then to communicate to potential customers the availability and advantages of the products. Finally, there is the logistics of making the products continually and conveniently available.

Governance As the enterprise grows, more people and greater organization are needed to manage its operation. Managers will need to be hired. Eventually, a

---

[1] Expressed by Plautus in 200–100 B.C.E. as *necesse est facere sumptum qui quaerit lucrum*—you must spend money, if you wish to make money, (in Act 1, Scene 3, line 217 of *Asinaria*) [1].

full C-suite will be needed, but also specialist managers (for example, in HR, technical areas, and finance).

Legal Services   In addition to incorporating the enterprise, drafting contracts and other standard lawyering, startup lawyers perform several important roles including educating the entrepreneur on financial transactions, standardized forms, and community norms; serving as "reputational intermediaries" (that is, supporting the reputation of the enterprise by association); and acting as "regulatory compliance experts" (that is, understanding and communicating regulatory and legal obligations) [2] (see also Suchman and Cahill [3]).

The relative weight of these expenses varies according to the nature of the enterprise. Some of the providers of these services might agree to defer payment, or take equity in the enterprise as (part) payment (for example, employee equity incentive schemes [4], and some startup lawyers [2]). Nevertheless, these needs, among others, require the entrepreneur to raise capital in order to progress the enterprise.

There are, of course, many potential sources of such capital:

Self   Although an individual entrepreneur or team of founders usually gives the initial impetus to an enterprise, only wealthy individuals can afford to fund the further development of the enterprise from their own pockets.

Altruists   Sometimes described as "friends, family, and fools" or "the bank of Mum and Dad." These are people who know the founders and have a vested interest in their success. They usually invest in the very early stages of the enterprise.

Crowdfunding   In general, crowdfunding involves the payment of small sums by many people, often via specialist websites. In return, these people may receive a product, recognition, equity in the enterprise, a promise of equity, interest on their payment, or even no consideration (in the case of donations).

Banks   Banks may offer loans to businesses, but generally require collateral or other surety of being repaid. Startups have difficulties satisfying these requirements.

Angels   Angel investors provide money to early stage ventures in return for interest on the loan, equity, or the promise of equity. They often also provide valuable business advice to the startup.

Equity Investors   Equity investors provide money and in return receive shares of the enterprise. This is known as equity financing. Most commonly, equity investors in private companies are venture capital institutions (VCs).

Equity in the enterprise may be in the form of common or preferred stock. Common stock is simply shares in the enterprise of the same kind that the founders hold. Preferred stock is also shares in the enterprise, but preferred stock holders have priority over holders of common stock in claims for the earnings or assets of the enterprise. In the event of the liquidation of the business, holders of these shares receive payment before holders of common stock. Preferred shares may also have other attributes that differ from common stock; for example, they might have guaranteed dividends and/or not hold voting rights.

The idea of the company issuing a "promise of equity," rather than direct equity, addresses a specific difficulty in startup financing. Excepting bank loans and altruistic donations, the sources of capital generally expect to receive a share in the enterprise, should it be successful. However, doing this fairly depends on a valuation of the enterprise, a notoriously difficult task for startups [5]. Unlike more mature companies, startups have no historical data to project into the future, little to no revenue, few tangible assets, uncertainty in the business model, and few—if any—comparable firms. Furthermore, there is the fact that most startups do not survive. Consequently, methodologies for valuing mature firms are inapplicable or of little reliability. Methodologies specific to startups have substantial subjective components [6]. As a result, the current focus of funding has moved to providing money in return for the promise of equity in the future, when the company is more mature and it becomes possible to determine a realistic value of the company. This "promise" is established by a contract that expresses the circumstances in which equity will be given, and how much. The Simple Agreement for Future Equity (SAFE) is one such contract.

However, SAFEs are comparatively recent.

1.2 A Little History

Innovative startups in the last century typically required substantial capital to set up the business, in the millions of dollars. The typical lifecycle of venture finance for a startup at the end of the twentieth century has been described [7] as starting with friends, family, and angel investors providing relatively small amounts of money (seed funding) to the enterprise in the early stages, in exchange for common stock. Then, as the enterprise shows its feasibility, more substantial amounts of money are raised with equity financing from VCs, through the issuance of convertible preferred stock (convertible to common stock). Eventually (if successful) the enterprise will receive several rounds of equity financing (those rounds denoted Series A, Series B, ...). The enterprise might obtain bridging loans from existing investors through convertible notes if money is needed to survive until the next round of venture capital or the sale of the enterprise.

A handbook on venture capital for entrepreneurs [8] published in 2001 listed four ways to obtain funding from VCs:

- sale of common stock
- sale of preferred stock
- convertible debenture (a long-term loan convertible into common stock)
- loan with warrant (a warrant is like an option to buy stock of the company)

The first two provide equity to the investor; the last two are loans that also provide the opportunity of obtaining equity, but in different forms.

Equity rounds usually involve millions of dollars, the money required for the business to go to the next stage of its development. In general, multiple investors are needed. Because of the amounts of money involved, investors require the company to agree to conditions (covenants) on the behavior of the company, to protect the investment, and require preferred stock, which have privileges beyond that of common stock (for example, regularly scheduled dividends, and priority over common stock holders in case of liquidation). Thus, achieving a successful equity round requires convincing sufficient investors to invest the amount of money needed, and negotiating with them covenants and privileges, as well as a price per share (or, equivalently, a valuation of the company). This is a time-consuming process; Wallace [9] estimates 3–6 months. In addition, a substantial contract is needed to formalize the outcome of the negotiations. Legal costs might be in the thousands of dollars per hour [8].

Obtaining a convertible debenture is also a long-running process. Although there might be only one counter-party, instead of several, many of the negotiating points in equity finance also arise in obtaining a convertible loan. The loan agreement contract alone could run from 25 to 100 pages [8], with similar legal expenses.

Overall, at the turn of the century, financing a startup with VCs was a heavyweight process, congruent with the millions of dollars involved, but expensive and time-consuming.

Shortly after the turn of the century, the costs of startups, especially software-based startups, began to reduce [7].[2] Previously, upfront costs for office space, computing equipment, software, etc., required investments of up to several million dollars for an enterprise to get off the ground. The advent of cloud computing, cloud-based software, and open-source software substantially reduced these costs. Furthermore, the growing availability of high-speed Internet access and social media reduced some costs of marketing and distribution.

With the reduction of costs, the need for VCs with deep pockets could be deferred. Furthermore, by delaying equity financing, companies may have a higher valuation and surrender less equity for the same money. Startup incubators and accelerators formed and proliferated, occupying a niche between angels and VCs. They could provide startups with support services, networking opportunities, mentoring, and money. Prominent among accelerators has been Y Combinator, which was founded in 2005.

---

[2] Coyle and Green [7] suggest roughly 2005 as the time these changes began to have an effect.

1.3 Convertible Notes

The reduction in costs also led to a growing use of more lightweight funding methods.[3] In particular, convertible notes, which previously had been used mostly for bridging loans, began to be used for regular financing.[4] A convertible note is a loan until a fixed date (called the maturity date), bearing interest that is to be paid regularly, and that is convertible to shares upon the occurrence of trigger events (such as an equity financing or an IPO) at the choice of the investor. If the loan is not converted, the principal must be repaid at maturity.

Convertible notes have advantages, for both the investor and the company. For the investor:

- The investor receives an income stream through the interest payments. This balances the risk of company failure, where the principal might not be fully recovered.
- As a debt, the note gives the investor priority over shareholders in the distribution of company assets, should the company fail.
- If the company does not fail, the investor can recover the principal of the loan at maturity.
- If the company is very successful, the investor has the opportunity to share in the success, as a shareholder, possibly at lower-than-market rates.
- Valuation of the company, which can be inaccurate for a startup, is deferred until the triggering event.

For the company:

- Much of the legal costs of equity financing, which become disproportionate as the money needed is reduced, are avoided.
- It is not necessary to synchronize investors in a single investment round. Instead, the company can incentivize early investors with better deals, and so avoid a commitment stand-off among investors. This has been named high-resolution fundraising [10]. As a result, funding can be concluded quicker.
- There is a reduction in covenants that restrict the actions of the company.
- When the investors are friends and family, deferred conversion removes a potential point of conflict [7].

On the other hand, there are always drawbacks. For the investor:

- The convertible note is a loan, and so subject to usury laws and minimum interest taxation rules that constrain the choice of interest rate [11, 12].
- Assuming conversion, shares are acquired at the time of conversion, instead of the time of investment. Investors can lose tax breaks because of this difference in timing [13].

---

[3] A change in attitude by VCs following the bursting of the dot-com bubble might also have been influential.

[4] Coyle and Green [7] have a detailed discussion on this transition.

- The company likely has a lower valuation at investment time, compared to conversion time. Thus, the convertible note would result in a smaller share of the company than direct equity finance.
- Similarly, there may be a conflict of interest with shareholders in that an investor would prefer an early triggering event (with a lower valuation).

For the company:

- For a company with limited or uncertain cash flow, the payment of interest is a burden. It can force the company to prioritize short-term payoffs over long-term growth of the business. (On the other hand, in many cases the interest was allowed to accrue, and was converted to equity. But this is balanced by the resulting further dilution of the founders' holding.)
- If maturity is reached without a triggering event then the company must repay the principal (or re-negotiate the loan), which can create financial stress on the company.
- High-resolution fundraising in smaller amounts can result in a continual fundraising effort, which can be a distraction from developing the company.
- Convertible debt creates complications in accounting and valuation (as we will see, in the simpler case of SAFEs, in Part II), although accounting has been simplified [14].

Despite the drawbacks, convertible notes grew in popularity.[5] Variations of convertible notes were introduced to allow the conversion at better than market rates. Among these variations were a discount on the market price, and a valuation cap (which ensures a maximum price per share for the conversion) [15]. In 2010, Paul Graham, co-founder of Y Combinator, claimed that "Convertible notes have won" [16] in that all investments in their current batch of startups were via convertible notes.[6] Nevertheless, the debt element of convertible notes raised legal and practical issues in relation to the California Finance Lender's Law [17].[7]

In late 2013, Y Combinator introduced SAFEs[8] as their new investment instrument [7, 18], and in 2014 made them available on their website.[9]

---

[5] Following the rise of lightweight funding was the rise of a method of funding based on multiple seed rounds.

[6] However, a 2018 survey [15] estimated that about half of seed fundings were equity financings.

[7] The CFLL required lenders to be licensed, but made an exception for venture capital bridge loans for one year or less. The resulting yearly negotiations from a position of weakness were a distraction for founders, as was the necessity to keep track of accrued interest [7].

[8] SAFEs were designed by Carolynn Levy, a partner at Y Combinator.

[9] Similar instruments were introduced at about the same time, notably a convertible equity security drafted by a law firm in conjunction with the Founder Institute in 2012, and the KISS (Keep It Simple Security) note, which was created by the startup accelerator 500 Startups (now 500 Global [19]) in 2014. See Coyle and Green [7] for more details.

1.4 Simple Agreements for Future Equity

A Simple Agreements for Future Equity (SAFE) can be viewed as a simplified convertible note [7] that omits debt aspects of convertible notes[10]: they eliminate interest payments, maturity dates, and the obligation to repay the money if there is no conversion. Instead they appear to be purchases of a right to receive shares (or, in some cases, money) in the case of triggering events. The SAFEs convert to a similar kind of share as received by the equity investor (preferred shares), although not quite the same. In comparison to other methods of financing, the SAFEs are quite brief, consisting of six or fewer pages.[11]

Several forms of SAFE were introduced by Y Combinator in 2013, varying on the method by which the investor would receive better than market conversion rates. One used a valuation cap, one used a discount, and one used both (with the investor receiving the more favorable of the two methods). In addition, there was a SAFE that offered no improvement in conversion rate, but provided a Most Favored Nation (MFN) clause. This clause allows the investor to choose the terms of any later SAFE (or other convertible security) in place of their current terms.

All these SAFEs, in the event of an equity financing, assume that the equity financing is described in terms of the value of the company *before* that financing— what is known as the pre-money valuation.

Compared to convertible notes, SAFEs reduce the protections of the investment and relieve the company of the problems of debt. Coyle and Green [7, 22] suggest that investors have a binary view of investing in startups—they will succeed or go bust—which helps explain the acceptance of reduced protections. SAFEs also avoid the issue of rules and regulations about interest, while both parties benefit from a further reduction in legal costs.

Although the SAFEs created quite an impact among investors and entrepreneurs, they did receive some criticism. Green and Coyle [22] considered SAFEs unsuitable for crowdfunding, where the investors are not sophisticated investors. In particular, they drew attention to the possibility that the company might survive without needing equity financing. In that case, the investor never receives equity in the company and,

---

[10] SAFEs are generally not regarded as debt instruments; certainly Y Combinator does not consider them such: the Post-Money SAFE Users Guide [20] states "we've always intended and believed the safe (original safe or new safe) to be an equity security," while the first page of the SAFE Primer [21] states "**A safe is not a debt instrument**" [emphasis in the original]. The comparison with convertible notes seems to make this clear, but it is not so clear-cut: in some circumstances the purchase price is to be returned to the SAFE holder. That has a debt-like flavor. Taxation authorities and law enforcement might take varying views, depending on the jurisdiction.

[11] Levy's aim was a simple, balanced, and layperson-friendly contract. As a result, she deliberately did not add terms to the contract that would have clarified its legal meaning. See Coyle and Green [7], pp. 168–170.

because all the debt provisions of convertible notes were stripped, the investor does not receive any interest. Thus, the investor is "stuck," receiving no value for their investment.[12]

In 2018, Y Combinator released new versions of the SAFEs that varied significantly from the original. They called these Post-Money SAFEs.[13] The SAFEs were based on the value of the company after accounting for the conversion of other SAFEs and other convertible securities, and the existence of an option pool, but still *before* the equity financing.[14] Y Combinator argued that this change made it easier for the founders to calculate the proportion of the company that was committed to SAFE holders, and was more consistent with viewing the SAFEs as fundraising independent of the equity round [20].[15] This change required substantial modifications to some definitions.

There were several other notable changes. The treatment of liquidity events was modified slightly. (We discuss this further in Chap. 8.) The new versions also addressed the issue of the stuck investor. Holders of a post-money SAFE now participate in dividends issued by the company.

The original SAFEs, in the event of an equity financing, provided for a *pro rata* rights agreement. Such an agreement allows the investor to retain their proportion of shares by investing in the following equity round (Series B, if the equity financing was Series A). This was omitted from the post-money SAFEs, with *pro rata* rights to be treated, instead, through a side letter [20].

The post-money SAFEs, in clause 5(a), now allow the company and a majority-in-interest of SAFE holders to amend the SAFE. They also, in clause 5(g), commit the parties to consider the SAFE a stock for all income tax purposes. This might be an attempt avoid the second drawback for investors of convertible notes, listed above: if the SAFE is a kind of stock then the 5-year clock for the QSBS tax break [26] starts ticking from the time of purchase of the SAFE, instead of the time of conversion (see Carr [13]).

One of the advantages of SAFEs is that they provide standardized terms, and so legal costs are reduced. However, this is undermined by parties inserting, deleting, or modifying clauses. The new SAFEs are "expressly copyrighted and made available under the Creative Commons Attribution-No Derivatives 4.0 License" [20]. Furthermore, they include the statement

> This Safe is one of the forms available at http://ycombinator.com/documents and the Company and the Investor agree that neither one has modified the form, except to fill in blanks and bracketed terms.

---

[12] This possibility is not purely theoretical: Toptal went this route [23].

[13] We will call the original SAFEs "Pre-Money SAFEs" in contradistinction to the new Post-Money SAFEs.

[14] In this sense, the name "Post-Money SAFE" is a misnomer [24].

[15] On the other hand, Post-Money SAFEs have been criticized as overly dilutive for founders: since SAFEs are not diluted by later SAFEs, the dilution is borne by the stockholders [25].

If the contract departs from the standard terms, the lack of this statement is a signal that terms may have changed. Thus, legal costs are limited when the standard contract is used.

Y Combinator's standard deal for startups, which had used the pre-money SAFEs, was now formulated in terms of the new post-money SAFEs. In 2024, that deal involves a combination of two different forms of Post-Money SAFE, one of which is an MFN SAFE, protecting Y Combinator's interests in the event of later SAFE issuance by the company [27]. The other provides the SAFE holder with a fixed percentage of the company shares at an equity financing.[16] The deal also includes a *pro rata* rights agreement.

In summary, substantial changes to the SAFEs were introduced, most significantly in the way SAFEs are converted to shares. Later, some smaller changes were made [20]. In 2021, the SAFE with both cap and discount was withdrawn because, in practice, Y Combinator did not encounter situations where they recommended it [20]. At the same time, the definition of a "Liquidity Event" was expanded to include direct listing of the company shares on a stock exchange. In 2023, some terms of the MFN SAFE were also altered, including the Liquidity Event clause, the Liquidity Priority clause, and the MFN Amendment Provision. In addition, the definitions of "Safe Preferred Stock" and "Subsequent Convertible Securities" were revised to clarify the original intent.

There remain criticisms of SAFEs, especially that users do not understand the effect of SAFEs on dilution (although the post-money SAFEs address this), nor the mechanics of conversion [24]. These are issues we address in Parts II and III.

1.5 Current Use of SAFEs

SAFEs are now in widespread use to finance startups. They are used in many accelerators in addition to Y Combinator.[17] Alchemist Accelerator [30] reportedly uses a SAFE. The Startup Grind community has set up a DAO (decentralized autonomous organization) that acts as a startup accelerator and which uses SAFEs [31].

SAFEs are used by some crowdfunding sites (for example, WeFunder [32] and Republic [33]). They are also used by many universities, encouraging students and staff to commercialize university-developed intellectual property (for example, University of Chicago [34], University of Pittsburgh [35], Oklahoma State University [36], UNSW [37]). Despite arguments that pre-money SAFEs are not safe for crowdfunding [22], it seems that pre-money SAFEs are the contracts still used. Several

---

[16] It is not clear whether this is achieved through one of the published SAFEs or with a different, unpublished SAFE [28]. In any case, the deal specifies that this SAFE is converted *after* other SAFEs are converted and the creation/increase of an option pool [27].

[17] Among the startup accelerators listed by GrowthMentor [29], several use SAFEs including: Jumpstart Foundry, gener8tor, Seed Round Capital, and FounderFuel.

capitalization table[18] management services [38–40] and financing support organizations, such as AngelList [41], support SAFEs. SAFEs also inspired SAFTs (Simple
Agreement for Future Tokens) [42] in the cryptocurrrency sector.

One consequence of the growth of use of convertible notes and SAFEs (and one of
the motivations for the introduction of the Post-Money SAFEs) is that it has become
more common for companies to conduct several rounds of lightweight funding before
seeking equity financing.

SAFEs are available in a wide range of countries, although legal details may
differ slightly, depending on the legal jurisdiction. Y Combinator itself has versions
of SAFEs for the US, Canada, Singapore and the Cayman Islands. Law firms have
adapted the SAFE to other jurisdictions and/or stand ready to give advice on such
contracts, including in Mexico, the UK, Europe, Israel, India, South Korea, Australia,
and New Zealand. SAFEs seem to be available, but deprecated, in China.[19] In Japan,
although SAFEs seem to be available [46] a version of the KISS—the J-KISS—is
more widely used [47].

However, the advantages and disadvantages of SAFEs, compared to other funding
mechanisms, will vary by country, since commercial realities and surrounding law
and regulation—such as taxation—can affect the usefulness of SAFEs. Many taxation
authorities have issued advice on the status of SAFEs.

Not all countries permit the use of SAFEs. Convertible notes and SAFEs are not
currently allowed in Russia [48]. Brazilian regulations seem to be a barrier to SAFEs
but an analogous contract called MISTO has been developed; however, MISTO is
structured as a convertible interest-free loan [49], so there are still differences with
a SAFE. Finally, it is unclear whether a SAFE is compliant with Islamic law—there
are differing opinions [50–52].

1.6 Conclusion

In summary, SAFEs are an established method of financing startups. They have been
propounded by a major startup accelerator (Y Combinator) and have been adopted by
other accelerators, crowdfunders, and universities. They have spread to many parts
of the world; although the original SAFEs require modification to address the local
laws and commercial environment, the central mechanisms for conversion of SAFEs
that we study are usually unchanged.

---

[18] A capitalization table (or cap table) of a company is a listing of shareholders, their number of
shares and the type of those shares.

[19] See Dicker [43] for their presence in China. However, MiraclePlus, which grew out of Y Combinator's former branch in China, does not offer SAFEs in mainland China [44], and Lin [45] suggests
that the environment in China is not suitable for transplanting US contractual terms.

References

1. Titus Maccius Plautus. Asinaria. BCE 200–100. Translated by Paul Nixon
2. Coyle JF, Green JM (2017) Startup Lawyering 2.0. North Carolina Law Rev 95(5):1403–1432
3. Suchman MC, Cahill ML (1996) The hired gun as facilitator: lawyers and the suppression of business disputes in Silicon Valley. Law Soc Inquiry 21(3):679–712
4. Hall BJ (2000) What you need to know about stock options. Harv Bus Rev. March–April
5. Damodaran A (2009) Valuing young, start-up and growth companies: Estimation issues and valuation challenges. SSRN 1418687
6. Olsen MR (2019) An empirical study of startup valuation. Master's thesis, Copenhagen Business School
7. Coyle JF, Green JM (2015) Contractual innovation in venture capital. Hastings Law J 66(1):133–183
8. Gladstone D, Gladstone L (2001) Venture capital handbook: an entrepreneur's guide to raising venture capital. FT Press (Updated and revised edition)
9. Wallace Q (2024) The founder's dilemma: priced equity rounds vs SAFEs. https://www.archangel.vc/post/founders-dilemma-priced-equity-rounds-vs-safes. Accessed May 2024
10. Graham P (2010) High resolution fundraising. https://www.paulgraham.com/hirefund.html. Accessed May 2024
11. Kagan J. Minimum-interest rules: definition, how it works, and example. https://www.investopedia.com/terms/m/minimum-interest-rules.asp. Accessed May 2024
12. West GD (2021) Convertible debt: a New York usury refresher. https://privateequity.weil.com/glenn-west-musings/convertible-debt-a-new-york-usury-refresher. Accessed May 2024
13. Carr P (2019) How to build or invest in a startup without paying capital gains tax. https://techcrunch.com/2019/12/04/how-to-build-or-invest-in-a-startup-without-paying-capital-gains-tax/. Accessed May 2024
14. Financial Accounting Standards Board (2020) Accounting Standards Update 2020-06, Debt—Debt with Conversion and Other Options (Subtopic 470-20) and Derivatives and Hedging—Contracts in Entity's Own Equity (Subtopic 815-40): Accounting for convertible instruments and contracts in an entity's own equity. http://fasb.org. Accessed Dec 2024
15. Coyle JF, Green JM (2018) The SAFE, the KISS, and the note: a survey of startup seed financing contracts. Minn Law Rev: Headnotes 103(Fall):42–66
16. Graham P (2010) Twitter post. https://x.com/paulg/status/22319113993. Accessed May 2024
17. Lewis M (2015) The California finance lender law: venture capital bridge loan exemption and general licensing requirements. https://www.morganlewis.com/-/media/files/special-topics/vcpefdeskbook/fundoperation/vcpefdeskbook_californiafinancelenderlaw.pdf. Accessed May 2024
18. Levy C (2018) Safe financing documents. https://www.ycombinator.com/documents/#safe. Accessed Sep 2019
19. 500 Global. https://500.co. Accessed May 2024
20. Y Combinator (2018) Post money safe user guide. https://www.ycombinator.com/docs/Post%20Money%20Safe%20User%20Guide.pdf. Accessed Dec 2024
21. Y Combinator (2016) SAFE Primer. https://web.archive.org/web/20180831020232/http://www.ycombinator.com/docs/SAFE_Primer.rtf
22. Green JM, Coyle JF (2016) Crowdfunding and the not-so-safe Safe. Va Law Rev Online 102:168–182
23. Matican R, Efrati A (2019) At booming Toptal no stock for employees or investors. https://www.theinformation.com/articles/at-booming-toptal-no-stock-for-employees-or-investors. Accessed June 2024
24. Reid M, McDonald M, Rock B (2020) Demystifying SAFEs: the good, the bad, and the ugly. https://www.dlapiper.com/en-au/insights/publications/2020/07/demystifying-safes. Accessed May 2024
25. Ancer J (2019) Why startups shouldn't use YC's post-money SAFE. https://siliconhillslawyer.com/2019/05/01/startups-shouldnt-use-yc-post-money-safe/. Accessed Nov 2024

26. Weltman B (2017) Qualified small business stock: What is it and how to use it. https://www.sba.gov/blog/qualified-small-business-stock-what-it-how-use-it. Accessed Nov 2024
27. Y Combinator (2024) Standard deal. https://www.ycombinator.com/deal. Accessed May 2024
28. Harvey C (2024) YC's secret SAFE. https://www.linkedin.com/pulse/ycs-secret-safe-chris-harvey-rsffc/. Accessed Nov 2024
29. Panagiotakopoulos F. The 53 best startup accelerators in the world (sorted by country). https://www.growthmentor.com/blog/best-startup-accelerators/. Accessed May 2024
30. Alchemist. Alchemist Accelerator. https://www.alchemistaccelerator.com. Accessed May 2024
31. Startup Grind. SG DAO. https://sg-dao.gitbook.io/sg-dao/. Accessed Apr 2023
32. Wefunder. https://www.wefunder.com/. Accessed Apr 2023
33. Republic. https://www.republic.com/. Accessed Apr 2023
34. University of Chicago. Six graduating students receive Polsky Founders' fund fellowships to work on their ventures full-time. https://polsky.uchicago.edu/2022/07/06/six-graduating-students-receive-polsky-founders-fund-fellowships-to-work-on-their-ventures-full-time/. Accessed Apr 2023
35. University of Pittsburgh. Big idea advantage fund. https://www.bigidea.pitt.edu/programs/signature-programs/big-idea-fund/
36. Oklahoma State University. Investment funding. https://cowboyinnovations.okstate.edu/brightest-orange-ventures/investment-funding.html. Accessed Apr 2023
37. UNSW Founders. Investment funding. https://unswfounders.com/accelerators. Accessed May 2024
38. Clara. https://clara.co/. Accessed May 2024
39. Cake. https://www.cakeequity.com/. Accessed May 2024
40. Carta. https://carta.com. Accessed Aug 2024
41. AngelList. What is a SAFE? https://learn.angellist.com/articles/safe-note. Accessed Apr 2023
42. Batiz-Benet J, Santori M, Clayburgh J (2017) The SAFT project: toward a compliant token sale framework. https://saftproject.com/static/SAFT-Project-Whitepaper.pdf. Accessed Dec 2024
43. Dicker A. The evolution of China venture capital legal terms. https://www.linkedin.com/pulse/evolution-china-venture-capital-legal-terms-art-dicker. Accessed May 2024
44. MiraclePlus. FAQ. https://www.miracleplus.com/en/faq/. Accessed May 2024
45. Lin L (2020) Contractual innovation in China's venture capital market. Eur Bus Organ Law Rev 21:101–138
46. Morimoto B (2023) Introduction of the Japanese version of "SAFE" for venture investment. https://www.nishimura.com/sites/default/files/newsletters/file/mergers_and_acquisitions_230608_en.pdf. Accessed May 2024
47. Reid Monroe-Sheridan A (2023) Convertible equity in the Japanese startup ecosystem. Univ Pennsylvania Asian Law Rev 18:195–221
48. Bonartseva EA, Tvarkovskii DA (2023) Onyatie konvertiruemogo zaima i ego otgranichenie ot inykh grazhdansko-pravovykh instrumentov [The concept of a convertible loan and its delimitation from other civil law instruments]. Voprosy rossiiskogo i mezhdunarodnogo prava [Matters Russ Int Law] 13(3A):423–432 (In Russian)
49. Latitud Ventures (2023) MISTO: the Brazilian answer to the YC safe. https://www.latitud.com/blog/misto-brazilian-yc-safe-mutuo-conversivel. Accessed May 2024
50. Bradford J. Is SAFE document equity conversion Shariah compliant? https://joebradford.net/is-safe-document-equity-conversion-shariah-compliant/. Accessed May 2024
51. Sultan Y (2021) Are SAFE & convertible notes Shariah compliant? https://www.linkedin.com/pulse/safe-convertible-notes-shariah-compliant-yousuf-sultan-csaa-cife-. Accessed May 2024
52. Adam F (2021) Shariah dynamics of SAFE agreements. https://amanahadvisors.com/shariah-dynamics-of-safe-agreements/. Accessed May 2024

Chapter 2
SAFE Contract Terms

The aim of this chapter is to give an overview of the legal text of SAFE contracts. SAFE contracts come in a number of forms. As initially issued by Y Combinator, in a form we call "Pre-Money SAFE," to distinguish from the later "Post-Money" versions, these contracts have two parameters, a "Valuation Cap" and a "Discount", that may or may not be included, giving four distinct contracts. The Valuation Cap gives a maximum valuation used to calculate a price at which the SAFE principal will be converted to shares, and the Discount relates to a percentage discount given on the price.

We use the Pre-Money and Post-Money versions of the SAFE with Cap Only [1, 2] as the main focus of this book, and the majority of the analysis and discussion of smart contracts is concerned with this form of SAFE. However, we will also explain in this chapter how the other three SAFE forms (Discount Only, Cap and Discount, and Most Favored Nation) differ from this form of SAFE.

2.1 Summary of SAFE Contract Structure

The different forms of SAFE contracts have a common overall structure, consisting of the following sections:

- *Preface*: The contract is prefaced by a declaration that the SAFE and related securities have not been registered under the Securities Act of 1933 or under state Acts. It warns that the SAFE may not be sold or transferred except as permitted by the Act and state securities law.[1]

---

[1] The Post-Money SAFEs also provide the name and the version number for the SAFE.

© The Author(s), under exclusive license to Springer Nature Singapore Pte Ltd. 2025
R. van der Meyden and M. J. Maher, *Simple Agreements for Future Equity (SAFE)*,
Blockchain Technologies, https://doi.org/10.1007/978-981-96-3920-5_2

- *Certification*: This identifies the parties to the contract (the Company and the Investor), as well as the values of the parameters (Date, Purchase Amount, Valuation Cap, and/or Discount) that are included in the particular SAFE. One of the objectives in the development of SAFEs was to provide a simple standard form enabling financing deals to be closed without extensive legal negotiations of terms, simply by settling upon values of these contract parameters. The Certification section also states that the Company issues to the Investor rights to certain shares subject to terms in the following sections.
- *Events*: This lists events of relevance to the performance of the contract (Equity Financing, Liquidity Event, Dissolution Event, and Termination), and describes the consequences of each.
- *Definitions*: This defines a number of terms (Capital Stock, Change of Control, Company Capitalization, Distribution, Dissolution Event, Equity Financing, Initial Public Offering, Liquidity Capitalization, Liquidity Event, Liquidity Price, Pro Rata Rights Agreement, Safe, and Safe Preferred Stock) that are used elsewhere in the contract.
- *Company Representations*: This contains a number of assertions made by the company that underpin the company's capacity to enter into the contract and the legal validity thereof.
- *Investor Representations*: Similarly, this contains a number of assertions by the investor that underpin their legal capacity to enter into the contract.
- *Miscellaneous*: It describes a number of provisos relating to revisions to the contract, delivery of relevant notices, rights not implied by the contract, transfer of rights associated to the contract, treatment of invalidity of part of the contract, the legal jurisdiction governing the contract, and treatment for taxation purposes[2].
- *Signature page*: Where the contract is signed by the investor and a representative of the company.

We expand upon the content of the Events and Definitions sections in the remainder of this chapter. As our focus in this book is on the operational aspects of SAFE contracts for purposes of implementation as smart contracts, we will not go into depth on the Representation and Miscellaneous sections, which are primarily relevant for legal interpretation of the contract, protecting against potential misinterpretations, and establishment of its validity in law.

In the following sections of this chapter, we describe each of the three main Events and their associated Definitions, for both Pre-Money and Post-Money SAFE versions. We focus on SAFE forms with Cap only [2], and sketch variations made in the other SAFE forms.

Figure 2.1 shows how the three events fit into the typical lifecycle of a startup company. After foundation, a startup company typically raises early forms of "seed" funding from friends and family, and from angel investors, who use the SAFE contract form to describe shares to be issued at a later time. An Equity Financing event occurs when the company first raises more significant funding from a venture capital firm,

---

[2] Only Post-Money SAFEs address taxation.

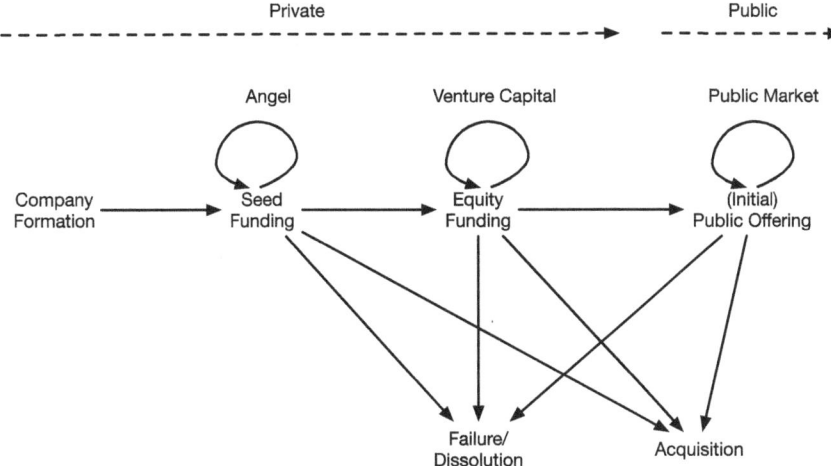

Fig. 2.1 Lifecycle of a startup company

which triggers the conversion of a SAFE to shares. The typical lifecycle requires multiple rounds of venture capital funding, but the ultimate aim in many cases is to list the company in public markets in an Initial Public Offering (IPO), at which point the early investors have an opportunity to profit from sale of their shares. However, an alternate possibility is that the company is acquired by another entity before it goes public. SAFE contracts classify IPO's and acquisitions that occur before termination of the SAFE as Liquidity Events. In the worst case, the company ultimately may not be sufficiently successful to raise a round of funding to sustain its operations, in which case it needs to be dissolved and the value of residual assets to be distributed to its investors. Dissolutions occurring before an Equity Financing or Liquidity Event are treated by the SAFE as a Dissolution event. Any of these event types terminate a SAFE contract.

2.2 Pre-Money SAFEs

2.2.1 Equity Financing

Equity Financing is defined in the SAFE as follows:

> **"Equity Financing"** means a bona fide transaction or series of transactions with the principal purpose of raising capital, pursuant to which the Company issues and sells Preferred Stock at a fixed pre-money valuation.

The key clause of the Pre-Money SAFE with Cap Only is the following, from the Events section:

1(a) **Equity Financing**. If there is an Equity Financing before the expiration or termination of this instrument, the Company will automatically issue to the Investor either: (1) a number of shares of Standard Preferred Stock equal to the Purchase Amount divided by the price per share of the Standard Preferred Stock, if the pre-money valuation is less than or equal to the Valuation Cap; or (2) a number of shares of Safe Preferred Stock equal to the Purchase Amount divided by the Safe Price, if the pre-money valuation is greater than the Valuation Cap.

This clause, effectively, describes how the number of shares to be issued to the SAFE investor in conversion of the SAFE are to be calculated at the time of an Equity Financing Event. We present this description as a mathematical formula in Chap. 4, where we discuss a number of subtleties in the interpretation of this clause.

The Equity Financing clause contains some further provisions which have some bearing on how we formalize the above:

In connection with the issuance of Standard Preferred Stock or Safe Preferred Stock, as applicable, by the Company to the Investor pursuant to this Section 1(a):

(i) The Investor will execute and deliver to the Company all transaction documents related to the Equity Financing; provided, that such documents are the same documents to be entered into with the purchasers of Standard Preferred Stock, with appropriate variations for the Safe Preferred Stock if applicable, and provided further, that such documents have customary exceptions to any drag-along applicable to the Investor, including, without limitation, limited representations and warranties and limited liability and indemnification obligations on the part of the Investor; and

(ii) The Investor and the Company will execute a Pro Rata Rights Agreement, unless the Investor is already included in such rights in the transaction documents related to the Equity Financing.

The clause refers to two kinds of stock. Standard Preferred Stock is simply the kind of shares obtained by a new investor as a result of the equity financing.

"Standard Preferred Stock" means the shares of a series of Preferred Stock issued to the investors investing new money in the Company in connection with the initial closing of the Equity Financing.

Safe Preferred Stock is intended to be as alike as possible to Standard Preferred Stock, except for using the Safe Price.

"Safe Preferred Stock" means the shares of a series of Preferred Stock issued to the Investor in an Equity Financing, having the identical rights, privileges, preferences and restrictions as the shares of Standard Preferred Stock, other than with respect to: (i) the per share liquidation preference and the conversion price for purposes of price-based anti-dilution protection, which will equal the Safe Price; and (ii) the basis for any dividend rights, which will be based on the Safe Price.

The notion of "price per share of the Standard Preferred Stock" is treated as a primitive input available at the time of the equity round, and the "Valuation Cap" is a parameter in the Certification section that is filled in when instantiating the contract before signing. The Safe Price is defined by the following clauses in the Definitions section:

"Safe Price" means the price per share equal to the Valuation Cap divided by the Company Capitalization.

"Company Capitalization" means the sum, as of immediately prior to the Equity Financing, of: (1) all shares of Capital Stock (on an as-converted basis) issued and outstanding, assuming exercise or conversion of all outstanding vested and unvested options, warrants and other convertible securities, but excluding (A) this instrument, (B) all other Safes, and (C) convertible promissory notes; and (2) all shares of Common Stock reserved and available for future grant under any equity incentive or similar plan of the Company, and/or any equity incentive or similar plan to be created or increased in connection with the Equity Financing.

"Capital Stock" means the capital stock of the Company, including, without limitation, the **"Common Stock"** and the **"Preferred Stock"**.

Notice that Company Capitalization refers to the total number of shares, and not their total worth.

Other forms of SAFE differ in the parameters of the instrument and in the way that the number of shares issued to the SAFE investor is calculated in an Equity Financing Event.

- In the Pre-Money SAFE with Discount Only, the Discount Rate parameter replaces the Valuation Cap parameter, and in place of 1(a), the conversion calculation is described as

    If there is an Equity Financing before the expiration or termination of this instrument, the Company will automatically issue to the Investor a number of shares of Safe Preferred Stock equal to the Purchase Amount divided by the Discount Price.

  subject to the definition

    "Discount Price" means the price per share of the Standard Preferred Stock sold in the Equity Financing multiplied by the Discount Rate.

  where

    The **"Discount Rate"** is [*100 minus the discount*]%.

- The Pre-Money SAFE with Cap and Discount has both a Valuation Cap and a Discount Price among its parameters, and specifies the shares to be issued in conversion of the SAFE by a formula that uses both of these parameters:

    If there is an Equity Financing before the expiration or termination of this instrument, the Company will automatically issue to the Investor a number of shares of Safe Preferred Stock equal to the Purchase Amount divided by the Conversion Price.

  where the Definitions section now contains

    "Conversion Price" means the either: (1) the Safe Price or (2) the Discount Price, whichever calculation results in a greater number of shares of Safe Preferred Stock.

The SAFE price is defined in this version exactly as above for the Pre-Money SAFE with Cap Only, while the discount price is defined exactly as above for the Pre-Money SAFE with Discount.

- The final SAFE version is the "Most Favored Nation" (MFN) SAFE. It has neither a Valuation Cap nor a Discount Rate as its parameters, In this SAFE, the conversion conditions are stated as

> If there is an Equity Financing before the expiration or termination of this instrument, the Company will automatically issue to the Investor a number of shares of Preferred Stock sold in the Equity Financing equal to the Purchase Amount divided by the price per share of the Preferred Stock.

In effect, this states that the SAFE investor receives their shares at the same price as the equity round investors. However, there is also the MFN clause which enables this SAFE to be converted to another form of SAFE in case the company offers better conditions to any other early investor:

> **"MFN" Amendment Provision**. If the Company issues any Subsequent Convertible Securities prior to termination of this instrument, the Company will promptly provide the Investor with written notice thereof, together with a copy of all documentation relating to such Subsequent Convertible Securities and, upon written request of the Investor, any additional information related to such Subsequent Convertible Securities as may be reasonably requested by the Investor. In the event the Investor determines that the terms of the Subsequent Convertible Securities are preferable to the terms of this instrument, the Investor will notify the Company in writing. Promptly after receipt of such written notice from the Investor, the Company agrees to amend and restate this instrument to be identical to the instrument(s) evidencing the Subsequent Convertible Securities.

The term Subsequent Convertible Securities is also defined in this SAFE.
This SAFE also has an extra condition in the definition of Equity Financing: it requires the equity financing to involve investment of at least $250,000.

> **"Equity Financing"** means a bona fide transaction or series of transactions with the principal purpose of raising capital, pursuant to which the Company issues and sells shares of Preferred Stock at a fixed pre-money valuation *with an aggregate sales price of not less than $250,000 (excluding all Subsequent Convertible Securities)*.

(Emphasis added to identify the difference from other SAFEs.)

2.2.2 Liquidity Events

Liquidity Events are defined in the Pre-Money SAFEs as a Change of Control or an Initial Public Offering (IPO). Each of these is also defined more precisely:

> **"Change of Control"** means (i) a transaction or series of related transactions in which any "person" or "group" (within the meaning of Section 13(d) and 14(d) of the Securities Exchange Act of 1934, as amended), becomes the "beneficial owner" (as defined in Rule 13d-3 under the Securities Exchange Act of 1934, as amended), directly or indirectly, of more than 50% of the outstanding voting securities of the Company having the right to vote for the election of members of the Company's board of directors, (ii) any reorganization, merger or consolidation of the Company, other than a transaction or series of related transactions in which the holders of the voting securities of the Company outstanding immediately prior to

such transaction or series of related transactions retain, immediately after such transaction or series of related transactions, at least a majority of the total voting power represented by the outstanding voting securities of the Company or such other surviving or resulting entity or (iii) a sale, lease or other disposition of all or substantially all of the assets of the Company.

"Initial Public Offering" means the closing of the Company's first firm commitment under-written initial public offering of Common Stock pursuant to a registration statement filed under the Securities Act.

Changes of Control, in all three cases, are situations where the shareholders, after the event, effectively have less than 50% control over the company or its pre-event assets, and in which the company potentially undergoes a change of focus in its activities. Both types of events cover situations where venture capital firms would often seek to benefit from the sale of (part of) their equity in the company.

The Liquidity Event clauses describe the treatment of an active SAFE at the time of a Liquidity Event. The Pre-Money SAFE with Cap Only has as parameters the principal (Purchase Amount) and a Valuation Cap. In a liquidity event, this SAFE offers the investor the option to "Cash Out" or to "Convert". In the case the investor chooses to Convert, the SAFE is converted to shares, with the number of shares issued in conversion of the SAFE defined as the Principal Amount divided by the Liquidity Price. Liquidity Price is defined as "the price per share equal to the Valuation Cap divided by the Liquidity Capitalization". The Liquidity Capitalization, in turn is defined by

"Liquidity Capitalization" means the number, as of immediately prior to the Liquidity Event, of shares of Capital Stock (on an as-converted basis) outstanding, assuming exercise or conversion of all outstanding vested and unvested options, warrants and other convertible securities, but excluding: (i) shares of Common Stock reserved and available for future grant under any equity incentive or similar plan; (ii) this instrument; (iii) other Safes; and (iv) convertible promissory notes.

Based on these definitions, the Pre-Money SAFE with Cap Only then states that the SAFE is treated as follows in case of a Liquidity Event.

(b) Liquidity Event. If there is a Liquidity Event before the expiration or termination of this instrument, the Investor will, at its option, either (i) receive a cash payment equal to the Purchase Amount (subject to the following paragraph) or (ii) automatically receive from the Company a number of shares of Common Stock equal to the Purchase Amount divided by the Liquidity Price, if the Investor fails to select the cash option.

In connection with Section (b)(i), the Purchase Amount will be due and payable by the Company to the Investor immediately prior to, or concurrent with, the consummation of the Liquidity Event. If there are not enough funds to pay the Investor and holders of other Safes (collectively, the "Cash-Out Investors") in full, then all of the Company's available funds will be distributed with equal priority and pro rata among the Cash-Out Investors in proportion to their Purchase Amounts, and the Convert Investors will automatically receive the number of shares of Common Stock equal to the remaining unpaid Purchase Amount divided by the Liquidity Price. In connection with a Change of Control intended to qualify as a tax-free reorganization, the Company may reduce, pro rata, the Purchase Amounts payable to the Cash-Out Investors by the amount determined by its board of directors in good faith to be advisable for such Change of Control to qualify as a tax-free reorganization for U.S. federal income tax purposes, and in such case, the Cash-Out Investors will automatically receive the number of shares of Common Stock equal to the remaining unpaid Purchase Amount divided by the Liquidity Price.

The option to Cash Out gives the SAFE holder downside protection against Liquidity Events that have the effect of causing an effective loss on their initial investment, or a change of direction of the company that does not meet their initial investment hypothesis. Conceptually, both Cash Out and Conversions happen just before the Liquidity Event, so that SAFE holders choosing to Convert can then be treated on a par with other shareholders in the Liquidity Event itself. (This may involve, for example, a distribution of a mix of cash and shares of an acquirer in exchange for shares in the company, or, in an IPO, rights to participate on specified terms based on the amount of the shareholding.)

The same terms are used in the SAFE with Cap and Discount. The SAFE with Discount, treats Liquidity events similarly, except that the Liquidity Price is defined by

"Liquidity Price" means the price per share equal to: the fair market value of the Common Stock at the time of the Liquidity Event, as determined by reference to the purchase price payable in connection with such Liquidity Event, multiplied by the Discount Rate.

The MFN SAFE defines Liquidity Price similarly, but without multiplying by the Discount Rate, which does not exist in that SAFE. Evidently, in the SAFE with Discount, a price determination similar to that used for the SAFE Price has been incorporated into the treatment of Liquidity Events. It is not clear why this was not done in the case of the SAFE with both Cap and Discount.

In situations where multiple SAFEs are active, the amount of value received by a SAFE holder taking a particular Liquidity Event option may depend on the options taken by other SAFE holders. This means that there is, effectively, a game-theoretic aspect to Liquidity Events. In Chap. 8, we develop a game-theoretic model of Liquidity Events, which we analyze to determine whether there is a game-theoretic equilibrium that captures the expected "best" options for each of the SAFE holders.

2.2.3 Dissolution Events

The Pre-Money SAFEs define Dissolution Events using the following:

"Dissolution Event" means (i) a voluntary termination of operations, (ii) a general assignment for the benefit of the Company's creditors or (iii) any other liquidation, dissolution or winding up of the Company (excluding a Liquidity Event), whether voluntary or involuntary.

Treatment of the SAFE upon the occurrence of such events is covered in the following clause.

(c) Dissolution Event. If there is a Dissolution Event before this instrument expires or terminates, the Company will pay an amount equal to the Purchase Amount, due and payable to the Investor immediately prior to, or concurrent with, the consummation of the Dissolution Event. The Purchase Amount will be paid prior and in preference to any Distribution of any of the assets of the Company to holders of outstanding Capital Stock by reason of their ownership thereof. If immediately prior to the consummation of the Dissolution Event, the assets of the Company legally available for distribution to the Investor and all holders of all other Safes (the "Dissolving Investors"), as determined in good faith by the Company's board of directors, are insufficient to permit the payment to the Dissolving Investors of their respective Purchase Amounts, then the entire assets of the Company legally available for distribution will be distributed with equal priority and pro rata among the Dissolving Investors in proportion to the Purchase Amounts they would otherwise be entitled to receive pursuant to this Section 1(c).

All the Pre-Money SAFE forms treat Dissolution events in the same way. We may note that, unlike in Equity Financing and Liquidity Events, the value of the SAFE's Valuation Cap or Discount does not affect the outcome.

2.3 Post-Money SAFEs

Y Combinator varied their standard contracts in September 2018, introducing a "Post-Money" version of the SAFE also in four versions (although the version with both Cap and Discount has since been discontinued). Appendix III of a guide to the Post-Money SAFEs [3] discusses the motivation for the changes made in these new contracts.

The overall structure of Post-Money SAFEs is similar to that of the Pre-Money SAFEs, but the approach to calculation of share distributions is altered in these documents. Pre-Money SAFEs treat the issuance of shares issued in conversion to the SAFE investors as concurrent with the issuance of shares to the new investors, with both calculated using the pre-money valuation. The main difference in the Post-Money SAFEs, intuitively, is that the issuance of shares to the SAFE investors is, in effect, viewed as being prior to the issuance of the shares to the new investors, so that the price calculation for the new investors is effectively based on the cap table of the company after the issuance of shares to the new investors. "Post" in these SAFEs refers, according to Y Combinator, to the fact that the "safe holder ownership is measured after (post) all the safe money is accounted for .... but still before (pre) the new money in the priced round that converts and dilutes the safes" [4].

Another change made in these SAFEs is to provide for the SAFE investor to receive a share of dividends distributed to shareholders prior to the conversion of the SAFE. The dividend amount is calculated as if the SAFE had been converted to Common Stock at the Liquidity Price (which is also used in Liquidity Event, as discussed below). Pre-Money SAFEs did not grant this right, leaving SAFE investors disadvantaged in situations where a company was able to develop into a sustainable enterprise without conducting an Equity Financing round.

There is a significant change in the MFN amendment provision in a recent version of the MFN SAFE (Version 1.2).

3. **"MFN" Amendment Provision**. If the Company issues any Subsequent Convertible Securities with terms more favorable than those of this Safe (including, without limitation, a valuation cap and/or discount) prior to termination of this Safe, the Company will

promptly provide the Investor with written notice thereof, together with a copy of such Subsequent Convertible Securities (the "**MFN Notice**") and, upon written request of the Investor, any additional information related to such Subsequent Convertible Securities as may be reasonably requested by the Investor. In the event the Investor determines that the terms of the Subsequent Convertible Securities are preferable to the terms of this instrument, the Investor will notify the Company in writing within 10 days of the receipt of the MFN Notice. Promptly after receipt of such written notice from the Investor, the Company agrees to amend and restate this instrument to be identical to the instrument(s) evidencing the Subsequent Convertible Securities.

This removes the right of the MFN SAFE holder to "all documentation relating to such Subsequent Convertible Securities," as provided in earlier MFN SAFEs, and imposes a 10 day deadline for the SAFE holder to activate the MFN amendment provision.

2.3.1 Equity Financing

In the case of the Post-Money SAFE with Cap Only [1], the Cap parameter in the certification is called a "Post-Money Valuation Cap," and the Equity Financing clause is changed to

Equity Financing. If there is an Equity Financing before the termination of this Safe, on the initial closing of such Equity Financing, this Safe will automatically convert into the greater of: (1) the number of shares of Standard Preferred Stock equal to the Purchase Amount divided by the lowest price per share of the Standard Preferred Stock; or (2) the number of shares of Safe Preferred Stock equal to the Purchase Amount divided by the Safe Price.

and the related definitions state

"Safe Price" means the price per share equal to the Post-Money Valuation Cap divided by the Company Capitalization.

"Company Capitalization" is calculated as of immediately prior to the Equity Financing and (without double-counting, in each case calculated on an as-converted to Common Stock basis):

- Includes all shares of Capital Stock issued and outstanding;
- Includes all Converting Securities;
- Includes all (i) issued and outstanding Options and (ii) Promised Options;
- Includes the Unissued Option Pool; and
- Excludes, notwithstanding the foregoing, any increases to the Unissued Option Pool (except to the extent necessary to cover Promised Options that exceed the Unissued Option Pool) in connection with the Equity Financing.

"Converting Securities" includes this Safe and other convertible securities issued by the Company, including but not limited to: (i) other Safes; (ii) convertible promissory notes and other convertible debt instruments; and (iii) convertible securities that have the right to convert into shares of Capital Stock.

There are two changes that are central to the understanding of the Equity Financing clause. First, the conditioning on the "pre-money valuation" to select the cases in the Equity Financing clause is replaced by a maximization over the number of shares. (This change resolves what is arguably a deficiency of the Pre-Money SAFE: we discuss this point in detail in Chap. 4.) Second, whereas "Company Capitalization" in the Pre-Money SAFE excludes the number of shares issued in conversion of SAFE contracts, these shares are included in the corresponding definition in the Post-Money SAFE. This in turn affects the calculation of the Safe Price, and the number of shares issued in conversion of the SAFE. A significant problem that this raises, taken up below, is that whereas the Pre-Money SAFE gives an explicit definition of the number of shares to be issued (depending on some undefined terms), the Post-Money SAFE has a circular definition, that can be understood as stating a set of constraints that need to be solved in order to determine the shares to be issued to a SAFE investor. We give a detailed discussion of the constraints and their solution in Chaps. 5 and 7.

There is a further change from the Pre-Money SAFE in that it is acknowledged that Preferred Stock may be offered in the Equity Financing at different prices to different investors, and the minimum such price is used in the calculation. Evidently, in the common case of all new investors in the Equity Financing event paying the same price, the minimum is just this shared price.

In the Post-Money SAFE with Cap and Discount, the Conversion Price is again the better of the Safe Price and the Discount Price. This SAFE was later withdrawn by Y Combinator.

The Post-Money SAFE with Discount converts at the Discount Price, defined as "the lowest price per share of the Standard Preferred Stock sold in the Equity Financing multiplied by the Discount Rate." The Post-Money MFN SAFE takes the Conversion Price to be "the lowest price per share of the Standard Preferred Stock." It also removes the threshold in the definition of Equity Financing.

2.3.2 Liquidity Events

Starting from version 1.1, in 2021, the Post-Money SAFE adds a third liquidity event:

> "**Liquidity Event**" means a Change of Control, a Direct Listing or an Initial Public Offering.

A Direct Listing is a listing on a national securities exchange.

> "**Direct Listing**" means the Company's initial listing of its Common Stock (other than shares of Common Stock not eligible for resale under Rule 144 under the Securities Act) on a national securities exchange by means of an effective registration statement on Form S-1 filed by the Company with the SEC that registers shares of existing capital stock of the Company for resale, as approved by the Company's board of directors. For the avoidance of doubt, a Direct Listing will not be deemed to be an underwritten offering and will not involve any underwriting services.

The definitions of Change of Control and Initial Public Offering are unchanged.

The treatment of Liquidity Events in the Post-Money SAFE with Cap Only differs from corresponding Pre-Money SAFEs in that the option for the SAFE holder to choose whether to Cash Out or Convert is replaced by a deterministic selection of an amount of value associated to each option, although an option remains if any other investor is granted a choice. In the Post-Money SAFE with Cap Only, this is stated as

(b) **Liquidity Event**. If there is a Liquidity Event before the termination of this Safe, the Investor will automatically be entitled (subject to the liquidation priority set forth in Section 1(d) below) to receive a portion of Proceeds, due and payable to the Investor immediately prior to, or concurrent with, the consummation of such Liquidity Event, equal to the greater of (i) the Purchase Amount (the "Cash-Out Amount") or (ii) the amount payable on the number of shares of Common Stock equal to the Purchase Amount divided by the Liquidity Price (the "Conversion Amount"). If any of the Company's securityholders are given a choice as to the form and amount of Proceeds to be received in a Liquidity Event, the Investor will be given the same choice, provided that the Investor may not choose to receive a form of consideration that the Investor would be ineligible to receive as a result of the Investor's failure to satisfy any requirement or limitation generally applicable to the Company's securityholders, or under any applicable laws.

The related definitions are

"**Liquidity Capitalization**" is calculated as of immediately prior to the Liquidity Event, and (without double-counting):

- Includes all shares of Capital Stock issued and outstanding;
- Includes all (i) issued and outstanding Options and (ii) to the extent receiving Proceeds, Promised Options;
- Includes all Converting Securities, other than any Safes and other convertible securities (including without limitation shares of Preferred Stock) where the holders of such securities are receiving Cash-Out Amounts or similar liquidation preference payments in lieu of Conversion Amounts or similar "as-converted" payments; and
- Excludes the Unissued Option Pool.

"**Liquidity Price**" means the price per share equal to the Post-Money Valuation Cap divided by the Liquidity Capitalization.

We may note here that the third item in the definition of Liquidity Capitalization *does* include SAFEs where the holders are receiving Conversion Amounts. In this regard, this clause of this Post-Money SAFE operates in the same way as the Equity Financing clause of this SAFE, and differently from the Pre-Money SAFE.

Moreover, a new clause is added that states a priority over these rights:

(d) **Liquidation Priority**. In a Liquidity Event or Dissolution Event, this Safe is intended to operate like standard non-participating Preferred Stock. The Investor's right to receive its Cash-Out Amount is:

(i) Junior to payment of outstanding indebtedness and creditor claims, including contractual claims for payment and convertible promissory notes (to the extent such convertible promissory notes are not actually or notionally converted into Capital Stock);

(ii) On par with payments for other Safes and/or Preferred Stock, and if the applicable Proceeds are insufficient to permit full payments to the Investor and such other Safes and/or Preferred Stock, the applicable Proceeds will be distributed pro rata to the Investor and such other Safes and/or Preferred Stock in proportion to the full payments that would otherwise be due; and

(iii) Senior to payments for Common Stock. The Investor's right to receive its Conversion Amount is (A) on par with payments for Common Stock and other Safes and/or Preferred Stock who are also receiving Conversion Amounts or Proceeds on a similar as-converted to Common Stock basis, and (B) junior to payments described in clauses (i) and (ii) above (in the latter case, to the extent such payments are Cash-Out Amounts or similar liquidation preferences).

There is not a corresponding clause in any of the Pre-Money SAFEs, though it might be argued that this priority order is implicit in these SAFEs and legal precedent, inasmuch as (i) creditors usually have priority over shareholders, and (ii) SAFEs are intended to convert into a type of Preferred Stock, so should rank equally with such stock, and have higher priority than Common Stock.

The Post-Money SAFE with Discount varies these terms by defining the Liquidity Price as "the price per share equal to the fair market value of the Common Stock at the time of the Liquidity Event, as determined by reference to the purchase price payable in connection with such Liquidity Event, multiplied by the Discount Rate."

Initially, the Post-Money MFN SAFE followed the SAFE with Discount, but with (in essence) a discount rate of 1.0. Version 1.2 in 2023 altered this: the Liquidity Event clause states simply that the SAFE holder is entitled to the "Purchase Amount (the 'Cash-Out Amount')." Nevertheless, if there are outstanding SAFEs that provide a choice between Cashout and Convert, the MFN SAFE holder is given the same choice. In any case, the Post-Money MFN SAFE defines the Liquidity Price in the same way as the Post-Money SAFE with Discount.

2.3.3 Dissolution Events

All Post-Money SAFEs have the same treatment of Dissolution Events:

(c) Dissolution Event. If there is a Dissolution Event before the termination of this Safe, the Investor will automatically be entitled (subject to the liquidation priority set forth in Section 1(d) below) to receive a portion of Proceeds equal to the Cash-Out Amount, due and payable to the Investor immediately prior to the consummation of the Dissolution Event.

Section 1(d) is the Liquidation Priority clause discussed in the previous section. Both Liquidity and Dissolution events refer to "Proceeds."

"Proceeds" means cash and other assets (including without limitation stock consideration) that are proceeds from the Liquidity Event or the Dissolution Event, as applicable, and legally available for distribution.

2.4 Termination

In brief, a SAFE contract terminates after an occurrence of an Equity Financing, Liquidity event, or Dissolution. The wording of the definition differs slightly between the Pre-Money and Post-Money SAFEs, due to slightly different contract structures. The Post-Money SAFEs define termination as:

> (d) **Termination**. This Safe will automatically terminate (without relieving the Company of any obligations arising from a prior breach of or non-compliance with this Safe) immediately following the earliest to occur of: (i) the issuance of Capital Stock to the Investor pursuant to the automatic conversion of this Safe under Section 1(a); or (ii) the payment, or setting aside for payment, of amounts due the Investor pursuant to Sections 1(b) or Section 1(c).

It is worth noting that it is possible that a SAFE does not terminate: that the company does not obtain equity financing, does not perform a liquidity event, and does not undergo dissolution. It simply carries on its business. In this case, the SAFE holder may receive no return on the investment.

To address this possibility, the Post-Money SAFEs require that SAFE holders participate in any dividend distribution to holders of common stock. In 5(c) it states:

> … if the Company pays a dividend on outstanding shares of Common Stock (that is not payable in shares of Common Stock) while this Safe is outstanding, the Company will pay the Dividend Amount to the Investor at the same time.

The dividend amount is defined as follows:

> **"Dividend Amount"** means, with respect to any date on which the Company pays a dividend on its outstanding Common Stock, the amount of such dividend that is paid per share of Common Stock multiplied by (x) the Purchase Amount divided by (y) the Liquidity Price (treating the dividend date as a Liquidity Event solely for purposes of calculating such Liquidity Price).

This completes our presentation of the terms of the Y Combinator SAFE contracts. (We have concentrated on the financial terms for the events that can terminate a SAFE, which are our focus for much of this book. Chapter 19 will discuss other aspects of SAFE contracts.) We now turn to a technical analysis of the consequences of these terms.

References

1. Y Combinator (2018) Safe: valuation cap, no discount. https://web.archive.org/web/20190626002912/https://www.ycombinator.com/docs/Postmoney%20Safe%20-%20Valuation%20Cap%20-%20v1.0.docx
2. Y Combinator (2016) Safe: Cap, no discount. https://web.archive.org/web/20180831020232/http://www.ycombinator.com/docs/SAFE_Cap.rtf
3. Y Combinator (2018) Post money safe user guide. https://www.ycombinator.com/docs/Post%20Money%20Safe%20User%20Guide.pdf. Accessed Dec 2024
4. Levy C (2018) Safe financing documents. https://www.ycombinator.com/documents/#safe. Accessed Sep 2019

Part II
Analysis: Single SAFE Scenarios

Chapter 3
Accounting Views of Convertible Instruments

Before embarking on a detailed analysis of the consequences of SAFEs in Equity Financing events, we first consider equations widely used to describe equity rounds that do not involve convertible instruments such as SAFEs. We show that these equations are inconsistent in the presence of (one or more) SAFEs. We resolve this inconsistency by showing that, depending on how SAFEs are accounted for (as liabilities or as potential shares in the cap table), different versions of these equations are valid.

3.1 Standard Equity Raise

Consider an equity financing round where the company has not issued any convertible instruments, so that the number of pre-existing shares S_{pre} is exactly the number of founders' shares s_f.[1] Suppose that the company raises extra capital m_{new} at a pre-money valuation of v_{pre}. Let p_{new} be the price of the newly issued shares, and s_{new} the number of newly issued shares. After the round, the number of shares in the company S_{post} is clearly $s_f + s_{new}$. The new money, price per share paid and number of new shares issued are related by the equation

$$m_{new} = s_{new} p_{new}. \qquad \text{(msp}_{new})$$

One issue for the new investor is to determine the price per share that they are prepared to pay. There are multiple ways this price might be determined in practice, e.g., by negotiation with the company, taking into the account the price last paid

---

[1] In the following discussion, we use subscripts *pre* and *post* as mnemonics to indicate whether the variables (other than p_{new} and m_{new}) refer to the pre-money or the post-money state.

© The Author(s), under exclusive license to Springer Nature Singapore Pte Ltd. 2025
R. van der Meyden and M. J. Maher, *Simple Agreements for Future Equity (SAFE)*,
Blockchain Technologies, https://doi.org/10.1007/978-981-96-3920-5_3

by other investors, and the investor's estimates of their expected returns, given their investment horizon. Another way is for the company and the investor to agree upon the pre-money valuation of the company. From this, we can determine a value per share for the existing shares in the company. Reasoning that the price of the newly issued shares should be the same as the value per share of the existing shares, at the pre-money valuation, we get that the transaction implies the following equation relating the pre-money valuation, outstanding shares, and the price per share:

$$v_{\text{pre}} = s_f\, p_{\text{new}}. \qquad\qquad (\text{vsp}_{\text{pre}})$$

Having determined a price per share p_{new} using this equation, we get, using Eq. $(\text{msp}_{\text{new}})$ that

$$s_{\text{new}} = m_{\text{new}}/p_{\text{new}} = m_{\text{new}} s_f / v_{\text{pre}}. \qquad\qquad (3.1)$$

One can take several views on the valuation of the company after the equity round. One is to value the company based on the revised number of outstanding shares, each share valued at the price per share paid in the equity round. This leads to a post-money valuation v_{post} given by the equation

$$v_{\text{post}} = S_{\text{post}}\, p_{\text{new}}. \qquad\qquad (\text{vsp}_{\text{post}})$$

An alternate view is to consider the way that the transaction changes the company's assets and liabilities. All that has changed is that the company now has an additional amount m_{new} of money in the bank; all its other assets and liabilities, valued at v_{pre}, are unchanged. On this view, the pre-money valuation and post-money valuation are related by the equation

$$v_{\text{post}} = v_{\text{pre}} + m_{\text{new}}. \qquad\qquad (\text{vm}_{\text{pre,post}})$$

The number of post-money shares is $S_{\text{post}} = s_f + s_{\text{new}}$, so we get that the amount of this post-money value per share after the raise is

$$
\begin{aligned}
v_{\text{post}}/S_{\text{post}} &= (v_{\text{pre}} + m_{\text{new}})/(s_f + s_{\text{new}}) && \text{by } (\text{vm}_{\text{pre, post}}) \\
&= (v_{\text{pre}} + m_{\text{new}})/(s_f + m_{\text{new}} s_f / v_{\text{pre}}) && \text{by } (3.1) \\
&= v_{\text{pre}}/s_f \\
&= p_{\text{new}} && \text{by } (\text{vsp}_{\text{pre}})
\end{aligned}
$$

i.e., the same as the pre-money value and price per share, as expected.[2] Note that Eq. (vsp_{post}) also gives $v_{\text{post}}/S_{\text{post}} = p_{\text{new}}$, so the two distinct ways of obtaining a post-money valuation expressed in Eqs. (vsp_{post}) and ($\text{vm}_{\text{pre, post}}$) are consistent.

We may also conclude from the fact that the post-money value per share is p_{new} that the new investor's share of the post-money valuation is $s_{\text{new}} p_{\text{new}} = m_{\text{new}}$. Thus, we also see that the investor neither gains nor loses by engaging in the transaction, they merely convert the form of their holding from cash to shares. In a similar sense, the founders also do not lose from the transaction: the value of their shares after the equity round is $s_f p_{\text{new}} = s_f(v_{\text{pre}}/s_f) = v_{\text{pre}}$, i.e., the pre-money valuation.

More generally, we will say that a transaction in which shares are issued for money is *conservative* for the investor if (ignoring transaction costs), immediately after the transaction, the value of the shares issued is at least equal to the money invested. The previous paragraph then shows that a standard equity round is conservative for the investor.

Stated in terms of monetary value, the effect of the equity round on the proportional shareholdings in the company is that after the raise, the founder holds a fraction

$$\frac{s_f}{s_f + s_{\text{new}}} = \frac{s_f}{s_f + m_{\text{new}} s_f / v_{\text{pre}}} = \frac{v_{\text{pre}}}{v_{\text{pre}} + m_{\text{new}}}$$

of the company, and the new investor holds

$$\frac{m_{\text{new}}}{v_{\text{pre}} + m_{\text{new}}}.$$

For purposes of later comparison with the effects of SAFEs, we note the effect of a sequence of two standard equity rounds (neither with convertible instruments in place), the first to an early investor for money m_{safe} at valuation c, and the second to a new investor for money m_{new} at valuation v_{pre}. Repeating the calculations of the present section, we obtain proportional shareholdings for the founders, early investor and new investor, respectively, of

$$\frac{c}{c + m_{\text{safe}}} \cdot \frac{v_{\text{pre}}}{v_{\text{pre}} + m_{\text{new}}}, \quad \frac{m_{\text{safe}}}{c + m_{\text{safe}}} \cdot \frac{v_{\text{pre}}}{v_{\text{pre}} + m_{\text{new}}}, \quad \frac{m_{\text{new}}}{v_{\text{pre}} + m_{\text{new}}}. \tag{3.2}$$

---

[2] It would be reasonable to take the view that the post-money value per share should be higher on the grounds that the new money enables the company to proceed with its development plans, whereas, without the new money, it may have to be liquidated for a lower return. However, if one takes the view that the pre-money valuation is based on the assumption that the equity round will proceed, the conclusion that the equity round should not change the value per share is reasonable.

For the new investor, this shareholding is worth m_{new}, as above, and the remaining value v_{pre} after the second round is divided between the founders and the early investor in the proportions

$$\frac{c}{c + m_{\text{safe}}}, \quad \frac{m_{\text{safe}}}{c + m_{\text{safe}}}$$

that result from the first equity round.

3.2 An Inconsistency

When convertible instruments, such as the SAFE contract, or other forms of convertible bond, are in place at the time of an equity round, they cause some fundamental changes to the properties of the round. Individually, each of Eqs. (msp$_{\text{new}}$), (vsp$_{\text{pre}}$), (vsp$_{\text{post}}$), and (vm$_{\text{pre,post}}$) governing a standard equity round is reasonable. However, taken together, they lead to an inconsistency when one of the consequences of the equity round is that convertible instruments convert into equity. In these situations, after the equity round, new shares have been issued not just to the new investor but also to the holder(s) of the convertible instruments, but no additional money is paid for these shares. (Money was paid for the convertible instruments at an earlier time, but it may have already been spent by the time of the equity round.) This leads to a contradiction between the above equations, stated in the following proposition.

Proposition 3.1 *Consider an equity round in which the pre-money state of the company consists of the founders holding shares s_f, and in which the company has issued a convertible instrument C to an early investor. Suppose that after the equity round, there are three shareholders, the founders, the early investor and the new investor, each holding a nonzero number of shares s_f, s_C and s_{new}, respectively, and new shares are issued in the round to the new investor at a non-zero price p_{new} and at no cost to the early investor. Then Eqs. (msp$_{\text{new}}$), (vsp$_{\text{pre}}$), (vsp$_{\text{post}}$), and (vm$_{\text{pre,post}}$) are inconsistent.*

Proof From the assumption about the three post-money shareholders, we get $S_{\text{post}} = s_f + s_C + s_{\text{new}}$. Thus, using (vsp$_{\text{post}}$), we have

$$v_{\text{post}} = p_{\text{new}} s_f + p_{\text{new}} s_C + p_{\text{new}} s_{\text{new}}.$$

By (vm$_{\text{pre, post}}$), the left-hand side of this equation equals $v_{\text{pre}} + m_{\text{new}}$, by (vsp$_{\text{pre}}$), the first term on the right equals v_{pre}, and by (msp$_{\text{new}}$), the rightmost term equals m_{new}. Thus, we have

$$v_{\text{pre}} + m_{\text{new}} = v_{\text{pre}} + p_{\text{new}} s_C + m_{\text{new}},$$

which is equivalent to $p_{\text{new}} s_C = 0$. Since both p_{new} and s_C have been assumed to be non-zero, this is a contradiction.   □

It follows that any attempt to make sense of such a situation needs, for consistency, to clarify which of the above equations is abandoned in understanding the equity round. As we will see, various approaches exist which resolve this issue by making different choices about which equation to abandon.

A related issue is that after the equity round, according to Eq. $(\mathrm{vm}_{\text{pre, post}})$, the post-money value per share is $(v_{\text{pre}} + m_{\text{new}})/(s_f + s_C + s_{\text{new}})$, which is less than the amount $(v_{\text{pre}} + m_{\text{new}})/(s_f + s_{\text{new}})$ it would be in the standard equity raise. If the new investor receives the same number of shares for their money, as in a standard equity round, this means that the new investor's shares are worth not m_{new}, as in a standard equity round, but a lesser amount. Whereas a standard equity round preserves value for the new investor, issuance of shares to the early investor *dilutes* the value of the new investor's shares, so that the new investor is left holding an immediate unrealized loss!

On the other hand, Eq. $(\mathrm{vsp}_{\text{post}})$ can be understood as stating that the post-money value per share is exactly equal to the price paid by the new investor, and it then follows by Eq. $(\mathrm{msp}_{\text{new}})$ that the post-money value of the new investor's shares is m_{new}. This directly contradicts the conclusion from the previous paragraph.

The following chapter considers the ramifications of these issues with respect to the Pre-Money SAFE with Cap Only, which adds some further complexity because of its recursive nature.

3.3 Two Accounting Views

The previous section indicates that when convertible instruments are involved in an equity round, the usual equations concerning the notions of pre-money and post-money valuation may be in conflict. This suggests that these notions are not well-defined in such a situation, and additional information may be required to determine which characterization best captures the situation. One way to do so is to consider the accounting status of the convertible instrument. There are two ways that the convertible instrument might be accounted for in the pre-money state. One is to view it as akin to a loan, which makes it a liability. Another is to account for it on the capitalization table of the company, as an obligation to issue shares. In what follows, we consider these alternatives from first principles.

3.3.1 Liability View

We consider first the consequences of an accounting model where the instrument C being converted is treated as a liability. Suppose the initial state of the company consists of assets A, and liabilities L and C, and the cap table consists of shares s_f (that is, $S_{\text{pre}} = s_f$). Let the valuation of the assets and liabilities be v_A, v_L and v_C, respectively. (If C is a loan, then it is easily valued: v_C is the outstanding principal and interest due. We will see in Chap. 4 that the recursive nature of SAFEs adds some complexities to the question of their valuation.) Thus, prior to the transaction, the company has a valuation

$$v_{\text{pre}} = v_A - v_L - v_C.$$

Consider an equity round where C is converted to shares at value equal to v_C and a new investor buys new shares for money m_{new}. The current share price is $p_{\text{new}} = v_{\text{pre}}/s_f$, and the number of new shares issued for the convertible instrument would therefore be $s_C = v_C/p_{\text{new}} = v_C s_f/v_{\text{pre}}$. (We emphasize that here v_C represents the value of the shares received by the convertible instrument holder at price p_{new}. This may be different from the value and price used for the actual conversion; case 2 of the Equity Financing clause of the SAFE is an example of this.) The number of shares issued to the new investor is given by Eq. (msp$_{\text{new}}$). The state of the company after the transaction consists of assets A plus new money m_{new}, liabilities L, and the cap table contains s_f, together with newly issued shares s_C and s_{new}. Valuing the assets and liabilities, we have $v_{\text{post}} = v_A + m_{\text{new}} - v_L$. Thus, we have that

$$v_{\text{post}} = v_{\text{pre}} + v_C + m_{\text{new}} . \tag{vmC$_{\text{pre,post}}$}$$

Note that the price per share after the transaction is therefore

$$\frac{v_{\text{post}}}{s_f + s_C + s_{\text{new}}} = \frac{v_{\text{pre}} + v_C + m_{\text{new}}}{s_f + (v_C s_f/v_{\text{pre}}) + (m_{\text{new}} s_f/v_{\text{pre}})} = \frac{v_{\text{pre}}}{s_f}$$

which is identical to the price before the transaction. In terms of assets and liabilities, the share price is $(v_A - v_L - v_C)/s_f$. This analysis shows that Eq. (vm$_{\text{pre, post}}$) does not correctly state the relationship between the pre-money and post-money valuation of the company when the convertible instrument is treated as a liability. Instead, Eq. (vmC$_{\text{pre, post}}$) captures this relationship. However, we do have

$$v_{\text{post}} = v_{\text{pre}} + v_C + m_{\text{new}} = p_{\text{new}} s_f + p_{\text{new}} s_C + p_{\text{new}} s_{\text{new}} = p_{\text{new}} S_{\text{post}},$$

so Eq. (vsp$_{\text{post}}$) does hold.

3.3.2 Cap Table View

However, particularly when C is not debt, but a convertible instrument like the SAFE, there is another way to do the accounting: we could consider the instrument C that will be converted to shares as already existing on the cap table of the company, rather than treated as a liability. On this view, the initial state of the company is given by assets A, liabilities L and cap table consisting of s_f and C. That is, $v_{\text{pre}} = v_A - v_L$ and $S_{\text{pre}} = s_f + s_C$.

The state after the equity round that converts C to an actual number of shares consists of assets A plus new money m_{new}, liabilities L and cap table comprised shares s_f, s_C, and s_{new}. Thus, by contrast with the situation above, we have, from an accounting point of view, that $v_{\text{pre}} = v_A - v_L$ and $v_{\text{post}} = v_A - v_L + m_{\text{new}} = v_{\text{pre}} + m_{\text{new}}$, so Eq. (vm$_{\text{pre, post}}$) *does* hold on this view.

It remains to determine a share price. Note that if we take the price to be v_{pre}/s_f, as in a standard equity round, and also assume that the transaction does not change the share price, we obtain the contradiction that the entirety of the final valuation $v_{\text{post}} = v_{\text{pre}} + m_{\text{new}}$ corresponds to the holder(s) of shares s_f and s_{new}, leaving no part of the valuation for the shares s_C. To escape this contradiction, we need to either assume that the share price decreases as a result of the transaction, or start with a different share price. A decrease in share price would imply that the transaction is unsatisfactory for the new investor, since it is not conservative. The price v_{pre}/s_f also does not take into account that we have recorded C on the cap table. We therefore consider just a different way to determine the share price.

To compute a share price that avoids the above paradox, we calculate this as if the shares s_C to be issued in conversion have already been issued. Thus, we have a share price

$$p_{\text{new}} = \frac{v_{\text{pre}}}{s_f + s_C}. \qquad \text{(vsCp}_{\text{pre}})$$

Take the price p_{new} at which shares are issued to the new investor to satisfy Eq. (vsCp$_{\text{pre}}$). As usual, the number of shares issued to the new investor, by Eq. (msp$_{\text{new}}$), satisfies $p_{\text{new}}s_{\text{new}} = m_{\text{new}}$. Thus,

$$p_{\text{new}} S_{\text{post}} = p_{\text{new}}(s_f + s_C) + p_{\text{new}}s_{\text{new}} = v_{\text{pre}} + m_{\text{new}} = v_{\text{post}},$$

so Eq. (vsp$_{\text{post}}$) holds. We can write this as

$$p_{\text{new}} = \frac{v_{\text{post}}}{S_{\text{post}}},$$

that is, the share price at which the transaction is conducted is equal to the value per share after the transaction.

3.3.3 Cap Table View (Alternate)

We can invert this reasoning, to show the following: under the assumptions that the values per share before and after the equity round are identical, that Eq. (vm$_{\text{pre, post}}$) captures the relation between pre- and post-money valuations, and that (msp$_{\text{new}}$) describes the number of shares issued to the new investor, Eq. (vsCp$_{\text{pre}}$) captures the share price at which the equity round should be conducted.

Treating v_C as the value of the shares s_C received in exchange for C, at price p_{new}, the number s_C satisfies

$$s_C = \frac{v_C}{p_{new}} = \frac{v_C}{v_{pre}} \cdot (s_f + s_C).$$

We can solve this equation for s_C to get

$$s_C = \frac{v_C s_f}{v_{pre} - v_C}.$$

Thus, we get the following further characterizations of the price per share:

$$p_{new} = \frac{v_{pre}}{s_f + s_C} = \frac{v_{pre}}{s_f + v_C s_f/(v_{pre} - v_C)} = \frac{v_{pre} - v_C}{s_f}$$

From the last of these, we see that the value of shares s_f at price p_{new} is $v_{pre} - v_C$, where the value of shares s_C at this price is v_C. Thus, this conversion price gives a coherent account of the impact of the conversion. The conclusion from this accounting view is that it is incorrect, pre-transaction, to view the valuation v_{pre} of the company as the value of the shares s_f. Instead, the valuation v_{pre} should be discounted by the value v_C to get the valuation of these shares. This discounted valuation is then preserved by the transaction, and the holder of C receives shares worth v_C.

3.3.4 Accounting Views: Conclusions

It is worth noting that on both Cap Table accounting views, the share price is

$$(v_{pre} - v_C)/s_f = (v_A - v_L - v_C)/s_f, \tag{3.3}$$

which is exactly the same as the share price calculated for the Liability view above. Although the two views disagree on the meaning of the term "pre-money valuation," they agree on the share price and yield the same number of shares issued. This suggests that share price is a more robust notion than "pre-money valuation." On both views, we obtained $v_{post} = v_A - v_L + m_{new}$, so "post-money valuation" is also a robust notion.

These considerations suggest that there are two distinct coherent interpretations of the terms "pre-money valuation" in an equity round in which shares are issued to a new investor for money m_{new}, and the instrument C is converted to shares. If we view C as a liability, then we can calculate a share price using Eq. (vsp$_{pre}$), but (vm$_{pre, post}$) is false and instead the correct relationship is given by $v_{post} = v_{pre} + v_C + m_{new}$. On the other hand, if we view C not as a liability, but as already represented on the cap table, then it is reasonable that (vm$_{pre, post}$) holds, but this view requires that rather than

Table 3.1 Accounting views related to equity round equations

View	Equations used	Equations derivable	Equations false
Liability	(msp_{new}), (vsp_{pre})	(vsp_{post}), $(vmC_{pre,\,post})$	$(vm_{pre,\,post})$, $(vsCp_{pre})$
Cap table	(msp_{new}), $(vsCp_{pre})$	(vsp_{post}), $(vm_{pre,\,post})$	(vsp_{pre}), $(vmC_{pre,\,post})$
Cap table (alternate)	(msp_{new}), (vsp_{post}), $(vm_{pre,\,post})$	$(vsCp_{pre})$	(vsp_{pre}), $(vmC_{pre,\,post})$

$$msp_{new} \qquad m_{new} = s_{new}p_{new}$$
$$vsp_{pre} \qquad v_{pre} = s_f\,p_{new}$$
$$vsp_{post} \qquad v_{post} = S_{post}p_{new}$$
$$vm_{pre,post} \qquad v_{post} = v_{pre} + m_{new}$$
$$vmC_{pre,post} \qquad v_{post} = v_{pre} + v_C + m_{new}$$
$$vsCp_{pre} \qquad p_{new} = \frac{v_{pre}}{s_f + s_C}$$
$$S_{post} = s_f + s_{safe} + s_{new}$$

Liability View $v_{pre} = v_A - v_L - v_C$; $S_{pre} = s_f$

Cap Table View $v_{pre} = v_A - v_L$; $S_{pre} = s_f + s_C$

Fig. 3.1 Equations related to conversion

using Eq. (vsp_{pre}), the pre-money valuation should be discounted when determining a share price. The discount can be represented either using Eqs. $(vsCp_{pre})$ or (3.3). Whichever view is adopted, the determination of valuation should bear in mind *what* is being valued. In the case of the Liability view, the valuation should include a deduction for the value of C, whereas in the Cap Table view, the valuation should ignore C or treat it as if already discharged. Table 3.1 summarizes these relationships. Figure 3.1 lists the related equations. We remark that using the variant definitions of v_{pre} and S_{pre} indicated for the Liability and Cap Table views, we can unify (vsp_{pre}) and $(vsCp_{pre})$ as

$$p_{new} = \frac{v_{pre}}{S_{pre}} \tag{3.4}$$

We should note that the accounting views discussed here are distinct from accounting as performed for standardized financial reporting and taxation. Accounting standards (such the Accounting Standards Codification (ASC)) may dictate whether a SAFE should be accounted for as a liability or within equity for such purposes. A recent study of accounting standards in relation to SAFEs [1] claims that it is common to classify a SAFE as a liability. Such standards may change [2]. The IRS treatment of SAFEs is said to be ambiguous [3, 4]. The post-money SAFEs explicitly commit the parties to treat the SAFE as stock for income tax purposes. Clause (g) of the Miscellaneous section [5] states:

> The parties acknowledge and agree that for United States federal and state income tax purposes this Safe is, and at all times has been, intended to be characterized as stock, and

more particularly as common stock for purposes of Sections 304, 305, 306, 354, 368, 1036 and 1202 of the Internal Revenue Code of 1986, as amended. Accordingly, the parties agree to treat this Safe consistent with the foregoing intent for all United States federal and state income tax purposes (including, without limitation, on their respective tax returns or other informational statements).

The point of our discussion is only to provide a framework in which to understand the different methods being used to convert SAFEs. The effect of accounting standards on the suitability of any accounting view for the purpose of conversion is unclear.

In the following chapter, we apply the general model developed in the present chapter to the specific case of the Pre-Money SAFE with Cap Only. In practice, a number of approaches are used in the operation of convertible instruments. We relate these to the two accounting views of this section.

References

1. Grant Thornton LLP (2024) Issuers' accounting for SAFEs. https://www.grantthornton.com/insights/newsletters/audit/2024/viewpoint/issuers-accounting-for-safe. Accessed June 2024
2. Financial Accounting Standards Board (2020) Accounting Standards Update 2020-06, Debt-Debt with Conversion and Other Options (Subtopic 470-20) and Derivatives and Hedging-Contracts in Entity's Own Equity (Subtopic 815-40): Accounting for convertible instruments and contracts in an entity's own equity. http://fasb.org. Accessed Dec 2024
3. Bell JR, Dorsey RW, Hamm J (2016) The simple agreement for future equity: advantages, disadvantages and tax treatment. Taxes: Tax Mag 94
4. Dolson SW (2024) Guide to the federal income tax treatment of SAFEs. https://frostbrowntodd.com/guide-to-the-federal-income-tax-treatment-of-safes/. Accessed Oct 2024
5. Y Combinator (2018) Safe: valuation cap, no discount. https://web.archive.org/web/20190626002912/https://www.ycombinator.com/docs/Postmoney%20Safe%20-%20Valuation%20Cap%20-%20v1.0.docx

Chapter 4
Methods for Converting a Pre-Money SAFE

In order to better understand the intentions underlying the definition of the SAFE, we analyze the impact that the definition of the SAFE has on the value of shares received in an equity round.

As discussed in Chap. 3, we can take different views on the pre-money and post-money valuation of the company. We consider these different views in order to determine which yields a coherent understanding of the way the SAFE has been constructed.

It appears that in practice, the views that investors take on the question of the meaning of "pre-money valuation" and "post-money valuation" results in a variety of methods being used to calculate the shares issued in an equity round in which a convertible instrument converts.

In this chapter, we analyze various methods used in practice in converting a SAFE in an equity raising, and some theoretically derived methods, using the insights drawn from the previous chapter. The methods analyzed are the "standard" method, the Percent-Ownership method, the Discounted Valuation method, the Dollars Invested method, and the Two-Stage methods. We focus on conversion of a single Pre-money SAFE with Cap Only (i.e., no discount).

One of the issues that arises in this, and subsequent chapters, is the question of whether the SAFE investor has any rights to negotiate in the terms of an equity round. On the one hand, the Equity Financing clause states a specific rule for conversion of the SAFE, and the contract does not explicitly give rights to negotiate its interpretation. On the other, the fact that part (i) of this clause requires the SAFE investor to execute certain documents at the time of the equity round potentially gives them some leverage to disrupt the round by withholding their signature. In practice, at least in the context of investor communities such as Silicon Valley, there is a social incentive for the parties to reach a compromise agreement: failure to come to agreement—or the enforcement of unfair terms—on a particular deal may tarnish the reputation of the equity investor, with consequences for their participation in future deals. A SAFE, in practice, may provide lesser protection for investors who are foreign to

R. van der Meyden and M. J. Maher, *Simple Agreements for Future Equity (SAFE)*, Blockchain Technologies, https://doi.org/10.1007/978-981-96-3920-5_4

such an investor community.[1] Our analysis will therefore consider, at various points, the alternatives of the SAFE investor participating or not participating in negotiation of the equity round terms.

4.1 Formalizing Conversion for the SAFE with Cap Only

Throughout this book, we consider only scenarios where the company has no outstanding convertible securities (including options and warrants) except for the SAFEs under consideration. This allows us to avoid the problem of accounting for the possible conversion of these securities.

We begin by reformulating key clauses of the Pre-Money SAFE with Cap Only in algebraic terms. As a result of the above assumption, the Company Capitalization before any equity financing S_{pre}, as defined in Pre-Money SAFEs, consists entirely of the shares of the existing shareholders (who, for convenience, we call the founders), that is,

$$S_{\mathrm{pre}} = s_f \tag{4.1}$$

Suppose an investor has paid m_{safe} for a SAFE with valuation cap c. We have already, in the previous chapter, introduced notation for describing the equity round and the state of the company before and after conversion. Now we can identify the Safe Price p_{safe} and the number of shares s_{safe} due to the SAFE investor as a result of the equity financing.

First, the Safe Price definition states that

$$p_{\mathrm{safe}} = \frac{c}{S_{\mathrm{pre}}} \tag{4.2}$$

We have assumed $S_{\mathrm{pre}} = s_f$, so we get a Safe Price of

$$p_{\mathrm{safe}} = \frac{c}{s_f}.$$

We can think of this as the price the SAFE investor would have paid had they purchased shares in a standard equity raise in which the company had a pre-money valuation equal to the cap c.

Second, we can write the Equity Financing clause (see Sect. 2.2.1) as follows:

1. If $v_{\mathrm{pre}} \leq c$, then $s_{\mathrm{safe}} = m_{\mathrm{safe}}/p_{\mathrm{new}}$.
2. If $v_{\mathrm{pre}} > c$, then $s_{\mathrm{safe}} = m_{\mathrm{safe}}/p_{\mathrm{safe}}$.

---

[1] Thanks to Alex Cook, Freehills, for helpful discussion on the question of SAFE investor negotiation rights.

This seems a simple and unproblematic expression of the Equity Financing clause, but we will see that there are multiple ways it can be interpreted. The variant interpretations hinge on the way that we determine v_{pre} and p_{new}.

4.2 Standard Method

We begin with an analysis of a literal interpretation of the SAFE contract, with the terms of the contract interpreted exactly as they would be in a standard equity round. We assume that the founders and the new investors agree upon a pre-money valuation v_{pre} for the company, use this and equation (vsp$_{pre}$) to derive a price per share paid by the new investor, and use equation (msp$_{new}$) to determine the number of shares received by the new investor. Using the Equity Financing clause, we can calculate the shareholding of the SAFE investor.

The use of equations (vsp$_{pre}$) and (msp$_{new}$) appears to be the method assumed by Y Combinator. For example, Example 1 in the Safe Primer [1] states the following (the numbers are approximate)[2]:

- Investor has purchased a safe for $100,000. The Valuation Cap is $5,000,000.
- The company negotiates with investors to sell $1,000,000 worth of Series A Preferred Stock at a $10,000,000 pre-money valuation. The company's fully-diluted outstanding capital stock immediately prior to the financing, including a 1,000,000 share option pool to be adopted in connection with the financing, is 11,000,000 shares.

The company will issue and sell 1,100,110 shares of Series A Preferred at $0.909 per share to the new investors. The company will issue and sell 220,022 shares of Series A-1 Preferred to the safe holder, at $0.4545 per share.

We now derive the proportional shareholdings of the parties, and their *values*. To avoid the contradiction from Sect. 3.2, we need to make a choice between Eqs. (vm$_{pre, post}$) and (vsp$_{post}$). We consider both options.

It appears from Colla [2] that investors frequently have the expectation that in setting a term sheet[3] with pre-money valuation v_{pre} and new money m_{new}, they are purchasing a proportion $m_{new}/(v_{pre} + m_{new})$ of the company. For a standard equity round, this does follow, using (vm$_{pre, post}$). However, as already noted, the conversion of SAFE shares dilutes the new investor, so that they hold less than this amount. We will argue that, moreover, if one adopts Eq. (vm$_{pre, post}$), the SAFE contract has some further unexpected consequences, and that the rationale for its design is not completely clear. Our analysis will show that this mode of operation is more coherent with respect to Eqs. (vsp$_{post}$) and (vmC$_{pre, post}$), corresponding to the Liability accounting view.

---

[2] The example contains a share option pool, which we have omitted in our formulation—the reader may treat the option pool shares as counted within founder shares s_f in our formulation.

[3] A term sheet is a non-binding agreement that outlines the terms for an investment. Once a term sheet is agreed upon, a formal contract is drawn up based on those terms.

For the analysis, we consider each of the cases of the pre-money valuation v_{pre}. We begin by calculating the proportional shareholding for each of the parties. We express this as a fraction of amounts of money, since this is informative for understanding the value of each shareholding. Note that, from (vsp$_{pre}$) and (msp$_{new}$), we have $p_{new} = v_{pre}/s_f$ and $s_{new} = m_{new}s_f/v_{pre}$, as we would in a standard equity round.

Case $v_{pre} \leq c$: In this case, we have $s_{safe} = m_{safe}/p_{new} = m_{safe}s_f/v_{pre}$. Thus,

$$S_{post} = s_f + s_{safe} + s_{new}$$
$$= s_f + (m_{safe}s_f/v_{pre}) + (m_{new}s_f/v_{pre})$$
$$= s_f(v_{pre} + m_{safe} + m_{new})/v_{pre}$$

We get that the proportional shareholding o_i for $i = f, safe, new$, of the founders, SAFE investor and new investor, respectively, can be expressed in the forms

$$\frac{v_{pre}}{v_{pre} + m_{safe} + m_{new}}, \quad \frac{m_{safe}}{v_{pre} + m_{safe} + m_{new}}, \quad \frac{m_{new}}{v_{pre} + m_{safe} + m_{new}}$$

after multiplying numerator and denominator by v_{pre}/s_f. Interestingly, these are precisely the same proportions of the company that they would hold had they raised both the SAFE investor's and the new investor's money at valuation v_{pre} in a single standard equity round.

If we were to assume that, as in the case of a standard equity raise, the price per share is the same immediately after the raise as the price at which the newly issued shares were sold, then, from the SAFE investor's point of view, the value of the shares they receive is $s_{safe}p_{new} = m_{safe}$. On this reasoning, the SAFE investor would be guaranteed in this case that, at the time of the equity round, they have not lost money on their investment. This calculation appears to be the motivation for the definition of this case. The assumption that the pre-money and post-money share price are the same underlies the characterization of post-money valuation given by Eq. (vsp$_{post}$), which gives $v_{post} = p_{new}S_{post} = v_{pre} + m_{safe} + m_{new}$ by the above calculations. On this view, the founders', SAFE investor's, and new investor's shares are worth $v_{pre}, m_{safe}, m_{new}$, respectively. This is a satisfactory outcome for each of the parties. It is conservative for both the SAFE investor and the new investor, and the founders retain the value v_{pre} that they have created.

However, the actual situation is more complex for the other (apparently more common [2]) characterization of post-money valuation. According to (vm$_{pre, post}$), the post-money valuation is $v_{pre} + m_{new}$. This means that the value of the holding of the SAFE investor after the round is

$$o_{safe}(v_{pre} + m_{new}) = m_{safe} \cdot \frac{v_{pre} + m_{new}}{v_{pre} + m_{safe} + m_{new}} < m_{safe}.$$

Thus, in fact, the SAFE investor does not get quite their money back in this case. Similarly, the value of the new investor's shares on this view of the post-money valuation is

$$o_{\text{new}}(v_{\text{pre}} + m_{\text{new}}) = m_{\text{new}} \cdot \frac{v_{\text{pre}} + m_{\text{new}}}{v_{\text{pre}} + m_{\text{safe}} + m_{\text{new}}} < m_{\text{new}}.$$

This is in agreement with the remarks in Sect. 3.2 concerning dilution of the new investor, since o_{new} is less than the proportion $m_{\text{new}}/(v_{\text{pre}} + m_{\text{new}})$ that the new investor would hold after a standard equity round. On this understanding, the Standard method is conservative for neither the SAFE investor nor the new investor.

Case $v_{\text{pre}} > c$: In this case $p_{\text{safe}} = c/s_f < v_{\text{pre}}/s_f = p_{\text{new}}$, and the number of shares held by the founders, SAFE investor and new investor after the equity round are, respectively,

$$s_f, \quad \frac{m_{\text{safe}} s_f}{c}, \quad \frac{m_{\text{new}} s_f}{v_{\text{pre}}}.$$

This implies that the proportional holdings can be written in terms of monetary value in the form

$$\frac{v_{\text{pre}}}{v_{\text{pre}} + \frac{v_{\text{pre}} m_{\text{safe}}}{c} + m_{\text{new}}}, \quad \frac{\frac{m_{\text{safe}} v_{\text{pre}}}{c}}{v_{\text{pre}} + \frac{v_{\text{pre}} m_{\text{safe}}}{c} + m_{\text{new}}}, \quad \frac{m_{\text{new}}}{v_{\text{pre}} + \frac{v_{\text{pre}} m_{\text{safe}}}{c} + m_{\text{new}}}.$$

As already noted above, and also in this Case, the new investor may receive shares of lesser post-money value than they paid for, if one takes Eq. ($v m_{\text{pre, post}}$) to describe the post-money valuation. On this understanding, the value of the SAFE holder's shares after the raise is

$$s_{\text{safe}} \frac{v_{\text{pre}} + m_{\text{new}}}{s_f + s_{\text{safe}} + s_{\text{new}}} = \frac{m_{\text{safe}}}{p_{\text{safe}}} \cdot \frac{v_{\text{pre}} + m_{\text{new}}}{s_f + s_{\text{safe}} + s_{\text{new}}}$$

$$= m_{\text{safe}} \cdot \frac{v_{\text{pre}} + m_{\text{new}}}{s_f p_{\text{safe}} + s_{\text{safe}} p_{\text{safe}} + s_{\text{new}} p_{\text{safe}}}$$

$$= m_{\text{safe}} \cdot \frac{v_{\text{pre}} + m_{\text{new}}}{c + m_{\text{safe}} + (c m_{\text{new}}/v_{\text{pre}})}$$

where the last step uses, respectively, the definition of p_{safe} in the SAFE contract, the definition of s_{safe} in the second case of the Equity Financing clause of the SAFE contract, and Eq. ($v s p_{\text{pre}}$). This means the SAFE investor makes an (unrealised) profit on the investment m_{safe} provided $v_{\text{pre}} + m_{\text{new}} > c + m_{\text{safe}} + (c m_{\text{new}}/v_{\text{pre}})$. This is a quadratic constraint on v_{pre}. Taking into account that we are in a case where $v_{\text{pre}} \geq c$, this constraint can be shown[4] to hold just when

---

[4] The polynomial has zeros at

$$v_{\text{pre}} = \frac{c + m_{\text{safe}} - m_{\text{new}} \pm \sqrt{(m_{\text{new}} - c - m_{\text{safe}})^2 + 4 m_{\text{new}} c}}{2}. \tag{4.3}$$

Prima facie, there may be two regions where the constraint is satisfied, to the left and right of these roots, but we have the further constraint that $v_{\text{pre}} \geq c$. It is therefore convenient to work with a translation: write $v_{\text{pre}} = c + x$, where $x \geq 0$. The constraint then becomes

$$v_{\text{pre}} \geq \frac{c + m_{\text{safe}} - m_{\text{new}} + \sqrt{(m_{\text{new}} - c - m_{\text{safe}})^2 + 4m_{\text{new}}c}}{2}.$$

This condition does not seem to have a strong rationale, particularly as it depends on the amount m_{new} of new money, whereas one might have expected a condition that depends just on the terms of the SAFE contract. A sufficient condition for this that does make some sense is $v_{\text{pre}} \geq c + m_{\text{safe}}$. (In a scenario where the SAFE investor had made their investment of m_{safe} in a standard round at valuation c, this would say that the company has not lost value from its post-money valuation $c + m_{\text{safe}}$ by the time of the new equity round at valuation v_{pre}.)

However, if we take the post-money valuation to be given by Eq. (vsp$_{\text{post}}$), we have that

$$
\begin{aligned}
v_{\text{post}} &= p_{\text{new}} S_{\text{post}} \\
&= p_{\text{new}} s_f + p_{\text{new}} S_{\text{safe}} + p_{\text{new}} s_{\text{new}} \\
&= v_{\text{pre}} + (v_{\text{pre}} m_{\text{safe}}/c) + m_{\text{new}}
\end{aligned}
$$

and $v_{\text{post}} o_{\text{safe}} = (v_{\text{pre}} + (v_{\text{pre}} m_{\text{safe}}/c) + m_{\text{new}}) o_{\text{safe}} = m_{\text{safe}} (v_{\text{pre}}/c)$. Since $v_{\text{pre}} > c$ in this case, this is always a gain on the SAFE investor's original investment. Additionally, $v_{\text{post}} o_{\text{new}} = m_{\text{new}}$, and $v_{\text{post}} o_f = v_{\text{pre}}$. On this understanding, therefore, the Standard method is conservative for both the SAFE investor and the new investor.

In summary, in both cases of the SAFE, we see that literal interpretation of the SAFE using (vsp$_{\text{pre}}$) and (msp$_{\text{new}}$) produces conservative outcomes for the parties under the assumption that the post-money valuation is given by (vsp$_{\text{post}}$), but is problematic when the post-money valuation is given by (vm$_{\text{pre, post}}$). This conclusion is consistent with the conclusions of Sect. 3.3: use of (vsp$_{\text{pre}}$) corresponds to the Liability view of the SAFE, which implies that (vsp$_{\text{post}}$) and (vmC$_{\text{pre, post}}$) describe the post-money valuation. From the above calculations, we see that the implied valuation v_{safe} of the SAFE is

$$v_{\text{safe}} = \begin{cases} m_{\text{safe}} & \text{if } v_{\text{pre}} \leq c \\ m_{\text{safe}} \cdot \frac{v_{\text{pre}}}{c} & \text{if } v_{\text{pre}} > c \end{cases}$$

on the Liability view.

$$x + \frac{x \cdot m_{\text{new}}}{c + x} > m_{\text{safe}}.$$

Note that in the case $x = 0$ this amounts to the practical absurdity $0 > m_{\text{safe}}$, so cannot be a solution. The value $x = 0$ therefore sits in the region where the polynomial is negative, and there is in fact only a single region satisfying both constraints, that to the right of the rightmost root.

4.3 Percent-Ownership and Discounted Valuation Methods

Notwithstanding the analysis of the previous section, on which the SAFE is coherent with respect to the Liability view of accounting, it appears from Colla [2], that equity round investors are interpreting term sheets stating pre-money valuation v_{pre} and new money m_{new} with the expectation (valid for standard equity rounds) that this delivers them a proportional shareholding of $m_{\mathrm{new}}/(v_{\mathrm{pre}} + m_{\mathrm{new}})$. As shown above, for both cases of the SAFE, the actual proportional shareholding is less than this when the Standard method is applied.

One of the approaches used in response to this perceived dilution suffered by a new investor as a consequence of shares issued in conversion of a convertible security is called the *percent-ownership* method [2]. The idea is to construct the equity round so as to guarantee to the new investor an agreed-upon share of the company on completion of the round. Having agreed upon this share, the other variables are calculated so as to deliver this outcome. This approach resolves the inconsistency of Sect. 3.2 by dropping equation (vsp$_{\mathrm{pre}}$) as a way to calculate p_{new} given v_{pre}, and instead calculates the value of p_{new} using Eqs. (vsp$_{\mathrm{post}}$) and (vm$_{\mathrm{pre, post}}$). We describe the working of this method for a SAFE in Sect. 4.3.1, and show that it is essentially equivalent to a another method that works with a discounted pre-money valuation, in Sect. 4.3.2.

4.3.1 Percent-Ownership Method

In our scenario, the Percent-Ownership method works in two steps, as follows.

Step 1: We fix the post-money proportional ownership o_{new} of the new investor to the agreed-upon value. To deliver this outcome, we need

$$o_{\mathrm{new}} = \frac{m_{\mathrm{new}}}{v_{\mathrm{post}}} \qquad \text{(omv}_{\mathrm{post}})$$

in terms of monetary value, so using Eq. (vm$_{\mathrm{pre, post}}$), we get

$$o_{\mathrm{new}}(v_{\mathrm{pre}} + m_{\mathrm{new}}) = m_{\mathrm{new}}. \qquad (4.4)$$

We can then calculate the pre-money valuation as $v_{\mathrm{pre}} = m_{\mathrm{new}}(1 - o_{\mathrm{new}})/o_{\mathrm{new}}$. We remark that, having determined v_{pre}, the remainder of the calculation does not need to treat o_{new} as an input, but can rederive it using Eqs. (omv$_{\mathrm{post}}$) and (vm$_{\mathrm{pre, post}}$). Thus, an equivalent starting point for Step 2 would be the assumption that m_{new} and v_{pre} are given, and that we seek to determine a price per share that yields a percentage ownership o_{new} whose post-money value is m_{new}.

Step 2: We now determine the values of the remaining variables. An alternate way to express the post-money proportional ownership attained for the new investors is to formulate it in terms of number of shares:

$$o_{\text{new}} = \frac{s_{\text{new}}}{s_f + s_{\text{safe}} + s_{\text{new}}}. \qquad (os_{\text{new}})$$

Combining this with (omv_{post}) and $(vm_{\text{pre, post}})$, we get

$$\frac{s_{\text{new}}}{s_f + s_{\text{safe}} + s_{\text{new}}} = \frac{m_{\text{new}}}{v_{\text{pre}} + m_{\text{new}}}. \qquad (4.5)$$

From this, we derive

$$s_{\text{new}} = \frac{m_{\text{new}}}{v_{\text{pre}}}(s_f + s_{\text{safe}}). \qquad (4.6)$$

It is worth noting that this formula effectively states that the new shares are issued at a price

$$p_{\text{new}} = v_{\text{pre}}/(s_f + s_{\text{safe}})$$

which suggests that the SAFE holder's shares s_{safe} have already been issued prior to the equity round at valuation v_{pre}. This corresponds precisely to the Cap Table accounting view, with the SAFE represented on the cap table as an obligation to issue shares. Equation (vsp_{pre}) fails on this method, but it is consistent to use $(vm_{\text{pre, post}})$ to determine the relationship between pre- and post-money valuations: since the SAFE is not represented as a liability, all that changes with respect to valuation is to add m_{new} to the assets of the company. (While equation (vsp_{pre}) does not express the relationship between p_{new} and v_{pre}, we note that we could use this equation to determine a valuation v_{pre}^* that does satisfy this equation, namely,

$$v_{\text{pre}}^* = p_{\text{new}}s_f = \frac{v_{\text{pre}}s_f}{s_f + s_{\text{safe}}} = v_{\text{pre}} - \frac{v_{\text{pre}}s_{\text{safe}}}{s_f + s_{\text{safe}}} = v_{\text{pre}} - p_{\text{new}}s_{\text{safe}}.$$

The term $v_S = p_{\text{new}}s_{\text{safe}}$ can be understood as the valuation of the SAFE contract, so we have $v_{\text{post}} = v_{\text{pre}} + m_{\text{new}} = v_{\text{pre}}^* + v_S + m_{\text{new}}$. Thus, the relationship between v_{pre}^* and v_{post} is precisely what we would expect on the Liability view accounting model, where the SAFE is represented as a liability that converts to shares in the course of the transaction at price p_{new}. Note moreover that

$$v_{\text{post}} = v_{\text{pre}}^* + v_S + m_{\text{new}} = p_{\text{new}}(s_f + s_{\text{safe}} + s_{\text{new}}) = p_{\text{new}}S_{\text{post}}$$

by Eq. (msp_{new}) and the definitions above, so that Eq. (vsp_{post}) is derivable.)

However, since we have not yet determined s_{safe}, which, according to the SAFE, depends on p_{new}, we cannot yet calculate p_{new}. Instead, we apply the definitions from the SAFE contract as additional constraints that together with Eq. (4.6) determine a value for p_{new}. The calculation is different in the two cases of the SAFE.

Case $v_{\text{pre}} \le c$: In this case the SAFE requires that $s_{\text{safe}} = m_{\text{safe}}/p_{\text{new}}$. Since $s_{\text{new}} = m_{\text{new}}/p_{\text{new}}$, by $(\text{msp}_{\text{new}})$, we get from (4.6) that

$$\frac{m_{\text{new}}}{p_{\text{new}}} = \frac{m_{\text{new}}}{v_{\text{pre}}}\left(s_f + \frac{m_{\text{safe}}}{p_{\text{new}}}\right).$$

We solve this for p_{new} to get

$$p_{\text{new}} = \frac{v_{\text{pre}} - m_{\text{safe}}}{s_f}.$$

(Obviously, this price makes sense only if $m_{\text{safe}} < v_{\text{pre}}$, so this case actually can occur only under the stronger condition $m_{\text{safe}} < v_{\text{pre}} \le c$.) This gives $s_{\text{safe}} = m_{\text{safe}}s_f/(v_{\text{pre}} - m_{\text{safe}})$. Since we have guaranteed that the post-money value per share is p_{new}, the SAFE holder's shares are worth $p_{\text{new}}s_{\text{safe}} = m_{\text{safe}}$, so the SAFE holder is guaranteed not to lose money. The proportional holding for the SAFE holder, expressed in terms of money, is

$$o_{\text{safe}} = \frac{m_{\text{safe}}}{v_{\text{pre}} + m_{\text{new}}}.$$

We already have o_{new} as given by Eq. (4.4), so the resulting founder share is

$$o_f = 1 - o_{\text{safe}} - o_{\text{new}}$$
$$= \frac{v_{\text{pre}} - m_{\text{safe}}}{v_{\text{pre}} + m_{\text{new}}}$$

According to the formulation $(\text{vm}_{\text{pre, post}})$ of post-money value that we have used in this derivation, the post-money value of the founder, SAFE investor and new investor's shares are, respectively, $v_{\text{pre}} - m_{\text{safe}}$, m_{safe} and m_{new}. This case is therefore conservative for the SAFE investor and the new investor. Note also that the sum of these post-money amounts is $v_{\text{pre}} + m_{\text{new}}$, so they satisfy $(\text{vm}_{\text{pre, post}})$.

Case $v_{\text{pre}} > c$: Here we have $s_{\text{safe}} = m_{\text{safe}}/p_{\text{safe}} = m_{\text{safe}}s_f/c$, and we get from (4.6) that

$$s_{\text{new}} = \frac{m_{\text{new}}}{v_{\text{pre}}}\left(s_f + \frac{m_{\text{safe}}s_f}{c}\right) = m_{\text{new}}\frac{s_f}{v_{\text{pre}}}\frac{c + m_{\text{safe}}}{c}.$$

Thus, the new investor pays a share price

$$p_{\text{new}} = \frac{v_{\text{pre}}}{s_f}\frac{c}{c + m_{\text{safe}}}$$

rather than the expected v_{pre}/s_f. The SAFE holder's share of the company in this case is

$$o_{\text{safe}} = \frac{\frac{m_{\text{safe}} s_f}{c}}{s_f + \frac{m_{\text{safe}} s_f}{c} + m_{\text{new}} \frac{s_f}{v_{\text{pre}}} \frac{c + m_{\text{safe}}}{c}}$$

$$= \frac{m_{\text{safe}}}{c + m_{\text{safe}}} \cdot \frac{v_{\text{pre}}}{v_{\text{pre}} + m_{\text{new}}}$$

and the founder share is the balance

$$o_f = \frac{c}{c + m_{\text{safe}}} \cdot \frac{v_{\text{pre}}}{v_{\text{pre}} + m_{\text{new}}}.$$

Comparing this outcome with (3.2), we see that this case operates in the same way as two equity rounds, in the first of which the SAFE holder purchases shares for money m_{safe} at pre-money valuation c, and in the second of which the new investor purchases shares at pre-money valuation v_{pre}. We may note that since the first of these rounds would have post-money valuation $c + m_{\text{safe}}$, the second of these rounds is a "down-round" in case $v_{\text{pre}} \in (c, c + m_{\text{safe}})$. Moreover, for valuations in this interval, we see that the value of the SAFE investors shares is $(v_{\text{pre}}/c + m_{\text{safe}}) \cdot m_{\text{safe}} < m_{\text{safe}}$, so this approach is not conservative for the SAFE investor. For the remaining case, where $v_{\text{pre}} \geq c + m_{\text{safe}}$, the approach is conservative.

4.3.2 Discounted Valuation Method

It is worth remarking that there is an alternate viewpoint that leads to essentially the same conclusions as just derived. One way that the new investor might respond to their perceived dilution is to "game" the situation, by conducting the equity round at a "discounted" pre-money valuation that differs from their actual valuation of the company, and which is designed to deliver an equity holding that better accords with this actual valuation, once the consequences of the SAFE are taking into account. (We emphasize that the discount we refer to here is for the *new* investor, and different from the discount on the conversion price provided to the SAFE investor in some versions of the SAFE—we are still concerned with the SAFE with Cap Only.)

Suppose that the new investor's actual pre-money valuation of the company (ignoring the existence of a SAFE contract) is v_{pre}. We write v_{pre}^- to denote the pre-money valuation at which the equity round is actually conducted—this variable is taken as an unknown, with its value to be derived. As in a standard equity round, we apply equation (vsp$_{\text{pre}}$) to determine a price per share from v_{pre}^-, and then determine the number of shares issued to the new investor using (msp$_{\text{new}}$) based on this price per share. For purposes of selecting the applicable case of the SAFE contract, we use v_{pre}^- rather than v_{pre}. We first use v_{pre}^- to calculate the number of shares issued to the SAFE investor and the new investor, but then calculate the value of those shares using v_{pre} and characterization (vm$_{\text{pre, post}}$) of the post-money valuation. Assuming that the post-money valuation of the new investor's shares should equal m_{new} gives a constraint on v_{pre}^- that we then use to derive a value for v_{pre}^-.

Since the equity round is conducted at valuation v_{pre}^-, based on existing shares s_f, we have $p_{\text{new}} s_f = v_{\text{pre}}^-$ by Eq. (vsp_{pre}) and $p_{\text{new}} s_{\text{new}} = m_{\text{new}}$ by equation (msp_{new}). After the equity round, using the actual valuation v_{pre} and Eq. ($\text{vm}_{\text{pre, post}}$), the post-money valuation is $v_{\text{pre}} + m_{\text{new}}$. Thus, the post-money value per share is

$$\frac{v_{\text{pre}} + m_{\text{new}}}{s_f + s_{\text{safe}} + s_{\text{new}}}.$$

In order for the new investor's shares to still be valued at m_{new} in the post-money state, this post-money value per share must equal p_{new}. Hence,

$$\begin{aligned} v_{\text{pre}} + m_{\text{new}} &= p_{\text{new}}(s_f + s_{\text{safe}} + s_{\text{new}}) \\ &= p_{\text{new}} s_f + p_{\text{new}} s_{\text{safe}} + p_{\text{new}} s_{\text{new}} \\ &= v_{\text{pre}}^- + p_{\text{new}} s_{\text{safe}} + m_{\text{new}} \end{aligned}$$

where we use Eq. (vsp_{pre}) for the first term and Eq. (msp_{new}) for the third term on the right. Thus, we conclude that $v_{\text{pre}}^- = v_{\text{pre}} - s_{\text{safe}} p_{\text{new}}$ is a valuation that guarantees that the post-money valuation of the new investor's shares will be m_{new}. Note also that the left and right hand sides of (first line of) the above equation are just the two formulations of v_{post} of equations ($\text{vm}_{\text{pre, post}}$) and ($\text{vsp}_{\text{post}}$), so this approach will guarantee that these formulations are equivalent (but with respect to different pre-money valuations). Also note that, by equation (msp_{new}), $p_{\text{new}} = m_{\text{new}}/s_{\text{new}}$ and so

$$\frac{v_{\text{pre}} + m_{\text{new}}}{s_f + s_{\text{safe}} + s_{\text{new}}} = \frac{m_{\text{new}}}{s_{\text{new}}} \tag{4.7}$$

which is equivalent to Eq. (4.5). Thus we have reached the same constraint as the Percent-Ownership method from a different perspective.

We do not yet have actual values for v_{pre}^-, s_{safe} or p_{new}, but we can obtain these by noting that the details of the SAFE contract provide additional constraints, and use these to derive the values of these variables. We obtain the same formulas for share price and proportional shareholding as we did with the Percent-Ownership method, and discounted valuations $v_{\text{pre}}^- = v_{\text{pre}} - m_{\text{safe}}$ when $v_{\text{pre}}^- \leq c$ and

$$v_{\text{pre}}^- = v_{\text{pre}} \cdot \frac{c}{c + m_{\text{safe}}}$$

otherwise.

4.3.3 Comparison of Percent-Ownership and Discounted Valuation, and Limitations

The Discounted Valuation method's perspective, of an equity round at the lower valuation v_{pre}^{-}, does result in one difference, however. Since the SAFE contract is invoked at valuation v_{pre}^{-} rather than v_{pre}, the conditions under which the cases of the SAFE contract apply are different. Instead of the conditions $v_{pre} \leq c$ and $v_{pre} > c$ for the cases given above, we have conditions $v_{pre}^{-} \leq c$ and $v_{pre}^{-} > c$, or equivalently, $v_{pre} - m_{safe} \leq c$ and $v_{pre} - m_{safe} > c$.

It is possible for these case conditions to be different in the two approaches of this section because, while the SAFE contract refers to both a pre-money valuation and a share price, it does not explicitly state how these should be related. One could take the view that it is an assumption of the contract that these variables are related to outstanding shares by Eq. (vsp$_{pre}$). The Percent-Ownership approach, uses a price which, by Eq. (vsp$_{pre}$), corresponds to the same discounted valuation $v_{pre} - s_{safe}p_{new}$ as just derived. Thus, there is a reasonable argument that the case conditions used in that approach should be the same as those used in the Discounted Valuation method, making the two approaches completely equivalent.

If either $v_{pre} \leq c$ (implying $v_{pre} - m_{safe} \leq c$) or $v_{pre} - m_{safe} > c$ (implying $v_{pre} > c$), then the two approaches agree with respect to selection of the SAFE case. However, when $c < v_{pre} \leq c + m_{safe}$, the Percent-Ownership and Discounted Valuation approaches yield for the SAFE investor proportional shareholdings of

$$\frac{m_{safe}}{c + m_{safe}} \cdot \frac{v_{pre}}{v_{pre} + m_{new}} = \frac{m_{safe}}{v_{pre} + m_{new}} \cdot \frac{v_{pre}}{c + m_{safe}} \text{ and } \frac{m_{safe}}{v_{pre} + m_{new}}$$

respectively. Under the condition $v_{pre} < c + m_{safe}$, the latter is the larger holding, so preferable to the SAFE investor. As already remarked above, the Percent-Ownership approach is not conservative for the SAFE investor when $v_{pre} \in (c, c + m_{safe})$. There is therefore a risk that, when $c < v_{pre} < c + m_{safe}$, the SAFE holder may legally challenge the Percent Ownership on the grounds that it does not satisfy equation (vsp$_{pre}$), and that this should be considered part of the implicit legal context of the SAFE contract, even if it is not explicitly stated. From the point of view of legal certainty, it may therefore be preferable to use the Discounted Valuation approach. (It is a weakness of the SAFE contract that it allows uncertainty on this point.)

A further reason to prefer Discounted Valuation approach is that, intuitively, one expects that the market should be able to set any price for the company's shares, but the Percent-Ownership approach does not support this. On that approach, in case $v_{pre} \leq c$, and $v_{pre} > m_{safe}$, we have that the price $p_{new} = (v_{pre} - m_{safe})/s_f$ takes on any value in the interval $(0, (c - m_{safe})/s_f]$. In case $v_{pre} > c$, the price $p_{new} = v_{pre}c/s_f(c + m_{safe})$ takes on any value in the interval $(c^2/s_f(c + m_{safe}), \infty)$. This leaves prices in the interval

$$((c - m_{safe})/s_f, \ c^2/s_f(c + m_{safe})]$$

unattainable.[5] On the other hand, under the Discounted Valuation approach, where the conditions for the two cases are $v_{pre} - m_{safe} \leq c$ and $v_{pre} - m_{safe} > c$, we obtain prices in the intervals $(0, \ c/s_f]$ and $(c/s_f, \ \infty)$, so all non-zero prices are attainable.

From an accounting perspective, since, in both the Percent-Ownership and Discounted Valuation methods, we have used Eq. (vm$_{pre, post}$), we should think of v_{pre} as a valuation of the company that does not include the SAFE as a liability. Consistent with this, as already noted, Eq. (4.6) effectively states that the price for the new investor is determined assuming that the SAFE shares are already represented on the cap table. However, the post-money valuation can also be written as $v_{pre}^- + p_{new}s_{safe} + m_{new}$. Taking $v_{safe} = p_{new}s_{safe}$ to be the valuation of the SAFE, we see that v_{pre}^- can be thought of as a valuation of the company on the assumption that the SAFE is accounted for as a liability, which is discharged in the course of the equity round, so that the correct equation relating this interpretation of the pre-money valuation to the post-money valuation is $v_{post} = v_{pre}^- + v_{safe} + m_{new}$. The analysis of both the Percent-Ownership and Discounted Valuation methods is therefore consistent with both accounting methods: we start with v_{pre} interpreted using the Cap Table view, and convert this to a different valuation with respect to the Liability View (v_{pre}^* for the Percent-Ownership method and v_{pre}^- for the Discounted Valuation method).

Using the conditions from the second (Discounted Valuation) method, we obtain a valuation of the SAFE given by

$$v_{safe} = \begin{cases} m_{safe} & \text{if } v_{pre} - m_{safe} \leq c \\ m_{safe} \cdot \frac{v_{pre}}{c + m_{safe}} & \text{if } v_{pre} - m_{safe} > c \end{cases}$$
$$= \begin{cases} m_{safe} & \text{if } v_{pre}^- \leq c \\ m_{safe} \cdot \frac{v_{pre}^-}{c} & \text{if } v_{pre}^- > c \end{cases}$$

Note that the second formulation, with case conditions expressed in terms of v_{pre}^-, corresponds directly to the valuation obtained from the Standard method with respect to the Liability view in Sect. 4.2.

4.4 Dollars Invested Method

Another approach that has been used in practice to deal with the perceived dilution of the new investor resulting from convertible bond conversion is the *Dollars Invested* method. According to Colla [2], this method is used to allay founder objections to the fact that the percent-ownership method results in them being diluted more than they expected when setting a term sheet with the new investors. As a compromise, the agreed pre-money valuation is used, but the post-money valuation is set to be equal

---

[5] Note that the statement that the left value is less than the right value is equivalent to $c^2 - m_{safe}^2 < c^2$, which holds provided $m_{safe} > 0$—we can safely assume that this is the case, else the SAFE is inconsequential.

to the pre-money valuation plus the new-money, plus the principal (and interest, if any) of the convertible bond. Based on this post-money valuation, a share price is calculated that results in the agreed ownership percentage for the new investor. That is, instead of Eq. (vm$_{pre, post}$), we use the following:

$$v_{post} = v_{pre} + m_{safe} + m_{new}. \qquad \text{(vmi}_{pre,post})$$

For the Pre-money SAFE with Cap Only, this might work as follows. In order to determine the price paid by the new investor, we equate the revised post-money valuation from equation (vmi$_{pre,post}$) with the product of the price and the number of post-money shares, as per Eq. (vsp$_{post}$), giving

$$v_{pre} + m_{safe} + m_{new} = p_{new}(s_f + s_{safe} + s_{new}) = p_{new}s_f + p_{new}s_{safe} + m_{new}. \quad (4.8)$$

where we use Eq. (msp$_{new}$) to get the rightmost term. There are two cases for s_{safe}, depending on v_{pre}, the pre-money valuation.[6]

Case $v_{pre} \leq c$: In this case $s_{safe}p_{new} = m_{safe}$, so Eq. (4.8) becomes $v_{pre} = p_{new}s_f$, and we deduce $p_{new} = v_{pre}/s_f$. (It is interesting to note that here we have *derived* rather than assumed equation (vsp$_{pre}$). However, it holds only in this case, and fails in the other case.) Dividing both sides of Eq. (4.8) by the left-hand side, we obtain the proportional shareholdings of the founders, SAFE investor and new investor, respectively, as

$$\frac{v_{pre}}{v_{pre} + m_{safe} + m_{new}}, \qquad \frac{m_{safe}}{v_{pre} + m_{safe} + m_{new}}, \qquad \frac{m_{new}}{v_{pre} + m_{safe} + m_{new}}.$$

This shareholding is the same as would have been produced had the SAFE holder and the new investor used their money to participate in a single equity round together, at pre-money valuation v_{pre}.

Case $v_{pre} > c$: In this case, we have $s_{safe} = m_{safe}s_f/c$. Thus Eq. (4.8) becomes

$$v_{pre} + m_{safe} = p_{new}s_f + p_{new}\frac{m_{safe}s_f}{c}$$

and we obtain

$$p_{new} = \frac{c}{c + m_{safe}}\frac{v_{pre} + m_{safe}}{s_f}.$$

---

[6] We remark that a question arises as to whether we should use the value v_{pre} agreed in the term sheet, or $v_{pre} + m_{safe}$, i.e., the post-money valuation being used in the calculation, minus the new money. The fact that we have $p_{new} = v_{pre}/s_f$, exactly as in a standard equity round suggests the former, but the SAFE holder may possibly have a legal case to argue for the latter. The latter approach would yield the same formulas for the two cases, but shift the intervals over which they hold, so we do not treat this as a separate method.

Dividing both sides of Eq. (4.8) by the left-hand side, we obtain the proportional shareholdings of the founders, SAFE investor and new investor, respectively, as

$$\frac{c}{c + m_{\text{safe}}} \cdot \frac{v_{\text{pre}} + m_{\text{safe}}}{v_{\text{pre}} + m_{\text{safe}} + m_{\text{new}}}, \quad \frac{m_{\text{safe}}}{c + m_{\text{safe}}} \cdot \frac{v_{\text{pre}} + m_{\text{safe}}}{v_{\text{pre}} + m_{\text{safe}} + m_{\text{new}}}, \quad \frac{m_{\text{new}}}{v_{\text{pre}} + m_{\text{safe}} + m_{\text{new}}}.$$

Comparing with (3.2), this is exactly the same outcome as would have been obtained in two equity rounds, in the first of which the SAFE holder purchased shares at valuation c, and in the second of which the new investor purchased shares at valuation $v_{\text{pre}} + m_{\text{safe}}$.

The coherence of this method depends on the view one takes of the "actual" post-money valuation, and the assumptions on which the pre-money valuation v_{pre} was determined. If v_{pre} was calculated on the assumption that the SAFE is not a liability, then the actual post-money valuation is $v_{\text{pre}} + m_{\text{new}}$, and the new investor suffers a dilution, as they do in the standard method, though to a smaller extent.

From an accounting perspective, perhaps the best rationale that can be given for this method is by viewing the transaction as a two-stage process. Suppose the new investor starts by valuing the company on the basis that the SAFE is represented as a liability with value m_{safe}, determining v_{pre} as the valuation on this assumption. In the first stage, the liability is transferred to an obligation on the cap table, increasing the valuation to $v_{\text{pre}} + m_{\text{safe}}$. The price per share p_{new} is then calculated on the assumption that the existing shares are s_f, plus the obligation to issue shares s_{safe}. This viewpoint yields the equation $v_{\text{pre}} + m_{\text{safe}} = p_{\text{new}}(s_f + s_{\text{safe}})$, which is equivalent to Eq. (4.8) used above, and supports the assumption that the post-money valuation is $v_{\text{pre}} + m_{\text{safe}} + m_{\text{new}}$.

Limitation: However, this method disagrees with the discounted valuation method on the proper valuation of the SAFE. The discounted valuation method takes this to be equal to $s_{\text{safe}} p_{\text{new}}$, which is equal to m_{safe} in case $v_{\text{pre}} \leq c + m_{\text{safe}}$, but which equals the larger amount

$$m_{\text{safe}} \cdot \frac{v_{\text{pre}}}{c + m_{\text{safe}}}$$

when $v_{\text{pre}} > c + m_{\text{safe}}$. Thus, the Dollars Invested method arguably undervalues the SAFE.

4.5 Two-Stage (Zero-Money Round) Approach

Some investors may respond to the perceived dilution they would suffer from the Standard method, with post-money defined by Eq. $(vm_{\text{pre, post}})$, by simply refusing to invest until the SAFE is extinguished by some means. One way this could be achieved is to structure the transaction in two stages. In the first stage, an artificial equity round is executed in which a minimal amount of new money is invested, simply in order to force conversion of the SAFE into shares. In the second stage, the new investor purchases shares at valuation v_{pre} for money m_{new}, exactly as in a standard equity round. This purchase is free from dilution by the issuance of SAFE shares, so, by the argument of Sect. 3.1, the new investor receives a share $m_{new}/(v_{pre} + m_{new})$, as expected.

Although it may not be legally feasible, in the limit, the "minimal amount" of new money invested in the first round is zero, so we assume for purposes of the analysis that no new money is invested in the first stage round. Consequently, we refer to this approach as the *Zero-Money Round* approach.

The SAFE investor may have legal grounds to challenge such an approach as being artificially constructed and prohibited by the terms of the SAFE. The text of the SAFE defines "Equity Financing" as

> a bona fide transaction or series of transactions with the principal purpose of raising capital,
> ...

and the SAFE investor might claim that the two stages constitute a single equity financing, that the first round does not have the principal purpose of raising capital, or that the transactions are not *bona fide*.[7] Nevertheless, this approach provides an interesting comparison with those already discussed.

We assume that the new investor is prepared to invest in stage 2 at valuation v_{pre}, and that this fact is known when constructing stage 1. In order to focus on the value of the SAFE investor's original investment, we will assume here that the SAFE investor does not take up, at additional cost, any *pro rata* rights that they might have in this second stage. The consequences of this two-stage approach depend on how the first stage is negotiated. There are two possibilities for the first stage, considered in the following subsections.

4.5.1 Zero-Money Round (Discounted Valuation Method)

The new investor could allow the conversion terms to be agreed between the company and the SAFE investor. Whereas, usually, the SAFE investor does not have rights to set the terms of the equity round, the negotiation between the company and SAFE investor in this situation gives some power to the SAFE investor to set terms for the conversion. They are likely to insist on avoidance of any loss, where possible. One way to achieve this is to use the Discounted Valuation method. Using the results of Sect. 4.3 with $m_{new} = 0$, we get the proportional shareholdings for the founders and SAFE investor after stage 1 of

$$\frac{v_{pre} - m_{safe}}{v_{pre}}, \quad \frac{m_{safe}}{v_{pre}}$$

---

[7] This definition also poses challenges to implementing SAFE contracts on a blockchain as a smart contract, as we discuss in Sect.16.3.

respectively, in case $m_{safe} < v_{pre} \leq c + m_{safe}$, and

$$\frac{c}{c + m_{safe}}, \quad \frac{m_{safe}}{c + m_{safe}}$$

in case $v_{pre} > c + m_{safe}$. In either case, with the company valued at v_{pre}, the value of the SAFE investor's shares is at least m_{safe}, so this step is conservative for the SAFE investor.

In stage 2, these holdings are diluted by the new money in the same way that the founders are diluted in a standard equity round as discussed in Sect. 3.1. Thus, the proportional holdings become, for the founders, SAFE investor and new investor, respectively:

1. In case $m_{safe} < v_{pre} \leq c + m_{safe}$:

$$\frac{v_{pre} - m_{safe}}{v_{pre} + m_{new}}, \quad \frac{m_{safe}}{v_{pre} + m_{new}}, \quad \frac{m_{new}}{v_{pre} + m_{new}}.$$

2. In case $v_{pre} > c + m_{safe}$:

$$\frac{c}{c + m_{safe}} \cdot \frac{v_{pre}}{v_{pre} + m_{new}}, \quad \frac{m_{safe}}{c + m_{safe}} \cdot \frac{v_{pre}}{v_{pre} + m_{new}}, \quad \frac{m_{new}}{v_{pre} + m_{new}}.$$

Since we have used the Discounted Valuation method in the first round, the same remarks concerning accounting model appropriate to the method apply as in Sect. 4.3. The first round is consistent with both the treatment of the SAFE as represented on the cap table, or (with respect to discounted valuation $v_{pre} - v_{safe}$) with the treatment of the SAFE as a liability valued at v_{safe}. The second round does not involve convertible instruments. Thus, at the post-money valuation $v_{pre} + m_{new}$, this method is always conservative for both the SAFE investor and the new investor.

In fact, we may note that the proportions we have calculated for the Zero-Money Round (Discounted Valuation) method, are exactly the same as for the Discounted Valuation method itself. That is, structuring the investment into two rounds, when using the Discounted Valuation method to discharge the SAFE, has the same effect as using this method for a single round in which the new investors money is included. We will therefore not consider this variant further.

4.5.2 Zero-Money Round (Standard Method)

The Discounted Valuation method results in a lower proportional shareholding for the founders. In the interests of maximizing founder incentives, the new investor may prefer to ensure that the founders retain a larger share. Clearly, the founders would prefer a better deal, and, arguably, the SAFE investor does not have bargaining rights in the construction of the equity round in which the SAFE converts. The first stage could therefore be constructed using the Standard method from Sect. 4.2, which delivers the founders a higher proportional shareholding.

In principle, there is the option for a more general type of collusion between the new investor and the founders, in which they conduct either of the two rounds at a valuation other than the new investor's actual valuation v_{pre}, and split the new investor's money across the two rounds, in such a way as to yield a desired result. We analyze this possibility in Sect. 6.4, but assume here that both rounds are conducted at valuation v_{pre}. In general, the founders get a greater benefit from a higher valuation (from the perspective of the share of the company that they retain, the implied value of their share, and their negotiation position with respect to future rounds), so they may be averse to rounds at valuations less than v_{pre}. Moreover, it could be argued that the new investor, to the extent that they are involved in the first stage, does not have an interest in this being conducted at a valuation greater than v_{pre}, since that may create a precedent for the founders to argue that the second round should also be conducted at such a higher valuation. We therefore conduct the analysis assuming that both the first round and second round are conducted at valuation v_{pre}.

Using the approach and results of Sect. 4.2 with $m_{new} = 0$, conducting the first round at valuation v_{pre} leads to the following consequences after stage 1.

1. If $v_{pre} \leq c$, then the founder and SAFE investor proportional shareholdings are

$$\frac{v_{pre}}{v_{pre} + m_{safe}}, \quad \frac{m_{safe}}{v_{pre} + m_{safe}}$$

   respectively.

2. If $v_{pre} > c$, then the founder and SAFE holder proportional shareholdings are

$$\frac{c}{c + m_{safe}}, \quad \frac{m_{safe}}{c + m_{safe}}$$

   respectively. The SAFE investor receives a fixed share of the company in this case, equivalent to having purchased their shares at valuation c in a standard equity round. They have made an unrealised gain in case this share of the valuation v_{pre} is greater than m_{safe}, which holds if $v_{pre} > c + m_{safe}$.

In stage 2, these holdings are diluted by the new money in the same way that the founders are diluted in a standard equity round as discussed in Sect. 3.1. Thus, the proportional holdings become, for the founders, SAFE investor and new investor, respectively:

1. In case $v_{pre} < c$:

$$\frac{v_{pre}}{v_{pre} + m_{safe}} \cdot \frac{v_{pre}}{v_{pre} + m_{new}}, \quad \frac{m_{safe}}{v_{pre} + m_{safe}} \cdot \frac{v_{pre}}{v_{pre} + m_{new}}, \quad \frac{m_{new}}{v_{pre} + m_{new}}.$$

2. In case $v_{\mathrm{pre}} \geq c$:

$$\frac{c}{c+m_{\mathrm{safe}}} \cdot \frac{v_{\mathrm{pre}}}{v_{\mathrm{pre}}+m_{\mathrm{new}}}, \quad \frac{m_{\mathrm{safe}}}{c+m_{\mathrm{safe}}} \cdot \frac{v_{\mathrm{pre}}}{v_{\mathrm{pre}}+m_{\mathrm{new}}}, \quad \frac{m_{\mathrm{new}}}{v_{\mathrm{pre}}+m_{\mathrm{new}}}.$$

Limitation: One expects that the appropriate accounting approach for this method is the same as for the Standard method applied in stage 1, namely, the Liability view. However, we note that this would yield a post-money valuation after stage 1 of $v_{\mathrm{pre}} + v_{\mathrm{safe}}$, which is greater than the pre-money valuation v_{pre} used for stage 2. In effect, the compromise that has been made in the interest of the founders is to conduct the first stage at a higher pre-money valuation than would be warranted by Liability accounting of the SAFE. This overvaluation is then corrected in a "down-round" in stage 2. This method is therefore not fully consistent with Liability accounting.

At post-money valuation $v_{\mathrm{pre}} + m_{\mathrm{new}}$, this method is always conservative for the new investor, and conservative for the SAFE investor in case $v_{\mathrm{pre}} \geq c$, but is not conservative for the SAFE investor in case $v_{\mathrm{pre}} < c$, since it yields the SAFE investor shares of value $m_{\mathrm{safe}} \cdot \frac{v_{\mathrm{pre}}}{v_{\mathrm{pre}}+m_{\mathrm{safe}}} < m_{\mathrm{safe}}$.

As already noted above, the SAFE investor may have have grounds to argue that the Zero-Money Round approach is prohibited by the terms of the SAFE contract. The legal risk in this approach is higher if the first round applies the Standard method, since the SAFE investor has a case that they were disadvantaged as a result of the artificial construction used, compared to the application of the Discounted Valuation method.

4.6 Summary

We have described and analyzed many different methods for converting a pre-money SAFE with Cap Only as a result of an equity financing. For each method, we have identified a valuation for the SAFE as a function of the company's pre-money valuation, and the proportions of the company owned by each of the founders, the SAFE investor, and the new investor. We also looked at the coherence of the methods in terms of the two accounting views defined in the previous chapter. We attempted to clarify the outcomes of these methods by describing them in terms of equity rounds.

Table 4.1 summarizes our conclusions, showing for each method an accounting stance on which it makes sense. Each accounting view is associated with an understanding of the meaning of the pre-money valuation v_{pre}. In some cases, the method is best understood as starting with a valuation v_{pre} in which the SAFE is treated with respect to one accounting view, but can also be understood under another accounting view at a different valuation with respect to which the terms of the SAFE are applied. The table captures this by indicating that one understanding can be "transformed"

Table 4.1 Accounting methods consistent with SAFE conversion methods for pre-money SAFEs with cap only

SAFE conversion method and abbreviation	Section	Consistent accounting view	Limitations
Standard (Std)	4.2	Liability (wrt v_{pre})	
Percent Ownership (PO)	4.3.1	Cap table (wrt v_{pre})	Price gap, case justification
Discounted Valuation (DV)	4.3.2	Cap table (wrt v_{pre}) transformed to Liability (wrt v_{pre}^-)	
Dollars Invested (DI)	4.4	Liability (wrt v_{pre}) transformed to Cap table (wrt $v_{pre} + m_{safe}$)	Undervalues SAFE
Zero-Money Round (Discounted) ZMR(DV)	4.5.1	Cap table (wrt v_{pre}) transformed to Liability (wrt v_{pre}^-) then standard round (wrt v_{pre})	
Zero-Money Round (Standard) (ZMR(Std))	4.5.2	Liability (wrt v_{pre}) then standard round (wrt v_{pre})	Down-round

to another. (We refer to the relevant sections for a statement of the precise sense in which this holds in each case.) Some of the methods have some limitations that makes their rationality questionable; we refer to the relevant sections for an explanation of the issues mentioned in the "Limitations" column.

References

1. Y Combinator (2016) SAFE Primer. https://web.archive.org/web/20180831020232/http://www.ycombinator.com/docs/SAFE_Primer.rtf
2. Colla D Calculating share price with outstanding convertible notes. https://www.cooleygo.com/calculating-share-price-outstanding-convertible-notes/. Accessed Sept 2019

Chapter 5
Conversion of Post-Money SAFEs

In Chap. 4, we focused on a Pre-Money SAFE (with Cap Only). In this chapter, we compare the Post-Money (with Cap Only [1]) SAFE's handling of Equity Financing with that of the Pre-Money SAFE, and analyze the Post-Money SAFE with respect to the same scenario as treated above, in which a single SAFE has been issued.

Terms of the Post-Money SAFE were introduced in Sect. 2.3. Like the Pre-Money SAFE, it has a Purchase Amount and a Cap, which is called the "Post-Money Cap." One apparent difference is that, whereas the Equity Financing clause in the Pre-Money SAFE makes explicit reference to both pre-money valuation and price, the Post-Money SAFE refers only to price. Thus, the issue we raised above, whether equation (vsp_{pre}) governs the relationship between pre-money valuation and price for purposes of the Equity Financing clause, has been obviated. The clauses defining "Safe Price" also contain modifications.

In the case of the Pre-Money SAFE, we argued that the contract implicitly introduces a circularity into the definition of pre-money valuation. There is similarly a circularity in the Post-Money SAFE: the number of SAFE shares issued depends on the Safe Price, which depends on the Company Capitalization, which in turn depends on the number of SAFE shares issued.[1] We show here that the recursion can be resolved for our simple scenario in which a single Post-Money SAFE has been issued. (The situation is somewhat more complex in situations where multiple SAFEs have been issued, where the SAFEs place more complex constraints on the conditions under which an equity round can be conducted. We discuss the more general case in Part III.)

---

[1] Given their name, perhaps it is not surprising that Y Combinator has a tendency to recursive contracts. "Y combinator" is a term from lambda calculus, concerned with recursion. See Wikipedia contributors [2], Barendregt [3].

© The Author(s), under exclusive license to Springer Nature Singapore Pte Ltd. 2025
R. van der Meyden and M. J. Maher, *Simple Agreements for Future Equity (SAFE)*,
Blockchain Technologies, https://doi.org/10.1007/978-981-96-3920-5_5

5.1 Conversion in Terms of Price

We first characterize the outcomes of the Post-Money SAFE in terms of price, and consider their relation to valuation below. We use the same variable names as for Pre-Money SAFEs. The Post-Money SAFE directly specifies the Purchase Amount of the SAFE m_{safe}, and the Post-money Valuation Cap c. We make the same simplifying assumptions about there being no options or other instruments, as above. The number of founder shares s_f is also fixed.

At the time of the equity round, the new money raised m_{new}, and the price per share p_{new} are given. As usual, we assume that equation (msp_{new}) is used to determine the number of shares s_{new} received by the new investor. (One of the differences between the Pre-Money SAFEs and the Post-Money SAFEs is that the latter allows that new equity round investors do not all pay the same price for their shares: there is an explicit reference to "the lowest price per share of the Standard Preferred Stock." It is unclear whether equity rounds with varying prices for new investors are much used in practice. In our analysis, we will for comparative purposes make the simplifying assumption that all new investors pay the same price p_{new}.)

As above, we assume that Common Stock, Safe Preferred Stock, and Standard Preferred Stock are essentially the same, but in fact there are differences (liquidation preference, anti-dilution protection, and dividend rights) that would need to be addressed in a more detailed analysis. We also assume a single equity investor, so that the "lowest price per share of the Standard Preferred Stock" unambiguously refers to p_{new}.

By the Equity Financing clause, we have that the number s_{safe} of shares issued to the SAFE investor satisfies

$$s_{safe} = \max\{m_{safe}/p_{new}, m_{safe}/p_{safe}\}.$$

By the definition of Safe Price,

$$p_{safe} = c/S_{pre}$$

where S_{pre} is the Company Capitalization. Since this has been defined to include all shares of capital stock and all SAFEs, in our simple scenario, we have

$$S_{pre} = s_f + s_{safe}.$$

Substituting, we get the equation

$$s_{safe} = \max\{m_{safe}/p_{new}, m_{safe}(s_f + s_{safe})/c\}$$

which explicitly displays that the SAFE gives a circular definition of s_{safe}. We have two cases, depending on which argument of the maximum is greater. (We include the case where the two terms are equal in both cases, in order to check consistency of the solutions when this equality holds.)

Case 1: $m_{safe}/p_{new} \geq m_{safe}(s_f + s_{safe})/c$. Equivalently, $s_{safe} \leq \frac{c}{p_{new}} - s_f$. In this case $s_{safe} = m_{safe}/p_{new}$. Consequently, the condition for this case is equivalent to

$$0 < p_{new} \leq \frac{c - m_{safe}}{s_f}.$$

Case 2: $m_{safe}/p_{new} \leq m_{safe}(s_f + s_{safe})/c$. In this case, s_{safe} is defined by the recursive equation $s_{safe} = m_{safe}(s_f + s_{safe})/c$. Solving for s_{safe}, we get

$$s_{safe} = \frac{m_{safe}s_f}{c - m_{safe}}. \tag{5.1}$$

Obviously, we need the constraint $m_{safe} < c$ for this solution to make sense, else we have an undefined or negative number of shares. This constraint is reasonable, if we treat the cap c as analogous to a post-money valuation for the SAFE investor's investment. (We will see below, when discussing the proportional shareholding outcomes, that this analogy makes sense.) It is also reasonable based on the intent of the contract: the rhetoric for the cap is that the investor gets a share of the upside over c.

Substituting for s_{safe} in the inequality and reorganizing, we get the following formulation of the condition for Case 2:

$$p_{new} \geq \frac{c - m_{safe}}{s_f}.$$

Note that this is the complement of the condition for Case 1.

When both cases apply, we have $p_{new} = (c - m_{safe})/s_f$ and the two cases agree on the value of s_{safe}. However, the SAFE does not specify whether the SAFE investor receives Standard Preferred or Safe Preferred stock, and it is also not made explicit that the investor has a choice. As the apparent intent of the clause is to deliver the maximum benefit to the SAFE investor, one expects that this question would be resolved in case of a legal dispute by the SAFE investor receiving whichever class of shares gives the greatest benefit. The details of the shareholder rights with respect to Liquidity, Dissolution, and Dividend events need to be taken into account to make a determination on this point. The definition of Safe Preferred shares states that these are identical to Standard Preferred shares except that the Safe Price is used when computing these rights. When the Safe Price and p_{new} are the same, then it follows that these two types of shares are actually equivalent, in which case it does not matter which is selected. Nevertheless, for uniformity with the pre-money SAFE, we assume that SAFE Preferred stock is preferable, so that the conditions for the two cases are $p_{new} \leq (c - m_{safe})/s_f$ and $p_{new} > (c - m_{safe})/s_f$.

Assuming that the transaction does not change the share price,[2] we can deduce monetary values for the resulting shareholdings of the founders, SAFE investor and new investor, respectively, by multiplying these by p_{new}, resulting in the following.

Case 1:

$$s_f\, p_{new}, \quad m_{safe}, \quad m_{new}$$

Case 2:

$$s_f\, p_{new}, \quad m_{safe}\, \frac{s_f\, p_{new}}{c - m_{safe}}, \quad m_{new}$$

On this assumption, the new investor is guaranteed a shareholding equivalent in value to their investment, as is the SAFE investor in Case 1. In Case 2, since we have $p_{new} > (c - m_{safe})/s_f$, the monetary value of the SAFE investor's shares is greater than m_{safe}. Thus, the contract guarantees that the equity round is conservative for both the SAFE investor and the new investor.

5.2   Conversion in Terms of Pre-Money Valuation

We now characterize these outcomes in terms of valuation. The pre- and post-money valuations are not defined in the Post-Money SAFE. Indeed, the text contains the following definition:

> **"Equity Financing"** means a bona fide transaction or series of transactions with the principal purpose of raising capital, pursuant to which the Company issues and sells Preferred Stock at a fixed valuation, including but not limited to, a pre-money or post-money valuation.

This suggests that, although the Safe Price is, effectively, calculated using the Cap Table accounting view at valuation c, the document has been deliberately designed to be noncommittal with respect to the conversion method and the approach used to determine a share price from a valuation. We therefore analyze the equity round transaction from the point of view of the two accounting methods, corresponding to different ways to relate price and valuation.

We note that by the analysis of the two accounting views in Chap. 3, for both the Liability and Cap Table views, the price of shares in the equity round is given by the equation

$$p_{new} = \frac{v_I - p_{new} s_{safe}}{s_f}$$

---

[2] As shown in Chap. 3, this assumption holds under both Cap Table and Liability accounting. However, as we explain below, some conditions must hold in order to obtain a non-zero price when applying these accounting methods to the Post-Money SAFE.

where $v_I = v_A - v_L$ is the inherent valuation of the company, based on the valuation of the assets v_A and liabilities v_L, but excluding convertible instruments. In the case of the Post-Money SAFE, the number of SAFE shares s_{safe} is defined using p_{new}, so to obtain p_{new}, we need to solve the equation

$$p_{\text{new}} = \frac{v_I - p_{\text{new}} s_{\text{safe}}(p_{\text{new}})}{s_f}.$$

This equation is the same for the Cap Table view as it is for the Liability view. We will therefore obtain the same share price in the two views and, consequently, the same issuance of shares to the SAFE investor.[3] It therefore suffices to consider just the Cap Table view.[4] Let the pre-money valuation of the company on Cap Table accounting be v_{pre}, so $v_{\text{pre}} = p_{\text{new}}(s_f + s_{\text{safe}})$.

In Case 1, we have

$$v_{\text{pre}} = (s_f + s_{\text{safe}}) p_{\text{new}} = (s_f + m_{\text{safe}}/p_{\text{new}}) p_{\text{new}} = s_f p_{\text{new}} + m_{\text{safe}}$$

so we can characterize p_{new} in terms of v_{pre} as $p_{\text{new}} = (v_{\text{pre}} - m_{\text{safe}})/s_f$. We note that this price only makes sense if $v_{\text{pre}} > m_{\text{safe}}$, so that this approach to determination of the price is applicable only when this condition holds.

In Case 2,

$$v_{\text{pre}} = (s_f + s_{\text{safe}}) p_{\text{new}} = \left(s_f + \frac{m_{\text{safe}} s_f}{c - m_{\text{safe}}} \right) p_{\text{new}} = \frac{c}{c - m_{\text{safe}}} \cdot s_f p_{\text{new}}$$

so

$$p_{\text{new}} = \frac{(c - m_{\text{safe}})}{c} \cdot \frac{v_{\text{pre}}}{s_f}.$$

This price is positive under the reasonable assumption that $c > m_{\text{safe}}$.

In both cases, $p_{\text{new}} \le (c - m_{\text{safe}})/s_f$ exactly when $v_{\text{pre}} \le c$, so the conditions for the two cases are $v_{\text{pre}} \le c$ and $v_{\text{pre}} > c$.

---

[3] We remark that this conclusion does not hold for the Pre-Money SAFEs. In that case, we have the SAFE issuance depends on both the price p_{new} and the pre-money valuation v_{pre}, so what needs to be solved is a simultaneous equation comprised of

$$p_{\text{new}} = \frac{v_I - p_{\text{new}} s_{\text{safe}}(p_{\text{new}}, v_{\text{pre}})}{s_f}$$

and $v_{\text{pre}} = v_I - p_{\text{new}} s_{\text{safe}}(p_{\text{new}}, v_{\text{pre}})$ in case of the Liability view and $v_{\text{pre}} = v_I$ in case of the Cap Table view. The Liability view corresponds to the Discounted Valuation method, and the Cap Table view corresponds to the Percent-Ownership method. We saw in Sect. 4.3 that these yield inequivalent solutions, with different case conditions.

[4] Note that the definition of Safe Price as $c/(s_f + s_{\text{safe}})$ suggests that in case this price is applied, the SAFE is being converted at valuation c using cap table accounting, so it is reasonable to treat the new investor's money in the same way (though at a different valuation).

On cap table accounting, the post-money valuation is $v_{\text{pre}} + m_{\text{new}}$, so the proportional shareholdings of the founders, SAFE investor and new investor, respectively, in the two cases are as follows.

Case 1: $(v_{\text{pre}} \leq c)$

$$\frac{v_{\text{pre}} - m_{\text{safe}}}{v_{\text{pre}}} \cdot \frac{v_{\text{pre}}}{v_{\text{pre}} + m_{\text{new}}}, \quad \frac{m_{\text{safe}}}{v_{\text{pre}}} \cdot \frac{v_{\text{pre}}}{v_{\text{pre}} + m_{\text{new}}}, \quad \frac{m_{\text{new}}}{v_{\text{pre}} + m_{\text{new}}}$$

Case 2: $(v_{\text{pre}} > c)$

$$\frac{c - m_{\text{safe}}}{c} \cdot \frac{v_{\text{pre}}}{v_{\text{pre}} + m_{\text{new}}}, \quad \frac{m_{\text{safe}}}{c} \cdot \frac{v_{\text{pre}}}{v_{\text{pre}} + m_{\text{new}}}, \quad \frac{m_{\text{new}}}{v_{\text{pre}} + m_{\text{new}}}$$

Intuitively, in both cases, the new investor is guaranteed a shareholding equivalent to their money. Case 1 is equivalent to a two stage process in which the SAFE investor first purchases shares worth m_{safe} in a standard equity round at valuation $v_{\text{pre}} - m_{\text{safe}}$, followed by the new investor buying shares worth m_{new} at valuation v_{pre}. The outcome of Case 2 is equivalent to that of a two stage process in which the SAFE investor first buys shares worth m_{safe} at valuation $c - m_{\text{safe}}$ in a standard equity round, followed by the new investor buying shares worth m_{new} at valuation v_{pre}. (As suggested above, on this view, c is analogous to the post-money valuation at which the SAFE investor's investment is made.)

We remark that the first stage of this process is always less advantageous to the founders, and more favorable to the SAFE investor, than a first stage investment of m_{safe} at pre-money valuation c, as was essentially involved in several cases of the analysis of the pre-money SAFE above.

Under the assumption that $v_{\text{pre}} > m_{\text{safe}}$, the valuation of the SAFE is

$$v_{safe} = \begin{cases} m_{\text{safe}} & v_{\text{pre}} \leq c \\ \frac{m_{\text{safe}}}{c} v_{\text{pre}} & v_{\text{pre}} > c \end{cases}.$$

5.3 Other Conversion Methods

The methods of conversion discussed in Chap. 4 for a Pre-Money SAFE (except the Standard method) are motivated by the perceived dilution of the new investor in the Standard method. However, as we have seen, under both the Liability and the Cap Table view Post-Money SAFEs are conservative for the new investor, so this motivation is lacking. Nevertheless, we consider the effect of other methods.

In the Zero-Money Round (Discounted Valuation) conversion method, the SAFE is first converted in an equity round with zero new money with price determined from the discounted valuation v_{pre}^- (interpreted using the Liability view), and we then run a standard round at valuation v_{pre}. As argued in the previous section, the price of the first round is the same using the Cap Table view as it is using the Liability view (all that differs is that now $m_{\text{new}} = 0$). As above, the cases are determined by whether $v_{\text{pre}} \leq c$. From the calculations above, we obtain proportions after the first round

$$\frac{v_{\text{pre}} - m_{\text{safe}}}{v_{\text{pre}}}, \quad \frac{m_{\text{safe}}}{v_{\text{pre}}}$$

in Case 1 and

$$\frac{c - m_{\text{safe}}}{c}, \quad \frac{m_{\text{safe}}}{c}$$

in case 2 for founders, SAFE investor respectively. Running the second round gives exactly the proportions for the single round Cap Table view as above.

This Zero-Money Round (Discounted Valuation) method and the Discounted Valuation method of Sect. 5.2 both calculate a price p_{new} that depends on the SAFE being converted, and, circularly, on whether this price yields a better outcome for the SAFE investor. These approaches both require the assumption that $v_{\text{pre}} > m_{\text{safe}}$. An alternate approach, more akin to the Zero-Money Round (Standard) method, would be to use the price $p_{\text{new}} = v_{\text{pre}}/s_f$ to convert the SAFE in the first round. For the Post-Money SAFE, this still requires an equation to be solved to obtain the number of shares s_{safe}, in case the SAFE Price gives the better outcome for the SAFE investor. Application of this method would not be conservative for the new investor, but we could use this approach as part of a "Zero-Money Round (Standard)" approach for the Post-Money SAFE.

The case condition for $p_{\text{new}} < p_{\text{safe}}$ in this case is

$$\frac{v_{\text{pre}}}{s_f} < \frac{c}{s_f + s_{\text{safe}}} = \frac{c}{s_f + \frac{m_{\text{safe}} s_f}{v_{\text{pre}}}}$$

which is equivalent to $v_{\text{pre}} < c - m_{\text{safe}}$. In this case, the first round gives proportions

$$\frac{v_{\text{pre}}}{v_{\text{pre}} + m_{\text{safe}}}, \quad \frac{m_{\text{safe}}}{v_{\text{pre}} + m_{\text{safe}}}$$

for the founders and SAFE investor, respectively. In the case $p_{\text{new}} \geq p_{\text{safe}}$, that is, $v_{\text{pre}} \geq c - m_{\text{safe}}$. we obtain proportions after the first round of

$$\frac{c - m_{\text{safe}}}{c}, \quad \frac{m_{\text{safe}}}{c}$$

These are exactly the same as the proportions obtained in the two cases in the methods above, except that the case condition differs. Note also that, while a Zero-Money Round method is always conservative for the new investor, this method is conservative for the SAFE investor only if $v_{\text{pre}} \geq c$. Since the second round would be priced at a lower amount $v_{\text{pre}}/(s_f + s_{\text{safe}})$, this is again a "down-round", similarly to the situation with this method in case of the Pre-Money SAFE. The main advantage to this approach is that it gives a somewhat grounded backup approach to dealing with situations where $v_{\text{pre}} \leq m_{\text{safe}}$.

5.4   Summary

Conversion for post-money SAFEs depends only on the (lowest) price of the equity round. The text of the SAFEs allows the use of any fixed valuation because the valuation is not used in the conversion of SAFEs. We have derived the appropriate price to use, given a pre-money valuation, interpreted with respect to either the Cap Table or Liability view. This price calculation yields the same price and share issuance for both of these views, under the assumption that $v_{\text{pre}} > m_{\text{safe}}$. The same results are also obtained using the Zero-Money Round (Discounted Valuation) method.

We see that the equity round conversion is conservative for the SAFE holder and the new investor. So the motivation for the proliferation of conversion methods that we saw for pre-money SAFEs is missing for post-money SAFEs and, therefore, we have not considered the application of the Dollars Invested method for post-money SAFEs.

The one exception is when $v_{\text{pre}} \leq m_{\text{safe}}$ in which case we can still use the Zero-Money Round (Standard) method. However, this has the same disadvantages as it does for pre-money SAFEs: it is non-conservative for the SAFE investor, and therefore caries legal risk.

References

1. Y Combinator (2018) Safe: valuation cap, no discount. https://web.archive.org/web/20190626002912/https://www.ycombinator.com/docs/Postmoney%20Safe%20-%20Valuation%20Cap%20-%20v1.0.docx
2. Wikipedia Contributors (2024) Fixed-point combinator. https://en.wikipedia.org/wiki/Fixed-point_combinator. Accessed Aug 2024
3. Barendregt H (1985) The Lambda calculus—its syntax and semantics, Volume 103 of Studies in logic and the foundations of mathematics. North-Holland

Chapter 6
Game-Theoretic Aspects of SAFE Conversion

We can view the different conversion methods as arising from a game involving the SAFE investor, the founders, and the equity investor. In this chapter, we analyze the methods in terms of this game, and identify optimal strategies for each of the players.

6.1 Introduction

In the previous chapters, we have discussed a variety of methods by which two forms of SAFE contract may be converted to shares at the time of the equity round. Factors distinguishing these methods are two distinct interpretations of the term "Pre-money Valuation," related to the accounting stance that one takes with respect to the SAFE.

We now compare the outcomes of these contracts and conversion methods to determine which is more favorable to each of the players (the founders, the SAFE investor, and the new equity round investor). There are two ways that one can undertake such an analysis.

1. **Conservative Methods**: One can suppose that it is common knowledge among the founders, SAFE investor, and new investors that all parties will negotiate with clarity about the meaning of "Pre-Money Valuation," so that only conversion methods appropriate to that interpretation will be considered, and the negotiation of a value for v_{pre} can take into account the selected meaning. A rational new investor will not accept an immediate loss, so only conservative methods are considered in this case.

2. **Deferred Interpretation**: One can suppose that the new investors, the SAFE investors, and the founders failed to distinguish between the two interpretations of "pre-money valuation." In particular, the interpretation had not been decided at the time the founders negotiated with the new investors a term sheet stating a pre-money valuation. (This seems to be not infrequently the case, according to

R. van der Meyden and M. J. Maher, *Simple Agreements for Future Equity (SAFE)*,
Blockchain Technologies, https://doi.org/10.1007/978-981-96-3920-5_6

[1]!) In such a situation v_{pre} is fixed, but the parties may still select a method for calculating p_{new}. All the methods should be compared in this case.

There is a game-theoretic dimension to the situation, in the form of a two-round game. In round 1, the founder and SAFE investor negotiate and choose a form for the SAFE investor's investment: either a standard equity round (to be followed later by a standard round with the new investor), a pre-money SAFE contract, or a post-money SAFE contract. In the second round, the founders and c the new investor negotiate and choose a method to be used to conduct the equity round. We analyze this game to determine the optimal strategy for the players.[1]

6.2 Conservative Methods

We consider first the cases where it is common knowledge that the participants have clarity about the interpretation of pre-money valuation and the accounting method, and use one of the conservative conversion methods that the analysis above shows to be appropriate for that interpretation. The game in this case proceeds in the following steps:

1. The founders and SAFE investor choose a form of contract: standard equity round for money m_{safe} at valuation c, Pre-Money SAFE for money m_{safe} and cap c, or a Post-Money SAFE for money m_{safe} and cap c'. (We argue below that when negotiating over the choice between a Pre-Money SAFE and a Post-Money SAFE, a particular value for c' is the expected outcome of this negotiation.)
2. The new investor chooses an inherent valuation of the company, i.e., a valuation $v_I = v_A - v_L$ where v_A and v_L are the valuations of the company assets A and liabilities L, *excluding* the SAFE contract.
3. The founders and new investor negotiate a conversion method and an associated interpretation and value of the pre-money valuation v_{pre}. The choices are the following: For the Pre-Money SAFE, the Standard method with Liability view, Discounted Valuation method with Cap Table view, or the Zero-Money Round (Standard) method with Liability view. For the Post-Money SAFE, either the

---

[1] Our analysis is by a discrete comparison of outcomes. We have not attempted to factor into the analysis any assumptions about the probability of v_{pre} falling into the above cases, since it is unclear what the appropriate distribution should be. Some data on ability of companies to meet their SAFE caps by the time of an equity round could in principle be obtained. A factor in such empirical results may be that founders will prefer to defer an equity round (possibly by raising additional SAFE investments) until they are able to obtain a valuation that is to their benefit, given the SAFE issued. Some weighting of the distribution towards the case $v_{pre} > c$ therefore seems likely. Against this, the failure rate of startups is high, so many SAFEs will ultimately be resolved by dissolution or sale of the company rather than an equity round.

Discounted Valuation method with Cap Table view (equivalently, the Standard method with Liability view), or the Zero-Money Round (Standard) method.[2]

We first determine a valuation v_{safe} for the SAFE contract at the time of the equity round for each of the conversion methods considered. Using this, we derive a value for v_{pre} for each method and compute the resulting distribution of shareholdings of the parties. We characterize this valuation in terms of the inherent valuation v_I to enable comparison of the various methods.

In case of the Standard method with the Liability view, we have the following valuation of the SAFE:

$$v_{safe} = \begin{cases} m_{safe} & v_{pre} \le c \\ m_{safe} \cdot \frac{v_{pre}}{c} & v_{pre} > c \end{cases}$$

On the Liability view, the SAFE is a liability that has not yet been accounted for in the inherent valuation v_I, so we have that the correct pre-money valuation is $v_{pre} = v_I - v_{safe}$. Thus, the above analysis has not yet escaped from the circularity of the SAFE. However, we may solve the equations resulting from the two cases to get

$$v_{safe} = \begin{cases} m_{safe} & v_I - m_{safe} \le c \\ m_{safe} \cdot \frac{v_I}{c + m_{safe}} & v_I - m_{safe} > c \end{cases} \tag{6.1}$$

which gives a formulation of the valuation of the SAFE in terms of the inherent valuation v_I of the company.[3]

In case of the two-stage Zero-Money Round (Standard) method, interpreted using the Liability view, we have

$$v_{safe} = \begin{cases} m_{safe} \cdot \frac{v_{pre}}{v_{pre} + m_{safe}} & v_{pre} \le c \\ m_{safe} \cdot \frac{v_{pre}}{c + m_{safe}} & v_{pre} > c \end{cases}$$

which *prima facie* seems different from the above case. However, we should note that in the two-stage methods, regardless of the approach used in the first stage, v_{pre} represents the valuation of the company at which the new investor is prepared to

---

[2] We omit the Dollars Invested method here because it is a non-conservative, less clearly rational compromise, and arguably motivated by the confusion concerning the meaning of "Pre-Money Valuation." As we will see, this motivation disappears as a result of the analysis that follows. The Percent-Ownership method differs from the Discounted Valuation method only in the case conditions. It also is omitted, since it has deficiencies including a price gap and not always being conservative for the SAFE investor: we assume that the Discounted Valuation method is the correct implementation of the Percent-Ownership conversion idea. The Zero-Money Round (Discounted Valuation) method with Cap Table view is also omitted, since it has been shown to be equivalent to the single round Discounted Valuation method.

[3] Note that the case condition for the second case $v_{pre} = v_I - v_{safe} > c$ is

$$v_I - m_{safe} \cdot \frac{v_I}{c + m_{safe}} > c,$$

which is equivalent to $v_I - m_{safe} > c$.

invest, *once the SAFE liability has been discharged*. Thus, the appropriate value for v_{pre} in this case is the inherent valuation $v_{\text{pre}} = v_I$. Substituting, we obtain

$$v_{\text{safe}} = \begin{cases} m_{\text{safe}} \cdot \frac{v_I}{v_I + m_{\text{safe}}} & v_I \leq c \\ m_{\text{safe}} \cdot \frac{v_I}{c + m_{\text{safe}}} & v_I > c \end{cases} \tag{6.2}$$

which is the same as Eq. (6.1) in case $v_I \geq c + m_{\text{safe}}$, but differs otherwise.

To compare pre-money and post-money SAFEs, we have the following result.

Proposition 6.1 *Consider the following scenarios:*

1. **Scenario 1**: *The company issues a Pre-Money SAFE with cap c for money m_{safe}.*
2. **Scenario 2**: *The company issues a Post-Money SAFE with cap $c' = c + m_{safe}$ for money m_{safe}.*

Then

1. *The SAFE is convertible in Scenario 1 using the Discounted Valuation method iff it is convertible in Scenario 2 using the Discounted Valuation method iff $v_I > m_{safe}$.*
2. *If $v_I > m_{safe}$, then the result of conversion using the Discounted Valuation method is the same in both scenarios, for all parties.*
3. *For all pre-money valuations, the result of using the Zero-Money Round (Standard) method is the same in both scenarios, for all parties.*

Proof The constraint $v_{\text{pre}} > m_{\text{safe}}$ on convertibility was noted above in the analysis of the Pre-Money SAFE and the Post-Money SAFE with respect to the Discounted Valuation method. Under the associated Cap Table view, we have $v_{\text{pre}} = v_I$.

The Post-Money SAFE, with Cap Table (equivalently, Liability with respect to v_{pre}^-) view, gives

$$v_{\text{safe}} = \begin{cases} m_{\text{safe}} & v_{\text{pre}} \leq c' \\ m_{\text{safe}} \cdot \frac{v_{\text{pre}}}{c'} & v_{\text{pre}} > c' \end{cases}$$

Substituting $c' = c + m_{\text{safe}}$ for c in the outcome for the Post-Money SAFE and simplifying, we get the value outcomes under the Cap Table view for the founders, SAFE investor, and new investor, respectively, of

$$v_I - m_{\text{safe}}, \quad m_{\text{safe}}, \quad m_{\text{new}}$$

in case $v_I \leq c + m_{\text{safe}}$, and

$$\frac{c}{c + m_{\text{safe}}} \cdot (v_I), \quad \frac{m_{\text{safe}}}{c + m_{\text{safe}}} \cdot (v_I), \quad m_{\text{new}}$$

in case $v_I > c + m_{\text{safe}}$. This is precisely the same outcome in all cases as the Pre-Money SAFE with Discounted Valuation method (as well as method Std(L)).

For the Zero-Money Round (Standard) method, note that the substituting $c + m_{safe}$ for c in the outcomes for the Zero-Money Round (Standard) method from Chap. 5, with $v_{pre} = v_I$ yields values

$$\frac{v_I}{v_I + m_{safe}} \cdot v_I, \quad \frac{m_{safe}}{v_I + m_{safe}} \cdot v_I, \quad m_{safe}$$

in case $v_I < m_{safe}$, and

$$\frac{c}{c + m_{safe}} \cdot v_I, \quad \frac{m_{safe}}{c + m_{safe}} \cdot v_I, \quad m_{safe}$$

otherwise. This is the same as the values obtained for the Zero-Money Round (Standard) method in Sect. 4.5.2. □

Thus, (assuming only one SAFE will be issued), if the founders and SAFE investor assume all players are fully rational, and anticipate use of either the Discounted Valuation method or the Zero-Money Round (Standard) method at the time of the equity round, we expect that a negotiation on the terms of a Pre-Money SAFE or a Post-Money SAFE will yield an instrument (with cap c or $c + m_{safe}$, respectively) with the same effective consequences at the time of the equity round in either case.[4] From this point of view, there is not a reason to prefer one over the other.[5] We may therefore narrow the analysis of the game to just the Pre-Money SAFE.

Using the above derivations of the values of v_{safe}, we obtain solutions for v_{pre} in terms of the inherent company valuation v_I for each approach (for the Cap Table view approaches, this is simply v_I). Substituting into the corresponding characterizations of proportional shareholdings for each of the methods, and noting that in all cases we have post-money valuation equal to $v_I + m_{new}$, we get the post-money values, in terms of v_I for each of the shareholders given in Table 6.1. (Dividing by $v_I + m_{new}$ gives the proportional holdings.) The table uses the following abbreviations for the approaches: $2R$ is two standard equity rounds, the first for money m_{safe} at valuation c, the second for money m_{new} at valuation v_I; $Std(L)$ is the Standard method with Liability view; $DV(C)$ is the Discounted Valuation method with Cap Table view, and $ZMR(Std, L)$ is the two-stage Zero-Money Round (Standard) method with Liability view.

From the point of view of the new investor, the outcome of each of these methods is the same. Any choice between these methods is therefore only of concern to the founders and the SAFE investor. We consider the ordering between the post-money values for the SAFE investor—since the total post-money valuation is the same in all cases, this is the inverse of the ordering for the founders.

---

[4] We ignore here other differences between the Pre-Money and Post-Money SAFEs, such as the ability to receive dividends in the Post-Money SAFE. Our comparison is only based on equity financing.

[5] We note, however, that our analysis concerns only a single SAFE—the situation is more complex when multiple SAFEs are issued.

Table 6.1 Value of shareholdings from standard equity rounds (2R) and, for rounds with a SAFE, conservative conversion approaches for founders, SAFE holder and new investor (ordered by increasing preference for SAFE holder and decreasing preference for founders)

Approach	Condition	Founders	SAFE	New
2R	True	$\frac{c}{c+m_{\text{safe}}} \cdot (v_I)$	$\frac{m_{\text{safe}}}{c+m_{\text{safe}}} \cdot (v_I)$	m_{new}
ZMR(Std, L)	$v_I \leq c$	$\frac{v_I}{v_I+m_{\text{safe}}} \cdot (v_I)$	$\frac{m_{\text{safe}}}{v_I+m_{\text{safe}}} \cdot (v_I)$	m_{new}
	$v_I > c$	$\frac{c}{c+m_{\text{safe}}} \cdot (v_I)$	$\frac{m_{\text{safe}}}{c+m_{\text{safe}}} \cdot (v_I)$	m_{new}
Std(L), DV(C)	$v_I - m_{\text{safe}} \leq c$	$v_I - m_{\text{safe}}$	m_{safe}	m_{new}
	$v_I - m_{\text{safe}} > c$	$\frac{c}{c+m_{\text{safe}}} \cdot (v_I)$	$\frac{m_{\text{safe}}}{c+m_{\text{safe}}} \cdot (v_I)$	m_{new}

In case $v_I \geq c + m_{\text{safe}}$, all the Pre-Money SAFE methods also produce exactly the same result as the Two-Round ($2R$) method. Note that the condition $v_I \geq c + m_{\text{safe}}$ essentially says that the inherent valuation of the company has not fallen below the post-money valuation of a standard round for money m_{safe} at valuation c. Thus, the differences between the Pre-Money SAFE approaches emerge only in case the equity round for the new investor's money is effectively a "down-round" from this post-money valuation.

The comparison between $2R$ and ZMR(Std, L) is straightforward. In case $v_I \leq c$, the ZMR(Std, L) value for the SAFE investor is larger because the term in the denominator is larger in case of $2R$. Thus, the SAFE investor always prefers ZMR(Std, L) over $2R$.

If $m_{\text{safe}} < v_I \leq c$, then approaches Std($L$) and DV($C$) agree that the SAFE investor effectively gets their money m_{safe} back in the form of shares. Approach ZMR(Std, L) gives the SAFE investor a lesser value in this case, and the founders a correspondingly greater value. Thus, the SAFE investor always prefers Std(L), DV(C) over ZMR(Std, L). As we noted above, ZMR(Std, L) might be selected by the new investor and founders in order to favor the founders, but if the SAFE investor holds the stronger bargaining position when in a negotiation with the founders concerning the conversion method,[6] ZMR(Std, L) is not likely to be selected.

However, we note that the Std(L) and DV(C) methods require that $m_{\text{safe}} < v_I$ in order to be applied. In case $v_I \leq m_{\text{safe}}$, approach ZMR(Std, L) might still be applied to resolve this impasse. This issue becomes more complicated in the case the company has issued multiple SAFEs, where there may be multiple valuation intervals in which the Post-Money SAFE cannot be converted, and Post-Money SAFEs are, in general, not equivalent to Pre-Money SAFEs. We discus this more general case in Chap. 7.

---

[6] The SAFE itself does not give the SAFE investor a right to negotiate, but this might be the case if they are contributing new money to the round, possibly because they hold pro rata rights.

In summary, it can be argued that when the Pre-Money SAFE and Post-Money SAFE are interpreted with clarity about the meaning of pre-money valuation, the values for c and m_{safe} that will be negotiated will be such that all conservative methods of execution, with an appropriate method of accounting, other than ZMR(Std, L), yield equivalent results. Since the SAFE conversion results are better for the SAFE investor in all cases, they will prefer a SAFE over a standard equity investment. In the negotiation concerning the conversion method, the founders would prefer to apply the Zero-Money Round (Standard) method. To the extent that they have bargaining power in the equity round, the SAFE investor will argue that this method not be applied, and they potentially have legal recourse if it is used. Given this legal risk, the new investor and founders may well prefer to apply the Discounted Valuation method instead.

6.3 Deferred Interpretation

Next, we consider the situation when the interpretation of "Pre-Money Valuation" is deferred until after the term sheet has been signed. The game in this case is played as follows:

1. The founders and the SAFE investor negotiate a form of contract.
2. The new investor chooses v_{pre} and m_{new}.
3. The founders and new investor choose a conversion method.

The choices of contract forms are "Two-Rounds" (i.e., two equity rounds, in the first of which the SAFE investor purchases shares at valuation c, and in the second of which the new investor purchases shares at valuation v_{pre}), a Pre-Money SAFE (with cap c and money m_{safe}) and a Post-Money SAFE.

While this order of events may not be fully rational, it appears that this situation is not just theoretical. Feld [2] writes

> Most notes are ambiguous as to whether they convert on a pre-money or a post-money basis. This can be especially confusing, and ambiguous, when there are multiple price caps. There are also some law firms whose standard documents are purposefully ambiguous to give the entrepreneur theoretical negotiating flexibility in the first priced round.
>
> If the entrepreneur knows this and is using it proactively so they get a higher post-money valuation, that's fair game. But if they don't know this, and they are negotiating terms with a VC who is expecting the notes to convert in the pre-money, it can create a mess after the terms are agreed to somewhere between the term sheet stage and the final definitives. This mess is especially yucky if the lawyers don't focus on the final cap table and the capitalization opinion until the last few days of the process. And, it gets even messier when some of the angels start suggesting that the ambiguity should work a certain way and the entrepreneur feels boxed in by the demands of his convertible note angels on one side and priced round VC on the other.

Table 6.2 Pre-Money SAFE: proportional shareholdings from different conversion methods (1). Proportional shareholdings, case of $v_{pre} \leq c$

Approach	Founders	SAFE investor	New investor
Two-Rounds	$\frac{c}{c+m_{safe}} \cdot \frac{v_{pre}}{v_{pre}+m_{new}}$ (1–4)	$\frac{m_{safe}}{c+m_{safe}} \cdot \frac{v_{pre}}{v_{pre}+m_{new}}$ (4)	$\frac{m_{new}}{v_{pre}+m_{new}}$ (1)
Standard	$\frac{v_{pre}}{v_{pre}+m_{safe}+m_{new}}$ (2)	$\frac{m_{safe}}{v_{pre}+m_{safe}+m_{new}}$ (2)	$\frac{m_{new}}{v_{pre}+m_{safe}+m_{new}}$ (2)
Discounted	$\frac{v_{pre}-m_{safe}}{v_{pre}+m_{new}}$ (5)	$\frac{m_{safe}}{v_{pre}+m_{new}}$ (1)	$\frac{m_{new}}{v_{pre}+m_{new}}$ (1)
Dollars Invested	$\frac{v_{pre}}{v_{pre}+m_{safe}+m_{new}}$ (2)	$\frac{m_{safe}}{v_{pre}+m_{safe}+m_{new}}$ (2)	$\frac{m_{new}}{v_{pre}+m_{safe}+m_{new}}$ (2)
ZMR (DV)	$\frac{v_{pre}-m_{safe}}{v_{pre}+m_{new}}$ (5)	$\frac{m_{safe}}{v_{pre}+m_{new}}$ (1)	$\frac{m_{new}}{v_{pre}+m_{new}}$ (1)
ZMR (Std)	$\frac{v_{pre}}{v_{pre}+m_{safe}} \cdot \frac{v_{pre}}{v_{pre}+m_{new}}$ (3)	$\frac{m_{safe}}{v_{pre}+m_{safe}} \cdot \frac{v_{pre}}{v_{pre}+m_{new}}$ (3)	$\frac{m_{new}}{v_{pre}+m_{new}}$ (1)

An issue with respect to the Post-Money SAFE is the choice of parameters. We argued above that a single Post-Money SAFE with cap $c + m_{safe}$ and money m_{safe} is equivalent, with respect to equity financing, to the Pre-Money SAFE with cap c and money m_{safe} for fully rational players. Thus, to avoid an unfair comparison, we should choose these equivalent parameter values for the Post-Money SAFE. However, the outcomes for each conversion method would then give the same result as the Pre-Money SAFE, making it redundant to include the Post-Money SAFE as an option. We therefore consider just the Pre-Money version of the SAFE.

As above, for the conservative methods, the choices of conversion method are the Standard method, Discounted Valuation method, the Two-Stage (Zero-Money Round) method based on either Standard or Discounted method. We also include the Dollars Invested method since we are now not dealing with fully rational players.

Tables 6.2, 6.3, and 6.4 summarize the relative shareholdings derived for each of the parties on the two forms of SAFE contracts and these models of their operation.[7] We have multiple tables here, since the outcomes of the SAFE contract have multiple cases, depending on the pre-money valuation v_{pre}. In each case, each column gives the outcomes on the different approaches for one of the three parties. The Standard, Dollars Invested, and Zero-Money Round (Standard) methods have outcomes that depend on the cases $v_{pre} \leq c$ and $v_{pre} > c$. In the case of the Discounted Valuation method, we use the given pre-money valuation v_{pre} to state the relative shareholdings, but which case of the "if" statement is used to calculate the shares issued in conversion of the SAFE is determined using the discounted valuation v_{pre}^-. This gives a case split on $v_{pre} \leq c + m_{safe}$ and $v_{pre} > c + m_{safe}$. Consequently, we need to split the analysis into three cases ($v_{pre} \leq c, c < v_{pre} \leq c + m_{safe}$ and $v_{pre} > c + m_{safe}$) rather than two.

As well as the proportional shareholding, we indicate the preference order of the outcome for that party, with (1) indicating the most preferred outcome (i.e., the

---

[7] Since we are concerned with relative shareholdings rather than valuations, and are dealing with parties who may not have clarity with respect to the accounting view they are using, we do not include the accounting view when naming a method in these tables.

Table 6.3 Pre-Money SAFE: proportional shareholdings from different conversion methods (2). Proportional shareholdings, case of $c < v_{pre} \leq c + m_{safe}$

Approach	Founders		SAFE investor		New investor	
Two-Rounds	$\frac{c}{c+m_{safe}} \cdot \frac{v_{pre}}{v_{pre}+m_{new}}$	(3)	$\frac{m_{safe}}{c+m_{safe}} \cdot \frac{v_{pre}}{v_{pre}+m_{new}}$	(6)	$\frac{m_{new}}{v_{pre}+m_{new}}$	(1)
Standard	$\frac{v_{pre}}{v_{pre}+\frac{v_{pre}m_{safe}}{c}+m_{new}}$	(1)	$\frac{\frac{v_{pre}m_{safe}}{c}}{v_{pre}+\frac{v_{pre}m_{safe}}{c}+m_{new}}$	(2)	$\frac{m_{new}}{v_{pre}+\frac{v_{pre}m_{safe}}{c}+m_{new}}$	(3)
Discounted	$\frac{v_{pre}-m_{safe}}{v_{pre}+m_{new}}$	(4)	$\frac{m_{safe}}{v_{pre}+m_{new}}$	(1-5)	$\frac{m_{new}}{v_{pre}+m_{new}}$	(1)
Dollars Invested	$\frac{c}{c+m_{safe}} \cdot \frac{v_{pre}+m_{safe}}{v_{pre}+m_{safe}+m_{new}}$	(2)	$\frac{m_{safe}}{c+m_{safe}} \cdot \frac{v_{pre}+m_{safe}}{v_{pre}+m_{safe}+m_{new}}$	(4)	$\frac{m_{new}}{v_{pre}+m_{safe}+m_{new}}$	(2)
ZMR (DV)	$\frac{v_{pre}-m_{safe}}{v_{pre}+m_{new}}$	(4)	$\frac{m_{safe}}{v_{pre}+m_{new}}$	(1-5)	$\frac{m_{new}}{v_{pre}+m_{new}}$	(1)
ZMR (Std)	$\frac{c}{c+m_{safe}} \cdot \frac{v_{pre}}{v_{pre}+m_{new}}$	(3)	$\frac{m_{safe}}{c+m_{safe}} \cdot \frac{v_{pre}}{v_{pre}+m_{new}}$	(6)	$\frac{m_{new}}{v_{pre}+m_{new}}$	(1)

Table 6.4 Pre-Money SAFE: proportional shareholdings from different conversion methods (3). Proportional shareholdings, case of $c + m_{safe} < v_{pre}$

Approach	Founders		SAFE investor		New investor	
Two-Rounds	$\frac{c}{c+m_{safe}} \cdot \frac{v_{pre}}{v_{pre}+m_{new}}$	(3)	$\frac{m_{safe}}{c+m_{safe}} \cdot \frac{v_{pre}}{v_{pre}+m_{new}}$	(3)	$\frac{m_{new}}{v_{pre}+m_{new}}$	(1)
Standard	$\frac{v_{pre}}{v_{pre}+\frac{v_{pre}m_{safe}}{c}+m_{new}}$	(1)	$\frac{\frac{v_{pre}m_{safe}}{c}}{v_{pre}+\frac{v_{pre}m_{safe}}{c}+m_{new}}$	(1)	$\frac{m_{new}}{v_{pre}+\frac{v_{pre}m_{safe}}{c}+m_{new}}$	(3)
Discounted	$\frac{c}{c+m_{safe}} \cdot \frac{v_{pre}}{v_{pre}+m_{new}}$	(3)	$\frac{m_{safe}}{c+m_{safe}} \cdot \frac{v_{pre}}{v_{pre}+m_{new}}$	(3)	$\frac{m_{new}}{v_{pre}+m_{new}}$	(1)
Dollars Invested	$\frac{c}{c+m_{safe}} \cdot \frac{v_{pre}+m_{safe}}{v_{pre}+m_{safe}+m_{new}}$	(2)	$\frac{m_{safe}}{c+m_{safe}} \cdot \frac{v_{pre}+m_{safe}}{v_{pre}+m_{safe}+m_{new}}$	(2)	$\frac{m_{new}}{v_{pre}+m_{safe}+m_{new}}$	(2)
ZMR (DV)	$\frac{c}{c+m_{safe}} \cdot \frac{v_{pre}}{v_{pre}+m_{new}}$	(3)	$\frac{m_{safe}}{c+m_{safe}} \cdot \frac{v_{pre}}{v_{pre}+m_{new}}$	(3)	$\frac{m_{new}}{v_{pre}+m_{new}}$	(1)
ZMR (Std)	$\frac{c}{c+m_{safe}} \cdot \frac{v_{pre}}{v_{pre}+m_{new}}$	(3)	$\frac{m_{safe}}{c+m_{safe}} \cdot \frac{v_{pre}}{v_{pre}+m_{new}}$	(3)	$\frac{m_{new}}{v_{pre}+m_{new}}$	(1)

largest proportional holding). Note that these rankings are valid only within each of the cases, they are not intended for comparisons across cases. In some cases (e.g., the case for the founders in the Two-Round method with $v_{pre} \leq c$), the preference order also depends on the value of m_{new}. Rather than further fragment the number of cases, we give a range of rankings in this case.

One immediate observation is that for the SAFE investor, the outcome for the Pre-Money SAFE is always at least as good as the outcome for the Two-Round method. The SAFE investor is likely to hold a stronger negotiating position than the founders, so the Pre-Money SAFE is likely to be the instrument selected in the first step of the game. We may therefore focus the analysis on the Pre-Money SAFE.

The three cases for v_{pre} give the following conclusions for the conversion method chosen at the final step of the game. In the analysis, we assume first that the final move is decided by negotiation between the founders and the new investor. The preferences of the founders and new investor turn out to be at odds in all cases, so we discuss two possible resolutions:

New holds: Here the new investor holds firm and insists on one of their first preferences. The founders yield and accept one of the new investor's first preferences but select one that is the founders' most preferred among that set, and

New yields: The new investors yield to one of their next best preferences. The founders select one of their best preferences among those.

SAFE Coalition: Strictly, the SAFE investor does not have legal standing in setting the terms of the equity round. However, we also consider the potential for them to form a coalition with one of the other players (either the new investor or the founders) to strengthen their position in the negotiation.

1. ($v_{pre} \leq c$): Here the founders and new investor do not have a common best choice: the founders prefer the Standard or Dollars Invested method, and the new investor prefers any of the other options.

   (a) If the new investor holds firm, the founders would select the Zero-Money Round (Standard) method, which is only their third preference. However, the Zero-Money Round methods do carry the risk of a legal dispute by the SAFE investor. To avoid this, the founders would select the Discounted method, which is their least preferred option.

   (b) If the new investor is prepared to yield, their second best options are the Standard or Dollars Invested methods, which are precisely the preferred options of the founders. Thus, the outcome is this case is either the Standard or Dollars Invested method.

   However, the SAFE investor would prefer, depending on m_{new}, either the Discounted, or Zero-Money Round (Discounted) options, which are among the most preferred options of the new investor. So they have an incentive to form a coalition with the new investor to argue for one of these options, most likely the Discounted method, in view of the legal questionability of the Zero-Money Round options.

   Thus, the outcome of this case depends on m_{new}, the negotiation strength of the parties, and their tolerance of legal risk, and could be any of the Standard, Discounted Valuation, Dollars Invested, or Zero-Money Round (Standard) methods.

2. ($c < v_{pre} \leq c + m_{safe}$): Again there is no common best option for the founders and new investor. The founders prefer the Standard method, and the new investor prefers the Discounted, Zero-Money Round (Discounted), or Zero-Money Round (Standard) options.

   (a) If the new investor holds firm, the founders would select the Zero-Money Round (Standard) option again, or their least preferred Discounted option in case they are averse to legal risk.

   (b) If the new investor yields, their second preference is the Dollars Invested method, which is also the second preference of the founders.

The stability of this case with respect to the SAFE investor is complex, since the SAFE investor's ranking is variable within this case. Depending on m_{new}, their interests may be aligned with the new investor by most preferring the Discounted or Zero-Money Round (Discounted) method, but their most preferred option may be also be aligned with the founders by preferring the Standard method. The Dollars Invested method is either the second or third preference of the SAFE investor.

Thus, in this case, again the selection of either the Standard, Discounted Valuation, Dollars Invested or Zero-Money Round (Standard) methods is conceivable.

3. ($c + m_{\text{safe}} < v_{\text{pre}}$): In this case, the rankings of the founders and new investor are once again in conflict, but the situation is somewhat simpler than the previous case. The founders and the SAFE investor both prefer the Standard method, which is the third and lowest preference of the new investor. Conversely, the new investor prefers the Discounted and the two Zero-Money Round methods, which are all the third and lowest preference of the founders and SAFE investor.

   (a) If the new investor holds firm, both the founders and the SAFE investor are indifferent among the new investor's first choices. The outcome is likely to be the Discounted method since it involves only a single equity round and is free from legal risk.

   (b) If the new investor is prepared to compromise, the Dollars Invested method, which is the unique second preference of all parties, is a stable compromise position.

In case the SAFE investor and the founders together have the strongest bargaining position, they may be able to get the new investor to accept the Standard method, but this is their least preferred option, so the case for this would need to be very strong (e.g., the prospects for the company are perceived to be so strong that the new investor is prepared to accept a much weaker deal than they might otherwise.) Thus, in this case, the Standard, Discounted or Dollars Invested methods, are conceivable selections.

In summary, as Feld suggests, the best answer to the question of what outcome can be expected for this version of the game is that "it's complicated." Depending on the parameters, negotiating strengths of the players, and attitude to legal risk, the ultimate selection of any of the conversions methods is conceivable. Table 6.5 summarizes the outcomes for the three possible assumptions with respect to negotiating position: new investor holds, new investor yields, and SAFE forms a dominant coalition. (Since the new investor holds the money, we do not expect the founders to hold the strongest position on their own.) The complexity of the negotiation scenarios resulting from a deferred interpretation of "Pre-Money Valuation" suggests that it would be more rational for all parties to set a clear policy for conversion of the SAFE, based on a clearly stated accounting status of the SAFE.

Table 6.5 Game outcomes by assumptions about v_{pre} and negotiation position

Case	New holds	New yields	SAFE coalition
$v_{pre} \leq c$	ZMR(Std) or DV	Std or DI	DV or ZMR(DV)
$c < v_{pre} \leq c + m_{safe}$	ZMR(Std) or DV	DI	DV or ZMR(DV) or Std
$c + m_{safe} < v_{pre}$	DV	DI	Std

6.4 Gaming the Two-Stage Process

As noted, inasmuch as the SAFE contract does not explicitly grant the SAFE investor negotiation rights, it leaves the terms of the conversion round open to negotiation between the founders and the new investor. This has the risk that they will collude against the interests of the SAFE investor. In the two-stage Zero-Money Round process of Sect. 4.5.2, we assumed that the first round is artificial in that zero new money is invested, and that the new investor invests all their money m_{new} in the second round at valuation v_{pre}, in order to obtain their desired share $m_{new}/(v_{pre} + m_{new})$ in the company, exactly as they would in a standard equity round. However, it is possible to generalize the two-stage process to allow more flexibility in the choice of valuation used in the two rounds, as well as to split the new investor's money across the two rounds.

We now conduct an analysis of a process in which the SAFE investor and the new investor collude as follows: they first negotiate on the proportion of shares to be owned by the new investor after the two-stage process is complete. This negotiation fixes a proportion o_{new} of shares for the new investor (which typically, would equal $m_{new}/(v_{pre} + m_{new})$). The two rounds are then constructed so as to deliver this proportion of shares to the new investor while maximizing the proportion of the company going to the founders (and therefore minimizing the proportion going to the SAFE holder). The new investor splits their money m_{new} into two portions m and $m_{new} - m$, with $0 \leq m \leq m_{new}$. In the first round, the new investor purchases shares for money m at pre-money valuation v_1, discharging the SAFE contract using the standard method. In the second round, the new investor purchases shares for money $m_{new} - m$ at pre-money valuation v_2. We seek constraints on these parameters that achieve a proportional shareholding of o_{new} for the new investor while optimizing the proportional shareholding of the founders. We note that this is equivalent to maximizing the ratio between the founders' share o_f and the SAFE investor's share o_{safe}. We again assume that the SAFE investor does not take up any *pro rata* rights that they may have in the second round. We conduct the analysis for the case that the SAFE is a Pre-Money SAFE with Cap and no Discount. Similar results can be obtained for other SAFE forms.

The outcome of the first round is a proportional shareholding for the founders, SAFE holder, and new investor, respectively, of

$$\frac{v_1}{v_1 + m_{safe} + m}, \quad \frac{m_{safe}}{v_1 + m_{safe} + m}, \quad \frac{m}{v_1 + m_{safe} + m}$$

in case $v_1 \le c$, and

$$\frac{v_1}{v_1 + m_{safe}\frac{v_1}{c} + m}, \quad \frac{m_{safe}\frac{v_1}{c}}{v_1 + m_{safe}\frac{v_1}{c} + m}, \quad \frac{m}{v_1 + m_{safe}\frac{v_1}{c} + m}$$

in case $v_1 > c$.

Determining the proportional shareholdings for the founders, SAFE investor, and new investor after the second round, we obtain proportions (respectively):

$$o_f = \frac{v_1}{v_1 + m_{safe} + m} \cdot \frac{v_2}{v_2 + m_{new} - m}$$

$$o_{safe} = \frac{m_{safe}}{v_1 + m_{safe} + m} \cdot \frac{v_2}{v_2 + m_{new} - m}$$

$$o_{new} = \frac{m}{v_1 + m_{safe} + m} \cdot \frac{v_2}{v_2 + m_{new} - m} + \frac{m_{new} - m}{v_2 + m_{new} - m}$$

in case $v_1 \le c$, and

$$o_f = \frac{v_1}{v_1 + m_{safe}\frac{v_1}{c} + m} \cdot \frac{v_2}{v_2 + m_{new} - m}$$

$$o_{safe} = \frac{m_{safe}\frac{v_1}{c}}{v_1 + m_{safe}\frac{v_1}{c} + m} \cdot \frac{v_2}{v_2 + m_{new} - m}$$

$$o_{new} = \frac{m}{v_1 + m_{safe}\frac{v_1}{c} + m} \cdot \frac{v_2}{v_2 + m_{new} - m} + \frac{m_{new} - m}{v_2 + m_{new} - m}$$

in case $v_1 > c$.

We conduct the analysis for the two cases, to determine the values of v_1, v_2 and m that give the optimal outcome for the founders, given that the new investor gets a share o_{new} at the end of the process.

Case 1: $v_1 \le c$. In this case, the ratio between the founders' share and the SAFE investor's share, is, from the above, $\frac{o_f}{o_{safe}} = \frac{v_1}{m_{safe}}$. Plainly this takes a maximum at $v_1 = c$. Noting that $o_f + o_{safe} + o_{new} = 1$ and substituting the value $o_{safe} = o_f m_{safe}/c$ obtained at the maximum, we derive that

$$o_f = \frac{c}{c + m_{safe}} \cdot (1 - o_{new}) \text{ and } o_{safe} = \frac{m_{safe}}{c + m_{safe}} \cdot (1 - o_{new}).$$

As this result was obtained without consideration of m or v_2, it remains to verify that the result can be obtained with feasible values for these variables. These variables are related because, when $v_1 = c$, the new investor proportion after the two rounds is

$$o_{\text{new}} = \frac{m}{c + m_{\text{safe}} + m} \cdot \frac{v_2}{v_2 + m_{\text{new}} - m} + \frac{m_{\text{new}} - m}{v_2 + m_{\text{new}} - m}$$

Solving for v_2, we get

$$v_2 = \frac{(m_{\text{new}} - m)(1 - o_{\text{new}})(c + m_{\text{safe}} + m)}{o_{\text{new}}(c + m_{\text{safe}} + m) - m} .$$

Note that all terms in the numerator are non-negative, so we have a feasible value $v_2 \geq 0$ satisfying the conditions for this case when $o_{\text{new}}(c + m_{\text{safe}} + m) - m > 0$, i.e., when

$$\frac{o_{\text{new}}(c + m_{\text{safe}})}{1 - o_{\text{new}}} > m .$$

This can always be satisfied by choosing m to be sufficiently small. (We note, however, that for the round to be considered *bona fide*, larger values are more credible).

Case 2: $v_1 > c$. In this case, note that the ratio between the founder share and the SAFE investor share is, from the above, $\frac{o_f}{o_{\text{safe}}} = \frac{c}{m_{\text{safe}}}$. This amount is independent of the choice of v_1, v_2 and m. Thus, after fixing the proportion going to the new investor after two rounds at o_{new}, this case does not provide any opportunity to manipulate the outcome by choosing parameters of the two-stage process.

As above, we derive that the corresponding founder share is

$$o_f = \frac{c}{c + m_{\text{safe}}} \cdot (1 - o_{\text{new}})$$

holding not just at the maximum, but *always*. This is identical to the *optimal* solution from Case 1, but this is to be expected, given that Case 2 gives a constant result and its definition is continuous with Case 1 at the boundary between the two cases, where the optimum of Case 1 occurs.

Again, we need to consider feasibility of the values for v_2 and m. Similarly to the above calculation, we get

$$v_2 = \frac{(m_{\text{new}} - m)(1 - o_{\text{new}})(v_1 + m_{\text{safe}}\frac{v_1}{c} + m)}{o_{\text{new}}(v_1 + m_{\text{safe}}\frac{v_1}{c} + m) - m} .$$

so that we require for feasibility that

$$o_{\text{new}}(v_1 + m_{\text{safe}}\frac{v_1}{c} + m) - m > 0$$

that is,

$$\frac{o_{new}(v_1 + m_{safe}\frac{v_1}{c})}{1 - o_{new}} > m .$$

Again, this is always possible by choosing m to be sufficiently small.

We conclude that the optimal share for the founders obtainable from the two-round process, given that the new investors receive share o_{new}, is always $c(1 - o_{new})/(c + m_{safe})$. There may be multiple choices of v_1, v_2 and m that yield this outcome.

The following example illustrates the extent of the advantage that can be obtained for the new investor and founders from this two-stage process in case $v_{pre} < c$.

Example 6.1 Consider a company whose founders have 1000 shares, and suppose an investor, Saffron, buys a Pre-Money SAFE with a valuation cap of $1,000,000 and no discount, for principal $10,000.

We suppose that later, a new investor Neville has $1,500 to invest, and that the development of the company has gone poorly, so that its valuation has dropped significantly to around $3,000, or $3 per share. If this investment were made using an equity round that discharges the SAFE according to the Standard method, the numbers of shares after the equity round would be Founders: 1000, Saffron: 3333, Neville: 500, and proportional shareholdings, roughly, Founders: 21%, Saffron: 69% and Neville: 10%. Saffron would have control of the company in this situation.

To avoid this outcome, Neville and the company collude to limit the influence of Saffron after an equity financing, by structuring Neville's investment into two rounds rather than one. It is agreed to structure the two rounds so as to give Neville a 1/3 stake at the end of the two stages. (Note that this is equivalent to the stake Neville would obtain if the SAFE were not present.)

Specifically, in spite of the low valuation, in the first round, Neville provides equity financing of $1000 at a Pre-Money Valuation of $1,000,000 and hence a price of $1000 per share. (Note that the valuation in this first round is equal to the cap. It was shown above that this gives the optimal share to the founders.) After this financing, the founders have 1000 shares, Saffron has 10, and Neville has 1, so the proportional shareholdings are, roughly, founders: 99%, Saffron: 0.99%, and Neville: 0.01%. The SAFE is terminated in this round. Saffron might well be satisfied with the deal since it implies a valuation of her shares roughly equal to her original investment, so she avoids a paper loss.

Later, in a second round of equity financing, Neville uses the remaining $500 to buy 504 shares at a price $0.99206 (a valuation of $1,002.976). The SAFE has terminated, so Saffron receives no shares in this round. The capitalization is now 1511 shares, of which the founders have 1000, Saffron has 10, and Neville has 505, or proportions of 66, 0.7 and 33.3%.

Thus, overall, rather than a 10% stake, Neville has obtained his desired 1/3 stake for the same money, and with a significantly different outcome for the founders and Saffron. The founders retain almost all of the controlling stake that Saffron would have obtained in a Standard SAFE round. If Neville had paid this price in a single round, Saffron would have received 3334 shares and become the majority shareholder.

If Saffron has a *pro rata* rights agreement with the company then she can obtain some shares at the same price as Neville, but that simply maintains her shareholding at about 1%. She still has arguably been defrauded of a majority share in the company.

It is of interest to compare this to an "honest" one-round process that guarantees a particular share to the new investor, namely, the Discounted Valuation method.

Recall from Sect. 4.3.1 that this involves a round conducted at valuation $v_{\text{pre}} = m_{\text{new}}(1 - o_{\text{new}})/o_{\text{new}}$ and yields founder share

$$\frac{v_{\text{pre}} - m_{\text{safe}}}{v_{\text{pre}} + m_{\text{new}}} = \frac{m_{\text{new}}\frac{1-o_{\text{new}}}{o_{\text{new}}} - m_{\text{safe}}}{m_{\text{new}}\frac{1-o_{\text{new}}}{o_{\text{new}}} + m_{\text{new}}}$$

$$= \frac{m_{\text{new}} - o_{\text{new}}(m_{\text{new}} + m_{\text{safe}})}{m_{\text{new}}}$$

in case $v_{\text{pre}} \leq c + m_{\text{safe}}$ and

$$\frac{c}{c + m_{\text{safe}}} \cdot \frac{v_{\text{pre}}}{v_{\text{pre}} + m_{\text{new}}} = \frac{c}{c + m_{\text{safe}}} \cdot \frac{m_{\text{new}}\frac{1-o_{\text{new}}}{o_{\text{new}}}}{m_{\text{new}}\frac{1-o_{\text{new}}}{o_{\text{new}}} + m_{\text{new}}}$$

$$= \frac{c(1 - o_{\text{new}})}{c + m_{\text{safe}}}$$

in case $v_{\text{pre}} > c + m_{\text{safe}}$. The latter is identical to the optimum attainable from the two-round process, showing that the two-round process gives the founders no advantage in this case.

In case $v_{\text{pre}} \leq c + m_{\text{safe}}$, we have $v_{\text{pre}} = \frac{m_{\text{new}}(1-o_{\text{new}})}{o_{\text{new}}} \leq c + m_{\text{safe}}$, that is,

$$o_{\text{new}} \geq \frac{m_{\text{new}}}{c + m_{\text{safe}} + m_{\text{new}}}.$$

The optimal two-round process is as good or better for the founders, compared to the Percent-Ownership method, if

$$\frac{c(1 - o_{\text{new}})}{c + m_{\text{safe}}} \geq \frac{m_{\text{new}} - o_{\text{new}}(m_{\text{new}} + m_{\text{safe}})}{m_{\text{new}}}$$

(and strictly better when the inequality is strict) or, equivalently,

$$o_{\text{new}} \geq \frac{m_{\text{new}}}{c + m_{\text{safe}} + m_{\text{new}}}. \tag{6.3}$$

Therefore, when $v_{\text{pre}} < c + m_{\text{safe}}$, the founders obtain an advantage from the two-round process, compared to the Percent-Ownership method.

The amount of advantage to the founders, compared to the Percent-Ownership method, is that they obtain an additional share of

$$\frac{m_{\text{safe}}}{c + m_{\text{safe}}} \cdot \frac{o_{\text{new}}(c + m_{\text{safe}} + m_{\text{new}}) - m_{\text{new}}}{m_{\text{new}}}$$

of the company. The share of the SAFE investor decreases by a corresponding amount.

As in the previous section, we have ignored the possibility that this loss causes the SAFE investor to litigate on the basis of the claim that the first round was not *bona fide*. It is likely that "gaming" of the SAFE along the lines contemplated in the present section is even less legally defensible than the use of a Zero-Money Round. For example, the SAFE investor may introduce an independent valuation of the company to argue that there has been a breach of contract. They may also argue that the intention of the contract was to provide them with downside protection in the event of a low de-facto valuation, and that they have been deprived of this protection by the artificial construction of the two rounds. It is not clear whether such arguments would be upheld, but the risk of costly litigation and the desire to keep the SAFE investor on good terms may inhibit the type of manipulation considered in the present section. Nevertheless, this type of manipulation has, reportedly, been seen in the wild.[8]

6.5 Summary

In this chapter, we have considered conversion of a SAFE from a game-theoretic viewpoint, in which the players are the founders, the SAFE investor, and the new investor. The diversity of conversion methods arises from a lack of clarity about the term "pre-money valuation" and the accounting view with respect to which the company is valued.

First, we have shown that if it is common knowledge that all players are rational in the sense of properly relating the accounting view to the understanding of the pre-money valuation at which the round is conducted, then all conversion processes have the same outcome, and the Discounted Valuation conversion method (starting with Cap table accounting), captures this canonical outcome. For this reason, we focus primarily on this conversion method in the following Part III, where we analyze conversion of multiple SAFEs.

Next, we have considered the conversion game under the assumption that the players leave the clarification of the meaning of "pre-money valuation" until after the term sheet with the new investor has been signed. While there is evidence that this occurs in practice, we have seen that this leads to a very complex negotiation scenario, in which the outcome of conversion depends on the relative negotiation strength of the players, and the coalitions that they form.

---

[8] Romain de Spoelberch, Polymorphic Capital, personal communication.

Finally, we have considered an aggressive (but legally questionable) form of "gaming" of SAFE conversion, in which the founders and new investor collude to artificially structure the equity round into two rounds, in order to deprive the SAFE investor of a proportional shareholding to which they would otherwise have been entitled.

References

1. Colla D Calculating share price with outstanding convertible notes. https://www.cooleygo.com/calculating-share-price-outstanding-convertible-notes/. Accessed Sep 2019
2. Feld B (2015) The pre-money vs. post-money confusion with convertible notes. https://feld.com/archives/2015/06/pre-money-vs-post-money-confusion-convertible-notes.html. Accessed Sep 2019

Part III
Analysis: Multiple SAFE Scenarios

Chapter 7
Equity Financing with Multiple SAFEs

We show in this chapter that many of the conclusions that apply for conversion of single SAFEs during equity financing extend to the scenario of multiple SAFEs, when they are Pre-Money SAFEs. However, for Post-Money SAFEs, we show that the situation is more complicated, and reveal a design flaw of these SAFE types: there can be multiple intervals of valuations at which SAFEs cannot be converted.

7.1 Introduction

In Part II, we considered several approaches to the conversion of a SAFE in the context of an Equity Financing event, and related them to Liability and Cap Table accounting of the SAFE. We argued that some of these approaches, like the Dollars Invested method, arise as a result of an incorrect understanding of the equations governing the equity round, the application of an inappropriate accounting method, or the equations being used. In considering the "rational" conservative methods in Sect. 6.1, where the players apply the correct equations for the conversion method being used, we found that these methods (the Standard method with Liability view, the Discounted Valuation method with Cap Table view, and the Zero-Round method with Cap Table view) all yield identical distributions of shares for the players subject to the constraint $v_{\mathrm{pre}} > m_{safe}$ on their applicability. The remaining conservative method, the Zero-Round method with Liability view does not require this constraint, so it is more broadly applicable, but it may be subject to legal challenge when the constraint $v_{\mathrm{pre}} > m_{safe}$ is satisfied. Thus, there is a case that the Discounted Valuation method with Cap Table view is, in general, the "correct" method to use when calculating the pre-money valuation and share price to be used in an equity round. We therefore narrow our focus to this method in what follows.

Our discussion in Part II dealt only with a simple scenario in which a *single* SAFE had been issued by the company by the time of an equity round. In this chapter, we describe how the Discounted Valuation method works in equity rounds where the company has issued *multiple* convertible instruments. In particular, we consider the application of the method when the company has issued SAFEs of the same two types (either uniformly Pre-Money SAFEs with Cap Only, or uniformly Post-Money SAFEs with Cap Only).[1] We also consider the relationship between these two types of SAFE when converted using the Discounted Valuation method.

The main conclusions of this chapter are that:

- The Discounted Valuation method extends to the case where multiple SAFEs have been issued, and remains conservative for both the new investors and the SAFE investors.
- Like the single SAFE case, there is a company valuation below which multiple Pre-Money SAFEs cannot be converted using the Discounted Valuation method. For Post-Money SAFEs, there are also valuations at which the SAFEs are not convertible using this method. However, the situation is significantly more complex for Post-Money SAFEs. The inability to convert occurs at higher valuations, and whereas the Discounted Valuation method is not applicable to Pre-Money SAFEs for valuations in a single interval, there may be multiple intervals over which Post-Money SAFEs are not convertible.
- In the situation where the company has issued a single SAFE contract, Proposition 6.1 established a correspondence between Pre-Money and Post-Money SAFEs, in the sense that for each Pre-Money SAFE, there is a Post-Money SAFE which, when converted using the Discounted Valuation method, yields the same share distribution as the Post-Money SAFE. In general, this correspondence does not hold with multiple SAFEs. However, we show that in scenarios where all Pre-Money SAFEs have been issued at the same Cap, there exists a scenario in which the company has issued instead a set of Post-Money SAFEs, with the same share distribution resulting under the Discounted Valuation method in these two scenarios. (There is, however, a limitation to this result in that the correspondence applies only at the level of the entire scenario, and not at the level of the individual SAFEs.)

The structure of the chapter is as follows. A description of the scenario with multiple SAFEs is given in Sect. 7.2. In Sect. 7.3 we give a general formulation of the Discounted Valuation method for a setting with multiple convertible instruments. The result of applying this method for conversion of multiple Pre-Money SAFEs with Cap Only is developed in Sect. 7.4. Conversion for multiple Post-Money SAFEs with Cap Only using the Discounted Valuation method is explained in Sect. 7.5. The two results from the application of the Discounted Valuation method for these two types of SAFE are compared in Sect. 7.6. Section 7.7 discusses scenarios with a mix of Pre-Money and Post-Money SAFEs. Finally, Sect. 7.8 makes some concluding remarks about conversion of multiple SAFEs.

---

[1] We believe that our analytic techniques apply to other forms of SAFE, with only minimal changes to the resulting formulas.

7.2 The Multiple SAFE Scenario

We now describe the scenario that we analyze in this chapter, in which multiple SAFEs have been issued by the time of the equity round, and the variables that we use in the analysis.

As before, we make the simplifying assumption that the company has not issued or promised options, and that there is no post-round option pool, so that the only pre-round non-converting equity that needs to be considered is the number founder shares, which, as above, we denote by s_f. (The other equity parameters, if desired, could be included in this variable.)

When dealing with multiple SAFEs, we assume the SAFEs are labeled i, for $i = 1, \ldots, k$. SAFE number i has the following associated variables, corresponding to the previous names in the single SAFE scenario:

1. a SAFE Cap c_i (previously c),
2. a Purchase Amount (money) m_i (previously m_{safe}),
3. a Safe Price p_i (previously, p_{safe}),
4. a number of shares issued in conversion s_i (previously s_{safe}),
5. a valuation v_i (previously v_{safe}).

We use these parameters for both Pre-Money and Post-Money SAFEs. As above, we denote the new investor's principal by m_{new}, the number of shares issued to the new investor by s_{new}, the Pre-Money Valuation by v_{pre}, and the share price paid by the new investor by p_{new}. (As above, we assume that all new investors pay the same price p_{new} for their Preferred shares.)

A number of the equations governing an equity round with a single convertible instrument, from Chap. 3 need to be generalized to the multi-SAFE scenario. In particular, we now have that the number of post-money shares S_{post} is given by

$$S_{\text{post}} = s_f + \left(\sum_{i=1}^{k} s_i \right) + s_{\text{new}}$$

and Eq. (3.3.1), relating the pre-money and post-money valuations is generalized to

$$v_{\text{post}} = v_{\text{pre}} + \left(\sum_{i=1}^{k} v_i \right) + m_{\text{new}} \qquad (vmC_{\text{pre,post}})$$

7.2.1 Pre-Money SAFE with Cap Only

For SAFE i, following the formalization in Sect. 4.1, the number of shares s_i to be issued in conversion of the SAFE is as follows:

1. If $v_{pre} \leq c_i$, then $s_i = m_i/p_{new}$
2. If $v_{pre} > c_i$, then $s_i = m_i/p_i$

where m_i is the purchase amount for SAFE i and the Safe Price p_i is defined by

$$p_i = \frac{c_i}{s_f} .$$

Recall that the variable v_{pre} here corresponds to the term "Pre-Money Valuation" in the natural language text of the contract. We have already noted (in Sect. 4.3), that some subtleties arise concerning the interpretation of this term in the context of the Discounted Valuation method.

We note that this contract has the following property with respect to the value of the shares received by the SAFE investor in an Equity Financing:

Proposition 7.1 *Suppose that the equity round is conducted at a valuation v_{pre}, with a share price p_{new} for the new investor's shares determined using Eq. (vsp_{pre}) (corresponding to the Liability view), and that the value per share after the equity round remains equal to p_{new}, that is, satisfies Eq. (vsp_{post}). Then the number shares issued to SAFE investor i satisfies*

$$s_i = \frac{m_i}{\min(p_{new}, p_i)}$$

and the value of these shares after the equity round is at least the money m_i that they invested.

Proof Note that if Eq. (vsp_{pre}) holds, then we have $v_{pre} \leq c_i$ iff

$$p_{new} = \frac{v_{pre}}{S_{pre}} \leq \frac{c_i}{S_{pre}} = p_i$$

where $S_{pre} = s_f$. If $v_{pre} \leq c_i$, then

$$s_i = \frac{m_i}{p_{new}} = \frac{m_i}{\min(p_{new}, p_i)}$$

and the value of the shares is

$$s_i p_{new} = \frac{m_i}{p_{new}} \cdot p_{new} = m_i .$$

If $v_{pre} > c_i$, then

$$s_i = \frac{m_i}{p_i} = \frac{m_i}{\min(p_{new}, p_i)}$$

and the value of the shares is

$$
\begin{aligned}
s_i p_{\text{new}} &= \frac{m_i}{p_i} \cdot p_{\text{new}} \\
&= \frac{m_i S_{\text{pre}}}{c_i} \cdot \frac{v_{\text{pre}}}{S_{\text{pre}}} \qquad\qquad \text{by } (\text{vsp}_{\text{pre}}) \\
&= m_i \cdot \frac{v_{\text{pre}}}{c_i} \\
&> m_i \qquad\qquad\qquad\qquad \text{by the case assumption.}
\end{aligned}
$$

In either case, the claim holds. $\qquad\qquad\qquad\qquad\qquad\qquad\qquad\qquad$ □

This result states that, under the assumptions, the equity round is conservative for the SAFE investors.

7.2.2 Post-Money SAFE with Cap Only

Rather than state an explicit conversion formula, the Post-Money SAFEs state a set of constraints that need to be solved in order to determine the number of shares issued to the SAFE investor. Whereas the number of shares issued for a Pre-Money SAFE is independent of other SAFEs issued, in the case of Post-Money SAFEs, these constraints include other SAFEs. In effect, the Post-Money SAFE conversion constraints view SAFEs as corresponding to a number of shares existing on the cap table at the time of the equity round.

With respect to our simplifying assumptions, the Company Capitalization S_{pre} is defined by the Post-Money SAFE not by Eq. (4.1) as in the Pre-Money SAFE, but by equation

$$
S_{\text{pre}} = s_f + \sum_{i=1}^{k} s_i \tag{7.1}
$$

where s_i is the number of shares that will be issued to SAFE investor i.

The Post-Money SAFE Equity Financing clause states that the number s_i of shares issued to the SAFE investor i satisfies

$$
s_i = \max\{m_i/p_{\text{new}}, m_i/p_i\} \tag{7.2}
$$

where p_{new} is the (minimum) price at which Standard Preferred shares are issued in the round and the SAFE Price p_i is defined as

$$
p_i = c_i/S_{\text{pre}} . \tag{7.3}
$$

As in the single SAFE scenario, since s_i depends on the SAFE Price p_i, which depends on S_{pre}, which in turn depends on s_i, these definitions are circular. To resolve the circularity, we need to take these equations as simultaneous constraints on these values, to be solved. We give the details of the solution in Sect. 7.5.

7.3 Discounted Valuation Method Revisited

Before investigating the specific consequences of the SAFE conversion formulas, it is helpful to recall the Discounted Valuation method for setting a share price in equity rounds involving a general notion of convertible contract (that is SAFE-like in the sense of being deterministic, so that the contract holder does not have a choice on whether to convert), and express its equations for the setting where there are multiple convertible contracts, rather than the single SAFE scenario for which it was introduced above.

The Discounted Valuation method can be understood as a response to the fact that issuance of shares s_i is dilutive for the new investor, which means that the round is not conservative. To compensate, this method artificially adjusts the pre-money valuation v_{pre} used to determine a share price to a value v_{pre}^- calculated to ensure that, immediately after the equity round, the new investor holds shares of value equal to the money they paid.

The assumption that the share price is determined from v_{pre}^- rather than v_{pre} is captured by equation

$$p_{new} = \frac{v_{pre}^-}{s_f} \tag{7.4}$$

Since shares are issued to the new investor at this price, we have equation (msp_{new}) with respect to v_{pre}^-. To obtain that the round is *minimally conservative* for the new investor (that is, to ensure that the post-money value of the new investors shares is *exactly* equal to their investment m_{new}), we need to have that the value per share after the equity round is equal to the share price paid by the new investor.

After the equity round, the capitalization of the company is $s_f + s_{new} + \sum_i s_i$. Moreover, assuming that the share price immediately after the equity round is equal to the share price before the equity round (as is the case for an equity round not involving convertible instruments), and that the post-money valuation is $v_{pre} + m_{new}$, we have

$$p_{new} = \frac{v_{pre} + m_{new}}{s_f + s_{new} + \sum_i s_i} \tag{7.5}$$

Reorganizing, this gives that

$$p_{new}s_f + p_{new}s_{new} + \sum_i p_{new}s_i = v_{pre} + m_{new}$$

so using equation (msp_{new}), we obtain

$$p_{new}s_f + \sum_i p_{new}s_i = v_{pre} \tag{7.6}$$

We can take two views of this equation. First, using Eq. (7.4), we can derive

$$v_{\text{pre}}^- = v_{\text{pre}} - \sum_i p_{\text{new}} s_i$$

Intuitively, each term $p_{\text{new}} s_i$ here can be understood as the valuation of the i-th convertible instrument, as determined from the number of shares s_i issued in exchange and the price per share p_{new}. Thus, we can understand v_{pre}^- as the result of discounting the pre-money valuation v_{pre} by the valuation of the convertible instruments. In effect, v_{pre}^- is the valuation of the company when the convertible instruments are accounted for as liabilities, assuming that v_{pre} is a valuation of all other assets and liabilities of the company (but excluding the convertible instruments). Note that, on this view, since these liabilities are discharged in the course the equity round, the valuation increases by an equivalent amount, so that the post-money valuation of the company is

$$v_{\text{pre}}^- + m_{\text{new}} + \sum_i p_{\text{new}} s_i = v_{\text{pre}} + m_{\text{new}}$$

exactly as expected. We also have that the generalized form of Eq. (vmC$_{\text{pre, post}}$) is satisfied, but with v_{pre}^- in place of v_{pre}, and with the valuation of SAFE i taken to be $v_i = p_{\text{new}} s_i$.

Alternately, note that Eq. (7.6) can be written as

$$p_{\text{new}} = \frac{v_{\text{pre}}}{s_f + \sum_i s_i} \tag{7.7}$$

This view can be understood as determining the share price on the understanding that the convertible instruments are represented not as liabilities, but as entries on the cap table of the company. That is, we calculate a share price from v_{pre} as if the shares s_i to be issued in conversion of the convertible instrument have already been issued before the equity round is conducted.

As with the single SAFE scenario, as discussed in Sect. 4.3, the alternative Percent-Ownership method also yields the same equations, based on the consideration that the new investor expects to receive for their money the same proportional shareholding in the company as they would in an equity round not involving convertible instruments. After the equity round, the new investor's actual share of the company is

$$o_{\text{new}} = \frac{s_{\text{new}}}{s_f + s_{\text{new}} + \sum_i s_i}$$

Multiplying numerator and denominator by p_{new}, we have

$$o_{\text{new}} = \frac{s_{\text{new}} p_{\text{new}}}{s_f p_{\text{new}} + s_{\text{new}} p_{\text{new}} + \sum_i s_i p_{\text{new}}}$$

$$= \frac{m_{\text{new}}}{s_f p_{\text{new}} + m_{\text{new}} + \sum_i s_i p_{\text{new}}} \qquad \text{by (msp}_{\text{new}})$$

Had the equity round been a standard equity round in which convertible instruments are not present, the new investor would have received a share

$$o_{\text{new}} = \frac{m_{\text{new}}}{v_{\text{pre}} + m_{\text{new}}}$$

of the company. Note that this share, plus the assumption that the post-money valuation of the company is $v_{\text{pre}} + m_{\text{new}}$, yields that the value of the shareholding is m_{new}, so that the round is minimally conservative for the new investor. Equating these two formulations of o_{new}, we again derive Eq. (7.6).

Thus, in fact, formula (7.6) underpins three different ways to understand the equity round:

- Decreasing the valuation used to determine the price of shares in the equity round so as to guarantee that the round is minimally conservative for the new investor. Equivalently, treating the convertible instruments as liabilities of the company (with valuation of these liabilities derived from the price at which shares are issued and the number of shares issued in conversion).
- Determining a price by treating the shares that will be issued in the equity round as already represented on the cap table.
- Setting the valuation of the company so as to ensure that the new investor receives an expected share of the company.

Ignoring any extraneous financial implications from these different viewpoints (e.g., from differing tax treatment), each yields that the round is minimally conservative for the new investor.

We note that, depending on the specifics of the convertible instruments, our characterizations of the discounted pre-money valuation v_{pre}^- and price p_{new} above may be insufficient to directly determine the values of these variables. The problem is that the numbers s_i of shares issued in conversion may be defined in terms of the pre-money valuation and price. As seen in Chap. 4, this circular dependency does in fact apply to the specific case of SAFE contracts, requiring further analysis. We take this up for the two types of SAFEs we consider in the following sections.

7.4   Discounted Valuation Method: Pre-Money SAFEs

We now consider the application of the Discounted Valuation method to our multi-SAFE scenario assuming multiple Pre-Money SAFEs (but no Post-Money SAFEs) have been issued. Our discussion in this section generalizes the treatment in Sect. 4.3 for equity rounds in the case where the company has issued a single Pre-Money SAFE. We suppose that the new investor's inherent pre-money valuation of the company (ignoring the existence of SAFE contracts) is v_{pre}.

The Discounted Valuation method is (minimally) conservative for the new investor by design, and also ensures that share price before the equity round is equal to the share price after the equity round. It follows using Proposition 7.1 that the equity round is also conservative for the SAFE investors.

We have not yet determined an actual value for v_{pre}^-. There is a circularity in the equation $v_{\text{pre}}^- = v_{\text{pre}} - \sum_i s_i \, p_{\text{new}}$, because, for the Pre-Money SAFE, the term $s_i \, p_{\text{new}}$ depends on the "Pre-Money Valuation" at which the equity round is conducted. We resolve this circularity by showing that the SAFE contract imposes additional constraints on these variables, that lead to a unique solution.

We note that, when using the Discounted Valuation method with Pre-Money SAFEs, a question arises concerning the term "Pre-Money Valuation" in this contract: should we interpret it as v_{pre}^- or as v_{pre}? In some cases, this choice leads to different values of s_i. See Sect. 4.3 for arguments leading to the conclusion that the Pre-Money SAFE is more coherent with the interpretation v_{pre}^-. We therefore use that interpretation in the following analysis. (As discussed in Sect. 4.3, although the analysis determines a price starting with a valuation v_{pre} according to the Cap Table view, the equity round is conducted at the calculated valuation v_{pre}^- corresponding to the Liability view.)

Using the interpretation v_{pre}^- of "Pre-Money Valuation", the conversion formula gives, using Proposition 7.1 and the definition of the Pre-Money SAFE,[2]

$$s_i \, p_{\text{new}} = \frac{m_i}{min(p_{\text{new}}, \, p_i)} \cdot \frac{v_{\text{pre}}^-}{s_f} = \frac{m_i}{min(v_{\text{pre}}^-, \, c_i)} \cdot v_{\text{pre}}^- \, . \tag{7.8}$$

Thus, we need to solve

$$v_{\text{pre}} = v_{\text{pre}}^- + \sum_{i:v_{\text{pre}}^- < c_i} m_i + \sum_{i:v_{\text{pre}}^- \geq c_i} \frac{m_i}{c_i} \cdot v_{\text{pre}}^- \tag{7.9}$$

for v_{pre}^-.

Since we need $v_{\text{pre}}^- > 0$ (else we get a zero or negative price per share) and $(m_i/c_i) \cdot v_{\text{pre}}^- \geq m_i$ when $v_{\text{pre}}^- \geq c_i$, we must have $v_{\text{pre}} > \sum_i m_i$ in order for this to be solvable. Conversely, we now show that we can solve for $v_{\text{pre}}^- > 0$ when $v_{\text{pre}} > \sum_i m_i$.

Note that the right-hand side of Eq. (7.9) is continuous, piecewise linear and increasing in v_{pre}^-, since $m_i = \frac{m_i}{c_i} \cdot v_{\text{pre}}^-$ when $v_{\text{pre}}^- = c_i$. Hence, in case $v_{\text{pre}} > \sum_i m_i$, we get a unique solution for v_{pre}^- in terms of v_{pre}. To express this, order the distinct SAFE caps among the c_i as $c_1' < c_2' < \ldots < c_k'$, and let $c_0' = 0$. The values $c_0', \ldots c_k'$ on the v_{pre}^- axis correspond to the points $C_0, \ldots, C_k$ on the v_{pre} axis, defined by

$$C_j = c_j' + \sum_{i:c_j' < c_i} m_i + \sum_{i:c_j' \geq c_i} \frac{m_i}{c_i} \cdot c_j' \, .$$

(See Fig. 7.1.) Note $C_0 = \sum_i m_i$.

---

[2] Note that if we used interpretation v_{pre} for "Pre-Money Valuation", but priced that round using Eq. (7.4), then Proposition 7.1 would not apply, because different values are being used for v_{pre} in the SAFE conversion conditions and the price calculation.

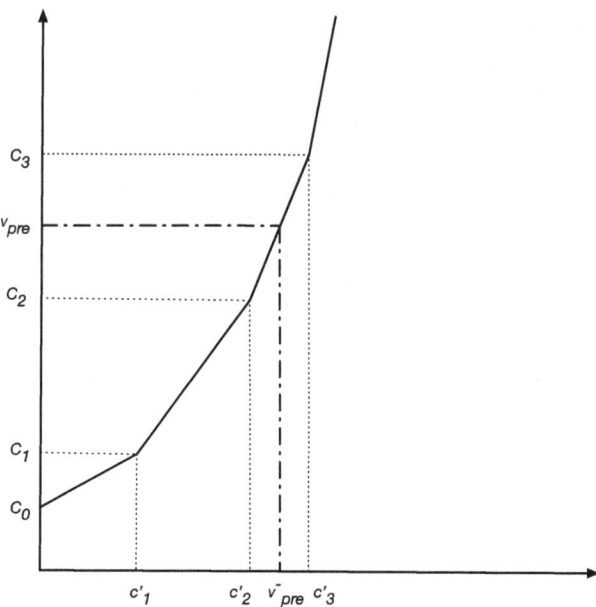

Fig. 7.1 v_{pre} as a function of v_{pre}^-

For $i = 1 \ldots k$ let $f(i)$ be the index such that $c_i = c'_{f(i)}$. Then $v_{\text{pre}}^- < c_i$ iff $v_{\text{pre}}^- < c'_{f(i)}$ iff $v_{\text{pre}} < C_{f(i)}$. Hence

$$v_{\text{pre}}^- = \frac{v_{\text{pre}} - \sum_{i:v_{\text{pre}} < C_{f(i)}} m_i}{1 + \sum_{i:v_{\text{pre}} \geq C_{f(i)}} \frac{m_i}{c_i}} \tag{7.10}$$

Note that we need

$$v_{\text{pre}} > \sum_{i:v_{\text{pre}} < C_{f(i)}} m_i$$

for this solution to yield $v_{\text{pre}}^- > 0$, but this holds when $v_{\text{pre}} > \sum_i m_i$.

We can now determine the number of shares issued in exchange for the i-th SAFE as

$$s_i = \frac{m_i s_f}{\min(v_{\text{pre}}^-, c_i)}$$

$$= \begin{cases} \frac{m_i s_f}{v_{\text{pre}}^-} & \text{if } v_{\text{pre}}^- < c_i \\[2mm] \frac{m_i s_f}{c_i} & \text{if } v_{\text{pre}}^- \geq c_i \end{cases}$$

$$= \begin{cases} \dfrac{m_i s_f \cdot \left(1 + \sum_{\ell: v_{\text{pre}} \geq C_{f(\ell)}} \frac{m_\ell}{c_\ell}\right)}{v_{\text{pre}} \cdot \left(1 - \sum_{\ell: v_{\text{pre}} < C_{f(\ell)}} m_\ell\right)} & \text{if } v_{\text{pre}} < C_{f(i)} \\[2em] \dfrac{m_i s_f}{c_i} & \text{if } v_{\text{pre}} \geq C_{f(i)} \end{cases}$$

The founder value after the equity round is

$$v_{\text{pre}} - \sum_i s_i p_{\text{new}} = v_{\text{pre}} - \left(\sum_{i: v_{\text{pre}}^- < c_i} m_i\right) - \left(\sum_{i: v_{\text{pre}}^- \geq c_i} m_i \frac{v_{\text{pre}}^-}{c_i}\right) \qquad (7.11)$$

7.5   Discounted Valuation Method: Post-Money SAFEs

As already noted, the Post-Money SAFE uses a calculation of the Safe Price from the Safe Cap in the case that $v_{\text{pre}} \geq c$, that is similar to that used in the Discounted Valuation method to obtain a discounted price for the equity round (Eq. (7.7)). However, it remains the case that the issuance of SAFE shares is dilutive for the new investor, irrespective of how the number of SAFE shares is calculated. It is therefore appropriate to determine the price of the round using the Discounted Valuation method to counteract this effect. We suppose that the company has issued post-money SAFEs to k investors, with investor $i = 1 \ldots k$ purchasing a SAFE for price m_i at post-money cap c_i.

The Company Capitalization, as defined by the Post-Money SAFE (pre the equity round, and inclusive of the SAFE shares), is

$$S_{\text{pre}} = s_f + \sum_{i=1}^{k} s_i$$

where s_i is the number of shares that will be issued to SAFE investor i. The Post-Money SAFE states that the number of shares issued to SAFE investor i in the equity round is given by

$$s_i = \max\left(\frac{m_i}{p_{\text{new}}}, \frac{m_i}{p_i}\right)$$

where p_i is the Safe Price, defined as $p_i = c_i / S_{\text{pre}}$.

To apply the Discounted Valuation method, we use a price p_{new} determined from the discounted valuation v_{pre}^- derived from v_{pre}. We have several equivalent formulations of this price (Eqs. (7.4) and (7.7)). In the present context, it is more convenient to use the formulation $p_{\text{new}} = v_{\text{pre}}^- / S_{\text{pre}}$ of Eq. (7.7), which yields that

$$s_i = \max\left(\frac{m_i S_{\text{pre}}}{v_{\text{pre}}^-}, \frac{m_i S_{\text{pre}}}{c_i}\right) = \frac{m_i S_{\text{pre}}}{\min(v_{\text{pre}}^-, c_i)} \qquad (7.12)$$

The valuation v_i of these shares is given by

$$v_i = s_i \, p_{\text{new}} = \frac{m_i S_{\text{pre}}}{\min(v_{\text{pre}}, c_i)} \cdot \frac{v_{\text{pre}}}{S_{\text{pre}}} = \frac{v_{\text{pre}} m_i}{\min(v_{\text{pre}}, c_i)}$$

from which we see that the round is conservative for the SAFE investors.

We also obtain the discounted valuation

$$v_{\text{pre}}^{-} = v_{\text{pre}} - \sum_i p_{\text{new}} s_i \tag{7.13}$$

$$= v_{\text{pre}} - \sum_i \frac{v_{\text{pre}} m_i}{\min(v_{\text{pre}}, c_i)} \tag{7.14}$$

$$= v_{\text{pre}} \left(1 - \sum_i \frac{m_i}{\min(v_{\text{pre}}, c_i)} \right) \tag{7.15}$$

Of course, S_{pre} depends on s_i, so Eq. (7.12) does not give a closed form solution for the number of shares s_i. To obtain a closed form solution, we first solve for S_{pre}. Substituting the value for the number of SAFE shares s_i for each investor into the equation for S_{pre}, we get

$$S_{\text{pre}} = s_f + \sum_{i=1}^{k} \frac{m_i S_{\text{pre}}}{w_i}$$

where $w_i = \min(v_{\text{pre}}, c_i)$ can be understood as the valuation at which SAFE i converts. This equation has a unique solution for S_{pre} as

$$S_{\text{pre}} = \frac{s_f}{1 - \sum_{i=1}^{k} \frac{m_i}{w_i}}$$

using which we can also solve uniquely for the number of shares for each SAFE investor as

$$s_i = \frac{m_i s_f}{w_i (1 - \sum_{j=1}^{k} \frac{m_j}{w_j})} .$$

Note that in order for these expressions for v_{pre}^{-} and s_i to be finite and positive, we need that

$$1 > \sum_{i=1}^{k} \frac{m_i}{w_i} . \tag{7.16}$$

This can be understood by noting that the proportion of the company that SAFE investor i receives, before the dilution by the new investor's money, is, by Eq. (7.12),

$$\frac{s_i}{S_{\text{pre}}} = \frac{m_i}{w_i} . \tag{7.17}$$

Hence inequality (7.16) says that the total share of the company controlled by the SAFE investors before the new investor's money should be less than 100%. (Some amount needs to be left over to account for the founder shares s_f.)

Define

$$\xi = \sum_{i=1}^{k} \frac{m_i}{c_i}$$

So, following this understanding of (7.16), ξ is the dilution (in proportional terms) that the founders would suffer (before the further dilution by the new investor) in the event that the pre-money valuation is greater than all valuation caps.

A significant point is that there are several ways in which this constraint (7.16) can be violated. We consider a number of cases:

Case 1: The company issues a set of SAFE notes with $\xi \geq 1$.

Since always $w_i \leq c_i$, this would imply that inequality (7.16) is violated. There is therefore an implicit obligation on the company not to issue excess SAFE notes in this way: doing so would render the company unable to meet its explicit obligations to the SAFE holders in the event of an Equity Financing.

Case 2: We have $\xi < 1$ and $v_{\text{pre}} \geq c_i$ for all $i = 1 \ldots k$.

Then for any valuation v_{pre} greater than or equal to the largest cap of any of the SAFE notes, we have $w_i = c_i$ for all i, so inequality (7.16) is satisfied and the numbers s_i are well defined.

Case 3: We have $\xi < 1$ and $\min(c_1, \ldots, c_k) \leq v_{\text{pre}} < c_i$ for some $i = 1 \ldots k$.

Let the SAFE caps be ordered as $c_1 \leq \ldots \leq c_k$, and suppose that $c_{j-1} \leq v_{\text{pre}} < c_j$. Then we have $w_i = c_i$ for $i < j$ and $w_i = v_{\text{pre}}$ for $i \geq j$, so inequality (7.16) amounts to

$$1 > \sum_{i=1}^{j-1} \frac{m_i}{c_i} + \sum_{i=j}^{k} \frac{m_i}{v_{\text{pre}}}$$

or, equivalently,

$$v_{\text{pre}} > \frac{\sum_{i=j}^{k} m_i}{1 - \sum_{i=1}^{j-1} \frac{m_i}{c_i}}. \tag{7.18}$$

Depending on the values of the SAFE caps c_i and prices m_i, this may not be possible for all values of v_{pre} in the interval $[c_{j-1}, c_j)$. Thus, even if the company has taken care not to exceed granting over a 100% share to the SAFE investors, as calculated using the SAFE caps, there may yet be pre-money valuations at which the SAFE contracts can never be converted. Note that both the numerator and denominator in the expression on the right-hand side of the inequality (7.18) increase as j descends from k to 1 (i.e., as v_{pre} decreases through the range of SAFE cap values). The value

SAFE Parameters:
$(m_1, c_1) = (1, 3),$
$(m_2, c_2) = (1, 4),$
$(m_3, c_3) = (2, 20),$
$v_{pre} = 4.4$

j	1	2	3
RHS (7.18)	4	4.5	4.8
(7.18) holds	T	F	F

SAFE Parameters:
$(m_1, c_1) = (1, 3),$
$(m_2, c_2) = (1, 5),$
$(m_3, c_3) = (2, 20),$
$v_{pre} = 4.4$

j	1	2	3
RHS (7.18)	4	4.5	4.3
(7.18) holds	T	F	T

Fig. 7.2 Summary of Example 7.1 showing Post-Money SAFE convertibility with v_{pre} in cap intervals $[c_{j-1}, c_j)$

of this expression behaves non-monotonically, so the set of intervals within which the SAFE cannot be converted is complex. The following example illustrates this.

Example 7.1 We give two examples of multiple Post-Money SAFE scenarios. Figure 7.2 summarizes the calculations in this example. Suppose a company issues three Post-Money SAFEs with Cap Only, with parameters as follows: $m_1 = 1, c_1 = 3; m_2 = 1, c_2 = 4; m_3 = 2, c_3 = 20$.

Then clearly, for $j = 1, 2, 3$, $\sum_{i=j}^k m_i$ are 4, 3, and 2, respectively, and $\frac{m_j}{c_j}$ are $\frac{1}{3}$, $\frac{1}{4}$, and $\frac{1}{10}$, respectively. Thus $\xi = \frac{41}{60} < 1$.

In the interval $[c_1, c_2)$ (i.e., $[3, 4)$), the right-hand side of (7.18) becomes $3/(2/3)$ = 4.5. No v_{pre} in the range $[3, 4)$ can exceed this. Thus, (7.18) cannot be satisfied anywhere in this interval. Hence, an equity financing using the Discounted Valuation method is not possible for values of v_{pre} in the range $[3, 4)$.

In the interval $[c_2, c_3)$ (i.e., $[4, 20)$), the right-hand side of (7.18) becomes $2/(5/12)$ = 4.8, Thus, an equity financing using the Discounted Valuation method is not possible for values of v_{pre} in the range $[4, 4.8)$.

In summary, with these three Post-Money SAFEs the Discounted Valuation method fails for company valuations in the range $[3, 4.8)$.

On the other hand, suppose SAFE 2 instead has a higher valuation cap, say the parameters are $m_2 = 1, c_2 = 5$. Then (7.18) becomes $v_{pre} > 2/(7/15) = 30/7 \approx 4.29$. Thus, an equity financing using the Discounted Valuation method is not possible for values of v_{pre} in the range $[3, 4.28)$.

In particular, a valuation of (say) 4.4 can be converted when SAFE 2 has the higher valuation cap, but not when it has the lower valuation cap.[3]

Case 4: The pre-money valuation satisfies $\sum_{l=1}^k m_i < v_{pre} \le \min(c_1, \ldots c_k)$. In this case, inequality (7.16) amounts to

---

[3] In terms of (7.16), $\frac{m_j}{w_j}$ are $\frac{1}{3}$, $\frac{1}{4}$, and $\frac{2}{4.4}$, respectively, so (7.16) is violated.

$$1 > \sum_{i=1}^{k} \frac{m_i}{v_{\text{pre}}}$$

which is necessarily satisfied in this case. Hence the SAFEs are always convertible in this case.

Case 5: The pre-money valuation satisfies $v_{\text{pre}} \leq \min(\sum_{i=1}^{k} m_i, c_1, \ldots c_k)$. In this case, we again have inequality (7.16) amounts to

$$1 > \sum_{i=1}^{k} \frac{m_i}{v_{\text{pre}}}$$

but this is false because $v_{\text{pre}} \leq \sum_{i=1}^{k} m_i$. Hence the SAFES are never convertible in this case.

We remark that whereas, for Pre-Money SAFEs, there is a single valuation interval over which the SAFE is not convertible using the Discounted Valuation method (namely, when $v_{\text{pre}} \leq \sum_i m_i$), for Post-Money SAFEs, there may be multiple, non-contiguous, such intervals. For example, for each of the two examples in Example 7.1, in addition to the interval of non-convertibility identified in the Example, we have non-convertibility over the interval $v_{\text{pre}} \in (0, 1)$, by Case 5.

It is also worth noting that in case all the Post-Money SAFEs are issued with the same cap c, that is, $c_i = c$ for all $i = 1 \ldots k$, then Case 3 cannot arise, as its condition is always false. This means that the set of Post-Money SAFEs is not convertible using the Discounted Valuation method only when $v_{\text{pre}} > \sum_i m_i$, exactly as with Pre-Money SAFEs. Indeed, we will see in the next section that there is a correspondence between Pre- and Post-Money SAFEs when all caps are identical.

7.6 Comparison

One way to compare Pre-Money SAFEs and Post-Money SAFEs is to consider contracts with identically valued parameters. In one scenario, suppose the company has issued k Pre-Money SAFEs with the i-th having Purchase Amount m_i and Cap c_i. Similarly, in a second scenario, suppose the company has issued k Post-Money SAFEs with the i-th having Purchase Amount m_i and Cap c_i. By Eqs. (7.8) and (7.15), we see that in the first scenario, SAFE investor i receives share valued at $v_{\text{pre}}^- m_i / \min(v_{\text{pre}}^-, c_i)$, and in the second scenario, $v_{\text{pre}} m_i / \min(v_{\text{pre}}, c_i)$. The value v_{pre}^- here is that calculated for Pre-Money SAFEs in Sect. 7.4, and satisfies $v_{\text{pre}}^- < v_{\text{pre}}$. (Note that a potentially different value for v_{pre}^- is calculated for the Post-Money SAFEs in Sect. 7.5.) Thus, the SAFE investor, in case $v_{\text{pre}} < c_i$, receives shares of a larger value (and therefore a larger share of the company) in the case of Post-Money SAFEs than they do in the case of Post-Money SAFEs.

However, knowing this, one expects that the company and the SAFE investor will negotiate on the value of the parameters, specifically, on the Cap amount. It is reasonable to consider the parameter settings that might result from such a negotiation. The following result shows that, provided that the Caps of different contracts are identical in each of the scenarios, an effectively equivalent set of contracts can be obtained in the Pre-Money and the Post-Money scenarios. This result generalizes to the multi-SAFE scenario the content of Proposition 6.1 concerned with the application of the Discounted Valuation method to scenarios with a single Pre- or Post-Money SAFE.

Proposition 7.2 *Consider the following scenarios, where* $m_1, \ldots, m_k, c > 0$:

1. **Scenario 1**: *The company issues a set of Pre-Money SAFEs* $i = 1 \ldots k$, *all with cap* c, *for Purchase Amounts* $m_1, \ldots, m_k$, *respectively.*
2. **Scenario 2**: *The company issues a set of Post-Money SAFEs* $i = 1 \ldots k$, *all with cap* $c + \sum_{i=1}^{k} m_i$, *for Purchase Amounts* $m_1, \ldots, m_k$, *respectively.*

Then

- *All contracts are convertible by the Discounted Valuation method in Scenario 1 iff all contracts are convertible by the Discounted Valuation method in Scenario 2 iff the pre-money valuation* v_{pre} *satisfies* $v_{pre} \geq \sum_i m_i$.
- *For all pre-money valuations* $v_{pre} \geq \sum_i m_i$, *and all* $i = 1 \ldots k$, *the post-money value of shares received by SAFE investor* i *in Scenario 1 converted using the Discounted Valuation method is the same as the post-money value of shares received by SAFE investor* i *in Scenario 2, converted using the Discounted Valuation method.*

Proof We first consider the conditions under which the SAFE contracts are convertible in the two scenarios. In the Pre-Money Scenario 1, the contracts are convertible by the Discounted Valuation method iff $v_{\text{pre}} > \sum_i m_i$, as shown in Sect. 7.4. In the Post-Money Scenario 2, the contracts are convertible iff condition (7.16) holds, with $c_i = c + \sum_j m_j$ for all $i = 1 \ldots k$. There are two cases.

- If $c + \sum_j m_j < v_{\text{pre}}$, then condition (7.16) is

$$\sum_i \frac{m_i}{c + \sum_j m_j} < 1 \qquad (7.19)$$

This is equivalent to $0 < c$, which always holds. Note that the condition $v_{\text{pre}} > \sum_i m_i$ for convertibility in Scenario 1 also always holds in this case.
- If $c + \sum_j m_j \geq v_{\text{pre}}$, then condition (7.16) is

$$\sum_i \frac{m_i}{v_{\text{pre}}} < 1$$

that is, $\sum_i m_i < v_{\text{pre}}$. This is equivalent to the condition under which we have convertibility in Scenario 1.

Thus, all contracts are convertible in Scenario 1 iff all contracts are convertible in Scenario 2 iff $v_{\text{pre}} > \sum_i m_i$.

In Scenario 1, using the analysis of Sect. 7.4, by Eq. (7.8), we have that

$$s_i p_{\text{new}} = \frac{m_i}{min(v_{\text{pre}}^-, c_i)} \cdot v_{\text{pre}}^-$$

$$= \begin{cases} m_i & v_{\text{pre}}^- \leq c \\ \frac{m_i v_{\text{pre}}^-}{c} & v_{\text{pre}}^- > c \end{cases}$$

since $c_i = c$ for all $i = 1 \ldots k$.[4] Also, we have that $c_0' = 0$ and $c_1' = c$ are the only distinct cap values to be considered, and we have

$$C_0 = \sum_i m_i$$

$$C_1 = c + \sum_{i:c_1' < c_i} m_i + \sum_{i:c_1' \geq c_i} \frac{m_i c_1'}{c_i}$$

$$= c + \sum_i m_i$$

since $c_i = c$ for all $i = 1 \ldots k$. We have $f(i) = 1$ for all i, since $c = c_1$. Hence, by Eq. (7.10),

$$v_{\text{pre}}^- = \frac{v_{\text{pre}} - \sum_{i:v_{\text{pre}} \leq C_{f(i)}} m_i}{1 + \sum_{i:v_{\text{pre}} > C_{f(i)}} \frac{m_i}{c_i}}$$

$$= \frac{v_{\text{pre}} - \sum_{i:v_{\text{pre}} \leq C_1} m_i}{1 + \sum_{i:v_{\text{pre}} > C_1} \frac{m_i}{c}}$$

In particular, in the case that $v_{\text{pre}}^- > c = c_1$, that is, $v_{\text{pre}} > C_1 = c + \sum_i m_i$, we have that

$$v_{\text{pre}}^- = \frac{v_{\text{pre}}}{1 + \sum_i \frac{m_i}{c}}$$

It follows that we can also characterize the valuation $s_i p_{\text{new}}$ as

$$s_i p_{\text{new}} = \begin{cases} m_i & v_{\text{pre}} \leq c + \sum_i m_i \\ \frac{m_i v_{\text{pre}}}{c + \sum_i m_i} & v_{\text{pre}} > c + \sum_i m_i \end{cases}$$

---

[4] By continuity, we could include the case of $v_{\text{pre}}^- = c_i$ in either the first or second case. We have used the given form here to simplify the reasoning using the closed form solution of v_{pre}^- that follows.

In the Post-Money Scenario 2, we have, by the analysis of Sect. 7.5 (with $c_i = c + \sum_i m_i$ for all $i = 1 \ldots k$), that

$$
\begin{aligned}
s_i \, p_{\text{new}} &= \frac{m_i \, v_{\text{pre}}}{\min(v_{\text{pre}}, c + \sum_i m_i)} \\
&= \begin{cases} m_i & v_{\text{pre}} \le c + \sum_i m_i \\ \frac{m_i \, v_{\text{pre}}}{c + \sum_i m_i} & v_{\text{pre}} > c + \sum_i m_i \end{cases}
\end{aligned}
$$

This is identical to the characterization of $s_i \, p_{\text{new}}$ in Scenario 1, so the value amounts of the shareholdings of the SAFE investors after the equity round are identical in the two scenarios. Since the Discounted Valuation method is always minimally conservative, it also follows that the new investor and the founders also have identically valued shareholdings after the equity round in the two scenarios.   □

It is worth noting, however, that because the transformation in Proposition 7.2 from the Pre-Money scenario to the Post-Money scenario includes the term $\sum_i m_i$, it needs to made globally, at the level of the scenarios, with knowledge of the total amount of money that will be raised by the company using SAFE contracts, rather than at the level of each individual SAFE. It is not possible to determine what would be the Post-Money SAFE equivalent to a given Pre-Money SAFE if the contracts were being issued one at a time, with the total money to be raised from SAFEs before the equity round unknown.

Furthermore, it is easy to see that, in general, it is not possible to replace multiple Post-Money SAFEs by equivalent—with respect to equity financing under the Discounted Valuation method—Pre-Money SAFEs. Any collection of Post-Money SAFEs that produce a gap in the valuations at which they convert cannot be imitated by Pre-Money SAFEs, since the latter do not produce such a gap.

7.7 Mixed SAFE Types

A mixture of Pre- and Post-Money SAFEs can be handled easily if the conversion method for Pre-Money SAFEs is compatible with the Liability view. In this case, the conversion can be done in two steps:

1. Convert Pre-Money SAFEs by any method compatible with the Liability view, ignoring the existence of Post-Money SAFEs.
2. Treating the shares created as now part of the company capitalization referred to by s_{common}, perform the conversion for the Post-Money SAFEs.

Note that, although in theory the methods chosen for Pre- and Post-Money SAFEs can differ in this two-step process, it may not be legally justifiable to treat the different types of SAFEs differently.

When the conversion method for Pre-Money SAFEs is based on the Cap Table view, the shares corresponding to the Post-Money SAFEs must appear in the cap table for the Pre-Money SAFEs. Consequently, we need to deal with interlinked circularities for each type of SAFE. This seems amenable to the techniques we have demonstrated in this book, although we have not pursued it.

Y Combinator advises against mixing Pre- and Post-Money SAFEs because it creates "cap table uncertainty" [1]. Presumably this refers to difficulty in understanding the degree of dilution of the founders. This is a problem that the Pre-Money SAFEs already presented and is addressed by the Post-Money SAFEs (but only if all SAFEs are Post-Money).

The alternatives Y Combinator suggests to companies that have outstanding Pre-Money SAFEs are to either (a) persuade existing Pre-Money SAFE holders to amend their Pre-Money SAFEs to Post-Money SAFEs, or (b) issue only Pre-Money SAFEs.

Approach (a) is sometimes possible, at least concerning equity financing, as we see from Proposition 7.2. However, when there are multiple SAFEs with different valuation caps, an exact equivalence is not, in general, possible, as we saw in the previous section. In addition, variations in other parts of the contracts mean that an equivalence of entire contracts is not possible, which means that individual negotiations with SAFE holders may be required. Approach (b) certainly simplifies managing the SAFEs, since it avoids dealing with two contracts with subtle differences. Furthermore, such management is hopefully simplified by our analyses in this book.

7.8 Conclusion

In this chapter, we have presented a general formulation of the Discounted Valuation method for dealing with equity rounds in which the company has issued multiple convertible instruments that convert to shareholdings in the course of the equity round, and developed the implications of this method in the case of two types of convertible instruments: Pre-Money SAFE contracts with Cap Only, and Post-Money SAFE contracts with Cap Only. We have also compared the outcomes and shown that there is an equivalence (with respect to Equity Financing events) between these contract types in certain situations.

The analysis shows that the Post-Money SAFEs, although arguably easier for investors to understand, are convertible using the Discounted Valuation method in a smaller set of circumstances than the Pre-Money SAFEs. To address this deficiency, we propose, in Chap. 10, an alternate definition of the conversion conditions for SAFEs that captures the intuitions underlying the Post-Money SAFEs, but are convertible in all situations where the Pre-Money SAFEs are convertible using the Discounted Valuation method, without requiring the application of that method to ensure conservatism.

SAFE contracts, both Pre- and Post-Money, come in three forms in addition to the versions that we have considered in this chapter. We expect that our general techniques apply to these forms and yield similar results, but we have not verified the details of the calculations for these other versions to confirm this. We also note that we have not fully addressed the effect of combining Pre- and Post-Money SAFEs in the same equity round, though we do not expect that there will be any significant difficulties with the calculations in this case. (It is unclear if such combinations will arise much in practice, especially considering Y Combinator's advice against it, but there might be some companies with outstanding Pre-Money SAFEs who, despite that advice, consider moving to the new Post-Money SAFEs.)

Reference

1. Y Combinator (2018) Post money safe user guide. https://www.ycombinator.com/docs/Post%20Money%20Safe%20User%20Guide.pdf. Accessed Dec 2024

Chapter 8
Liquidity Events with Multiple SAFEs

Upon a liquidity event (say, the company is acquired or conducts an IPO), SAFE holders must make a choice between receiving a payout based on a return of cash invested or on taking shares in the company. This results in a situation where the value a SAFE holder receives can be dependent on the choices of other SAFE holders. In this chapter, we analyze this situation with game-theoretic methods. Additionally, we discuss the distribution of dividends to Post-Money SAFEs, which is defined using terms used in the treatment of liquidity events.

8.1 Liquidity Events

The definition of the conditions under which a Liquidity Event occurs were stated in Chap. 2. Since these terms refer to legislation, their interpretation requires extra-contractual reasoning, which we will not pursue here. We note, moreover that there may be situations in which it might be argued that the conditions for an Equity Financing and a Liquidity event overlap. Specifically, Equity Financing events that result in a Change of Control have this property. In such situations, a determination needs to be made as to which of the consequences of these events applies. We will not attempt to argue the case for either resolution. Our analysis in this chapter will simply assume that it has already been determined that the circumstances do in fact constitute a Liquidity Event.

In a Liquidity Event, the Pre-Money SAFE holders have an option to receive either cash (the Cashout Amount), or to convert the SAFE to a certain number of shares, the value of which is called the Conversion Amount. In the latter case, the distribution of the proceeds of the Liquidity event is then based on the rights associated to this shareholding. (In acquisitions, the proceeds to be distributed to the investors may be money and/or acquirer shares, but we focus in this chapter on the corresponding monetary value.)

R. van der Meyden and M. J. Maher, *Simple Agreements for Future Equity (SAFE)*, Blockchain Technologies, https://doi.org/10.1007/978-981-96-3920-5_8

The Liquidity Event clause in most Post-Money SAFE contracts is slightly different, and states that the SAFE holder gets an amount of the "Proceeds" of the Liquidity event equal to the maximum of the Cashout Amount and the Conversion amount.

For both types of SAFEs, these amounts may depend on the Cashout and Conversion amounts of other investors, so it is unclear that there is a determinate maximum. This is so particularly as the Cashout rights are senior to the Conversion rights, so that moneys to be paid out to the holders of shares received in conversion first have the Cashout amounts deducted.

As a simple example of the way choices of SAFE holders affect other SAFE holders, consider the situation where there are two SAFE investors A and B, who buy identical contracts with purchase price m. Suppose the value remaining for SAFE holders V is equal to m. If A chooses Convert and B chooses Cashout, then B receives m and A receives 0, and vice versa. If they both choose Convert then A and B share m with holders of capital stock after conversion of preferred shares. On the other hand, if they both choose Cashout they receive $m/2$ each.

In general, the situation is more complicated: more SAFEs, of different types, with different parameters. In those situations, it can be difficult to know what choice to make upon a liquidity event, and whether to regard fellow SAFE holders as collaborators, with whom to strategize to increase your payout, or competitors, from whom to hide and/or mislead as to your intentions (or both).

SAFE contracts therefore create a game-theoretic scenario in which the players are the holders of SAFE contracts, and the moves are to choose to Cashout or to Convert. The details of the payouts depend on the specific type of SAFE, of which there are multiple versions, depending on a number of parameters.

In the following we define a general game model for liquidity events involving SAFEs and other convertible instruments. We will show that, for Pre-Money SAFEs, SAFE holders have an optimum choice of Cashout or Convert, while, for Post-Money SAFEs with Cap, the maximization required in the Liquidity Event clause is well-defined.

However, when there are both Pre- and Post-Money SAFEs with Cap outstanding these guarantees are lost.

8.2   The Liquidity Event Game

We consider a liquidity event, carried out under the following assumptions. An amount of value $V > 0$ is to be paid out to the investors in the company. We assume that all debts of the company and claims senior to the shareholders and holders of convertible instruments have already been extinguished, so that V represents the amount of value to be distributed to the investors with claims on equity in the company. We assume that V consists of monetary value, but the analysis would be essentially the same if it were denominated in some other form that can be uncontroversially valued in monetary terms: for example, publicly traded shares of the acquirer in case of an acquisition of the company. (In situations where different investors place widely different valuations on the proceeds to be distributed, the game would be much more general, to the extent that it becomes unclear that it is sufficiently specific to admit any interesting analysis beyond game-theoretic generalities, rather than the specific analysis we conduct here.)

We remark that the assumption that V represents a uniform type of asset is a simplification. In acquisitions, what is available for distribution could be a mix of cash and shares of the acquirer. In this case, investors choosing Cashout would have a first claim on the cash, but it may not be sufficient for a full payout to these investors. However, as the balance of their Cashout amounts could then be treated as a claim on the other type of assets, it seems reasonable to treat V as if it were uniform, for purposes of analysis.

We suppose that a non-empty finite set $\mathbf{C}$ of convertible instruments has been issued by the company, with instrument $i \in \mathbf{C}$ purchased for amount $m_i > 0$, the "principal". In a liquidity event, each instrument gives the investor a choice between "cashing out" or "converting". In case of a cashout, the investor recovers their purchase amount (or a *pro rata* amount, in case there is insufficient value to be distributed). In case of a conversion, the instrument is first converted to shares (of the company undergoing the Liquidity event), and the value remaining is distributed in proportion to the shareholdings after conversion. Cashouts are senior to conversions, so that the purchase amounts are repaid to investors choosing to Cashout, before a distribution is made based on existing shareholdings and shares issued in conversion. We assume that preferred stock, options, and other convertible instruments have already been converted to common stock, and that common stockholders are treated at the same priority as converting investors.

Thus, the set of players of the game is $\mathbf{C}$, and each $i \in \mathbf{C}$ has two moves Cashout and Convert. We may therefore represent a strategy profile $\sigma : \mathbf{C} \to$ {Cashout, Convert} (which describes the actions of all the players in one play of the game) by the set $K \subseteq \mathbf{C}$ of instruments that are cashed out, that is, the set of $i \in \mathbf{C}$ such that $\sigma(i) = $ Cashout. Conversely, $\overline{K} = \mathbf{C} \setminus K$ is the set of instruments that convert. We describe the payouts $U_i(K)$ for each player $i \in \mathbf{C}$ when K is the set of instruments that are cashed out.

For $i \in K$, the payouts are straightforwardly defined from the purchase amounts. Define

$$m(K) = \sum_{i \in K} m_i$$

to be the total amount paid for the instruments that are being cashed out. The payouts for $i \in K$ are then defined by

$$U_i(K) = \begin{cases} m_i & \text{if } m(K) \leq V \\ \frac{m_i}{m(K)} \cdot V & \text{otherwise} \end{cases}$$

That is, investors cashing out receive their purchase amount, if the funds V to be distributed are enough to cover such payments, or a *pro rata* share of V otherwise.

We write $\mathrm{cash}(K)$ for $\sum_{i \in K} U_i(K)$, the total payout to investors that take the cashout option.[1] This means that the amount to be distributed to shareholders after the cashouts is $V - \mathrm{cash}(K)$. This amount is greater than or equal to zero, and equal to zero if $m(K) \geq V$.

We assume that the payouts to investors $i \in \overline{K}$ taking the conversion option is defined by

$$U_i(K) = \frac{m_i}{\gamma_i f(\overline{K})} \cdot (V - \mathrm{cash}(K))$$

where

• $f : \mathcal{P}(\mathbf{C}) \to \mathbb{R}$ is a function that is linear in the sense that for $X \subseteq \mathbf{C}$, we have

$$f(X) = \beta + \sum_{i \in X} \alpha_i$$

where β and the α_i are positive constants, and
• γ_i is a positive constant.

Thus, the payout to investor i is proportional to the purchase amount m_i paid in by investor i, divided by a number $f(\overline{K})$ that depends on the set $\overline{K}$ of investors converting, and a constant factor γ_i that depends only on the investor i. Plainly, for this distribution to be well-defined, we require that for all non-empty $C = \overline{K} \subseteq \mathbf{C}$, we have

$$\sum_{i \in C} \frac{m_i}{\gamma_i f(C)} < 1 \tag{8.1}$$

Strictness of the inequality comes from taking into account that the company will have a non-zero number of shares before adding converted shares. The restriction to non-empty set C is because in the case $C = \emptyset$, there is no distribution to be made to holders of convertible instruments.

In particular, taking $C = \{i\}$, we see that

$$m_i < \gamma_i f(\{i\}) = \gamma_i(\beta + \alpha_i) \tag{8.2}$$

is a necessary condition for the distribution to be well-defined.

In summary, a Liquidity Event game consists of players $\mathbf{C}$, with moves Cashout or Convert, and payoffs as defined by U above, where constraint (8.1) (and, consequently, (8.2)) is satisfied by the parameters.

A further constraint, for technical reasons to be explained below, is the following inequality: for each $i \in \mathbf{C}$,

$$\gamma_i \alpha_i \leq m_i \tag{8.3}$$

---

[1] It is easy to see that $\mathrm{cash}(K) \leq m(K)$ and if $m(K) \leq V$ then $\mathrm{cash}(K) = m(K)$.

We will show in the next section that the above model is sufficiently general to cover the liquidity event provisions of all versions of the Pre-Money and Post-Money SAFE contracts.

8.3 SAFE Contract Variants

The different SAFEs treat liquidity differently. The SAFEs with Cap define conversion using the valuation cap, while conversion of the SAFE with Discount uses the discounted fair market price. The Pre-Money MFN SAFE acts like a SAFE with a discount of 0% (i.e., a discount rate of 1). The Pre-Money SAFEs and Post-Money SAFEs use different definitions of Liquidity Capitalization: the Pre-Money SAFEs exclude SAFEs (and other convertible instruments), while they are included in the Post-Money SAFEs. (In a sense, the Pre-Money SAFE takes a Liability view of Liquidity Capitalization, while the Post-Money SAFE takes a Cap Table view.)

We show in this section that the Liquidity Event provisions of all these SAFEs are instances of the general game model of Sect. 8.2. In all these variants, the distribution in the case of investors choosing to cash out is as described above for the case $i \in K$, so we focus on the case where $i \in \overline{K}$.

We suppose that all SAFEs issued by the company are of the same type, though they may differ in their parameter settings. (We discuss mixes of Pre- and Post-Money SAFEs in Sect. 8.7.)

The Liquidity Event clauses of these contracts have qualifications relating to the form of the distribution (e.g., a mix of cash and shares), and provisions relating to tax-free reorganizations, which we also omit for simplicity.

The treatments of liquidity in SAFEs fall into three main classes:

- Pre-Money SAFEs with Cap (both Cap Only and Cap with Discount), discussed in Sect. 8.3.1,
- SAFEs with Discount (both Pre-Money and Post-Money), and Pre-Money MFN SAFEs and early Post-Money MFN SAFEs, discussed in Sect. 8.3.2, and
- Post-Money SAFEs with Cap (both Cap Only and Cap with Discount), described in Sect. 8.3.3.

The more recent Post-Money MFN SAFEs (versions 1.2 and later) simply provide a Cashout.

8.3.1 Pre-Money SAFE with Cap

The Pre-Money SAFE with Cap Only has as parameters the purchase amount m_i and a Valuation Cap c_i. In a liquidity event, this SAFE offers the investor the option to cash out or to convert. The number of shares issued in conversion of the SAFE is defined as the purchase amount m_i divided by the Liquidity Price. Liquidity Price is defined as "the price per share equal to the Valuation Cap divided by the Liquidity Capitalization." The Liquidity Capitalization, in turn is defined by

"**Liquidity Capitalization**" means the number, as of immediately prior to the Liquidity Event, of shares of Capital Stock (on an as-converted basis) outstanding, assuming exercise or conversion of all outstanding vested and unvested options, warrants and other convertible securities, but excluding: (i) shares of Common Stock reserved and available for future grant under any equity incentive or similar plan; (ii) this instrument; (iii) other Safes; and (iv) convertible promissory notes.

We write S_{liq} for the liquidity capitalization. Thus, we obtain that the Liquidity Price for SAFE i is equal to c_i/S_{liq}, and the number of shares issued in conversion is equal to $m_i S_{liq}/c_i$. After conversion, the total number of shares in the company is therefore $S_{liq} + \sum_{i \in \overline{K}} m_i S_{liq}/c_i$, and we obtain that the share of investor $i \in \overline{K}$ of the distribution is

$$
\begin{aligned}
U_i(K) &= \frac{m_i S_{liq}/c_i}{S_{liq} + \sum_{j \in \overline{K}} m_j S_{liq}/c_j} \cdot (V - \texttt{cash}(K)) \\
&= \frac{m_i}{\gamma_i f(\overline{K})} \cdot (V - \texttt{cash}(K))
\end{aligned}
$$

where $\gamma_i = c_i$ and $f(X) = 1 + \sum_{j \in X} m_j/c_j$. This distribution therefore fits the Liquidity game model with $\beta = 1$ and $\alpha_i = m_i/c_i$.

It is immediate from the definitions that $\sum_{i \in \overline{K}} U_i(K) < V - \texttt{cash}(K)$, when $V > \texttt{cash}(K)$, so condition (8.1) holds. Note also that $\gamma_i \alpha_i = m_i$ in this case, so inequality (8.3) is satisfied.

The Pre-Money SAFE with Cap and Discount behaves identically to the Cap Only for liquidation events (it does not use the discount). Thus the above applies equally to that SAFE.

8.3.2 Pre and Post-Money SAFE with Discount Only

The treatment of liquidation is the same in both the Pre-Money and Post-Money SAFEs with Discount. These SAFEs have as parameters the purchase amount m_i and a Discount Rate $d_i \leq 1$. Using this, the Liquidity Price is defined by

"**Liquidity Price**" means the price per share equal to: the fair market value of the Common Stock at the time of the Liquidity Event, as determined by reference to the purchase price payable in connection with such Liquidity Event, multiplied by the Discount Rate.

Thus, the Liquidity Price is for SAFE i is $p_{fair} d_i$ where p_{fair} is provided as an input to the liquidity event. The number of shares issued in conversion is again the purchase amount divided by the Liquidity Price, giving a share of the distribution for SAFE investor i of

$$U_i(K) = \frac{m_i/p_{fair}d_i}{S_{liq} + \sum_{j \in \overline{K}} m_j/p_{fair}d_j} \cdot (V - \texttt{cash}(K))$$

$$= \frac{m_i}{\gamma_i f(\overline{K})} \cdot (V - \texttt{cash}(K))$$

where $\gamma_i = p_{fair}d_i$ and $f(X) = S_{liq} + \sum_{j \in X} m_j/p_{fair}d_j$. Here we have $\beta = S_{liq}$ and $\alpha_j = m_j/p_{fair}d_j$. Hence, again, $\gamma_i\alpha_i = m_i$ and inequality (8.3) is satisfied. It is immediate from the definitions that condition (8.1) holds.

As noted above, the Pre-Money MFN SAFE treats liquidity events like a Discount Only SAFE, but without a discount. Thus the above applies to that MFN SAFE with the additional simplification that $d_i = 1$. Early versions of the Post-Money MFN SAFE are the same. From version 1.2, the Post-Money MFN SAFE simply returns the purchase amount m_i (or a *pro rata* share in case of insufficient funds). These MFN SAFEs are not included in the Liquidity Event game. However, using the upcoming Theorem 8.1, it is straightforward to eliminate these SAFEs from consideration.[2]

8.3.3 Post-Money SAFE with Cap

We explain the derivation here for the Post-Money SAFE with Cap Only. The parameters of this contract, like the Pre-Money version, are the purchase amount m_i paid for SAFE i, and a cap c_i that is called the Post-Money Cap of the SAFE. (The now-withdrawn Post-Money SAFE with Cap and Discount behaves identically to the Post-Money SAFE with Cap Only, for liquidation events. Thus the analysis to follow applies equally to that SAFE.)

For a Post-Money SAFE to be meaningful and well-defined, we need $m_i < c_i$, because on conversion, the Post-Money SAFE is converted to a number of shares that gives the SAFE investor a m_i/c_i share of the company (in the case of a conversion triggered by an equity round, this share is prior to the new issuance for the new money). For the same reason, we need that the total conversion share of all SAFEs is less than 1, so that the founders are left with a non-zero share of the company. That is, we need $\sum_{i \in C} m_i/c_i < 1$.

The Liquidity Event clause of the SAFE says

Liquidity Event. If there is a Liquidity Event before the termination of this Safe, this Safe will automatically be entitled to receive a portion of Proceeds, due and payable to the Investor immediately prior to, or concurrent with, the consummation of such Liquidity Event, equal to the greater of (i) the Purchase Amount (the "Cash-Out Amount") or (ii) the amount payable on the number of shares of Common Stock equal to the Purchase Amount divided by the Liquidity Price (the "Conversion Amount").

Compared to the Pre-Money version, we might note that this assumes that the greater of (i) and (ii) is well-defined. Because of the dependence of these values on other

---

[2] If $V > m(\mathbf{C})$ then we simply remove these SAFEs from $\mathbf{C}$ and subtract all their purchase amounts from V. Otherwise, all players should choose to Cashout, by Theorem 8.1.

SAFEs, this is in fact not clear, so we will model this maximization instead as a choice of the investor, and derive the existence of an optimum choice.

The Liquidity Price and Liquidity Capitalization are defined by

"Liquidity Price" means the price per share equal to the Post-Money Valuation Cap divided by the Liquidity Capitalization.

"Liquidity Capitalization" is calculated as of immediately prior to the Liquidity Event, and (without double- counting):

- Includes all shares of Capital Stock issued and outstanding;
- Includes all (i) issued and outstanding Options and (ii) to the extent receiving Proceeds, Promised Options;
- Includes all Converting Securities, other than any Safes and other convertible securities (including without limitation shares of Preferred Stock) where the holders of such securities are receiving Cash-Out Amounts or similar liquidation preference payments in lieu of Conversion Amounts or similar "as-converted" payments; and
- Excludes the Unissued Option Pool.

SAFEs are a form of Converting Security. Thus, whereas the Liquidity Capitalization in the case of the Pre-Money SAFEs was a constant, this definition is a function of the set of SAFEs not already cashed out.

To formalize these definitions, suppose $C = \overline{K}$ is the set of SAFES that are being converted, and let $i \in C$. (We use C rather than $\overline{K}$ in the following in order to improve readability of expressions where $\overline{K}$ would appear in subscripts.) Relative to these parameters, let

- $s(C)_i$ be the number of shares issued to SAFE investor i in conversion of their SAFE,
- $S_{liq}(C)$ be the Liquidity Capitalization,
- $\texttt{cash}(C)_i$ be the Cashout Amount of the SAFE i,
- $\texttt{conv}(C)_i$ be the Conversion Amount of the SAFE i, and
- $P_{liq}(C)_i$ be the Liquidity Price of SAFE i.

Note that these values may depend on their parameters.

The conversion value is determined as follows. We assume that there are no convertible securities other than SAFEs, so write s_{common} for the (constant) number of common shares after all outstanding options and warrants have been vested. Then

$$S_{liq}(C) = s_{common} + \sum_{i \in C} s(C)_i$$

by the definition of Liquidity Capitalization. However, $s(C)_i$ in turn depends on $S_{liq}(C)$, so we have a circularity. Specifically, we have that, the Liquidity Price is defined as

$$P_{liq}(C)_i = c_i / S_{liq}(C)$$

and, when $i \in C$, that $s(C)_i$ is defined as the purchase amount divided by the Liquidity Price, i.e.,

$$s(C)_i = m_i / P_{liq}(C)_i$$

Combining these equations, we get

$$s(C)_i = (m_i/c_i) S_{liq}(C) \tag{8.4}$$

and hence

$$s(C)_i = (m_i/c_i)(s_{common} + \sum_{i \in C} s(C)_i)$$

which displays the circularity via $s(C)_i$ (we could also display it via $s(C)$).

However, we do not actually need to compute $s(C)_i$ in order to determine the Cashout and Conversion amounts for a given choice of C. Clause (d), "Liquidation Priority" states that the conversion amounts are junior to cashout amounts, that is, cashouts are paid first, and the remaining value is issued to common and converted shares. For $i \in K$, the Cashout Amounts are identical to the amounts $U_i(K)$ of Sect. 8.2.

The Conversion Value of SAFE i is the share of the total Conversion amount corresponding to the proportional shareholding after conversion, that is,

$$\mathrm{conv}(C)_i = \frac{s(C)_i}{s(C)} \cdot (V - \mathrm{cash}(K))$$

$$= \frac{m_i}{c_i} \cdot (V - \mathrm{cash}(K))$$

by Eq. (8.4). The payout to investor $i \in C = \overline{K}$ is $U_i(K) = \mathrm{conv}(C)_i$. This fits the game model of Sect. 8.2 with $\gamma_i = c_i$ and $f(X) = 1$, that is, $\beta = 1$ and $\alpha_j = 0$ for all $j \in C$. Note also that $\gamma_i \alpha_i - m_i = -m_i < 0$, so inequality (8.3) is satisfied. We have already stated that $\sum_{j \in C} m_i/c_i < 1$ as an assumption: it follows from this that condition (8.1) holds.

The above derivation does leave open the question of whether $s(C)_i$ has been properly defined, given the circularity. This can be addressed by noting that the circularity is easily resolved by treating these definitions as simultaneous equations. Summing Eq. (8.4) over $i \in C$, we get

$$\sum_{i \in C} s(C)_i = \left(\sum_{i \in C} m_i/c_i \right) \left(s_{common} + \sum_{i \in C} s(C)_i \right)$$

which we can reorganize to get a closed form solution for the total converted shares:

$$\sum_{i \in C} s(C)_i = \frac{\sum_{i \in C} m_i/c_i}{1 - \sum_{i \in C} m_i/c_i} \cdot s_{common}$$

Substituting this back into the above definition of the Liquidity Cap, we get

$$S_{liq}(C) = s_{common} + \frac{\sum_{i \in C} m_i/c_i}{1 - \sum_{j \in C} m_j/c_j} \cdot s_{common}$$

$$= \frac{s_{common}}{1 - \sum_{j \in C} m_j/c_j}$$

Also, substituting into Eq. (8.4), we get

$$s(C)_i = (m_i/c_i) S_{liq}(C)$$

$$= \frac{m_i/c_i}{1 - \sum_{j \in C} m_j/c_j} \cdot s_{common}$$

which shows that the notion of Conversion Payout is well-defined, given C.

8.4 Nash Equilibria in the General Game

The SAFE contracts' handling of Liquidity Events makes the most sense if it can be proved that there is always a choice of moves that is the best possible for all players.

Given a Liquidity Event game on a set of players C, which determines payouts $U_i(K)$ for $i \in C$, define the binary relation $\preceq$ on the set of strategy profiles $\mathcal{P}(C)$ by $K_1 \preceq K_2$ if for all $i \in C$ we have $U_i(K_1) \leq U_i(K_2)$. It can be seen that this relation is a quasi-order, i.e., it is reflexive and transitive. We write $K \approx K'$ when both $K \preceq K'$ and $K' \preceq K$. Intuitively, $K \approx K'$ when K and K' yield the same payouts for all players. (We will see below that this is possible even when $K \neq K'$.)

An *optimum* strategy profile is a strategy profile K^* such that $K \preceq K^*$ for all strategy profiles K. Optima, when they exist, are not necessarily unique, since we may have two distinct optima $K \approx K'$.

In general, it is not clear that an optimum exists. A weaker notion is the following. A strategy profile K is a *pure-strategy Nash equilibrium* if for all players i, we have $U_i(K') \leq U_i(K)$ for all strategy profiles K' that differ from K only in the move of player i, that is, for which $K' = K \cup \{i\}$ or $K' = K \setminus \{i\}$. We call these simply "Nash equilibria" henceforth, since we do not need to consider mixed Nash equilibria in this book.

Plainly, an optimum, if one exists, is also a Nash equilibrium. We may therefore approach the question of the existence of optima via an analysis of the Nash equilibria. A strategy profile that is not a Nash equilibrium is unstable, in the sense that some player has an incentive to change their move, possibly to the detriment of another player. Where an optimum does not exist, it is therefore preferable that the "solution" to the game be a Nash equilibrium. However, multiple Nash equilibria may exist. A reasonable criterion for choice among Nash equilibria is to choose one that is optimal among the Nash equilibria: that is, a Nash equilibrium K such that $K' \preceq K$ for all Nash equilibria K'. Again, there is no *a priori* guarantee that such an optimum exists, however. (We give an example of this below.)

In our application to Liquidity Event games, it is desirable that an optimum Nash equilibrium exists for all values of V, since this means that there is always a distribution to the holders of the convertible instruments that avoids inherent conflicts between investors. The contracts in question could be argued to have been poorly designed or poorly selected if there were values of V for which the Liquidity Event game lacks an optimum Nash equilibrium, since there could then potentially eventuate a liquidity event situation where any Nash equilibrium selected would leave at least one investor worse off than in some other Nash equilibrium. The injured party would be motivated to pursue legal action in this case, seeking redress by extra-contractual means. A well-designed contract would give greater legal certainty and reduce legal costs by ensuring that this possibility could not arise.

It is plausible that the investors would collaborate in choosing their moves so as to select an optimum Nash equilibrium when one exists, since it gives them their maximum payout among all stable alternatives. Such an agreement, once reached, also presents little risk that one of the parties to the agreement would maliciously change their move at the last minute.

The following result identifies one condition under which an optimum Nash equilibrium is guaranteed to exist: when the total amount V to be paid out is less than the total cashout amounts due to the players. Roughly, in this case, an investor who elects to convert rather than cash out runs the risk that the funds remaining to be distributed for the lower priority conversion amounts will be zero. This makes it a safer strategy to cash out.

Recall that, for a set $X \subseteq \mathbf{C}$, $m(X) = \sum_{i \in X} m_i$, that is, the total purchase amount for investors in the set X.

Theorem 8.1 *If $0 < V < m(\mathbf{C})$ then $K = \mathbf{C}$ is the unique Nash equilibrium.*

Proof Suppose that $V < m(\mathbf{C})$. To show that the strategy profile $K = \mathbf{C}$ in which all investors take cash is the unique Nash equilibrium, suppose that $K \neq \mathbf{C}$ is a Nash equilibrium. Note first that we must have $m(K) < V$. For, if $m(K) \geq V$ then, for any $i \in \overline{K}$, we have $U_i(K) = 0$, because all value is paid out to investors that cash out, but $U_i(K \cup \{i\}) > 0$ since in $K \cup \{i\}$, investor i participates (*pro rata*) in a cashout. This contradicts the assumption that K is a Nash equilibrium.

It follows that $\texttt{cash}(K) = m(K)$.

We next argue that we cannot have $i \in \overline{K}$ with $m(K) + m_i \geq V$. For, suppose that this is the case. Then we would have

$$U_i(K \cup \{i\}) = \frac{m_i}{m(K) + m_i} \cdot V$$

since i participates *pro rata* in the cashout. Also, as argued above, we have $m(K) < V$, so $U_i(K) = \frac{m_i}{\gamma_i f(\overline{K})}(V - m(K))$. From the assumption that K is a Nash equilibrium, we have

$$\frac{m_i}{m(K) + m_i} \cdot V \leq \frac{m_i}{\gamma_i f(\overline{K})}(V - m(K)) .$$

From this, we get that

$$(m(K) + m_i - \gamma_i f(\overline{K}))V \geq m(K)(m(K) + m_i)$$
$$\geq m(K)V$$

by using the assumption that $m(K) + m_i \geq V$. Since $V > 0$ it then follows that $m_i \geq \gamma_i f(\overline{K})$, which contradicts the condition (8.2) for the game to be well-defined. This shows that we must have $m(K) + m_i < V$ for all $i \in \overline{K}$.

In this case, we have, for $i \in \overline{K}$, that $U_i(K \cup \{i\}) = m_i$ and

$$U_i(K) = \frac{m_i}{\gamma_i f(\overline{K})}(V - m(K)) \, .$$

Since K is a Nash equilibrium, we have

$$m_i \leq \frac{m_i}{\gamma_i f(\overline{K})}(V - m(K)) \, .$$

Summing over $i \in \overline{K}$, and using condition (8.1) we get

$$m(\overline{K}) = \sum_{i \in \overline{K}} m_i \leq (\sum_{i \in \overline{K}} \tfrac{m_i}{\gamma_i f(\overline{K})})(V - m(K))$$
$$< V - m(K)$$

where, for the last inequality, we use the fact $V - m(K) > 0$, established above.

But this implies that $m(\mathbf{C}) = m(\overline{K}) + m(K) < V$, contradicting the assumption. This shows that the only possible Nash equilibrium in case $V < m(\mathbf{C})$ is $K = \mathbf{C}$. Indeed the arguments above also show that $K = \mathbf{C}$ is a Nash equilibrium, because it has been shown in the case of every strategy profile $K' = \mathbf{C} \setminus \{i\}$, that i gains by switching their move from conversion to cashout. $\square$

Another class of games that have the same unique Nash equilibrium is identified in the next result.

Theorem 8.2 *Let* $\Gamma = \min_{i \in \mathbf{C}} \gamma_i$.

- *If* $V < \gamma_i \beta$ *then Cashout is the better option for SAFE i, independent of the choices of other SAFEs.*
- *If* $V < \Gamma \beta$ *then* $K = \mathbf{C}$ *is the unique Nash equilibrium.*

Proof If $V < \gamma_i \beta$ then $V < \gamma_i f(\overline{K})$ for every K. Hence, if $i \notin K$,

$$U_i(K) = \frac{m_i}{\gamma_i f(\overline{K})}(V - m(K)) \leq m_i \frac{V}{\gamma_i f(\overline{K})} \leq m_i$$

That is, if $V < \gamma_i \beta$ then the conversion payout for SAFE i is less than the cashout amount, so SAFE i should choose Cashout, independent of the choices made by other SAFEs.

For the second part, if $V < \Gamma \beta$ then $V < \gamma_i \beta$, for every i. Thus every SAFE should choose to Cashout, and $\mathbf{C}$ is the unique Nash equilibrium. $\qquad \square$

At the other end of the valuation spectrum, if the value of the company is sufficiently large, the SAFE holders will choose to Convert. This gives a second condition under which there is a unique Nash equilibrium.

Theorem 8.3 *Consider the Liquidity Event game.*

- *If $V - m(\mathbf{C}) > \gamma_i f(\mathbf{C}) - m_i$, then Convert is the better option for SAFE i, independent of the choices of other SAFEs.*
- *If $V - m(\mathbf{C}) > \max_{i \in \mathbf{C}} (\gamma_i f(\mathbf{C}) - m_i)$ then $K = \emptyset$ is the unique Nash equilibrium.*

Proof Suppose $V - m(\mathbf{C}) > \gamma_i f(\mathbf{C}) - m_i$. Then for all $K \subseteq \mathbf{C}$, we have $V - m(\mathbf{C}) > \gamma_i f(\overline{K}) - m_i$, because $f(\mathbf{C}) \geq f(\overline{K})$ and $\gamma_i > 0$. For $i \notin K$, we have $\mathrm{cash}(K) \leq m(K) \leq m(\mathbf{C}) - m_i$, so

$$U_i(K) = \frac{m_i}{\gamma_i f(\overline{K})} \cdot (V - \mathrm{cash}(K))) \geq \frac{m_i}{\gamma_i f(\overline{K})} \cdot (V - m(\mathbf{C}) + m_i) > m_i \ .$$

In other words, the payout for i by Conversion is better than the Cashout amount, independent of the choices of other players.

If the condition is true for every i then every player will Convert. That is, if $V - m(\mathbf{C}) > \max_{i \in \mathbf{C}} (\gamma_i f(\mathbf{C}) - m_i)$ then every player will choose to Convert, because Cashout is a worse option. Hence, $K = \emptyset$ is the unique Nash equilibrium. $\qquad \square$

The simpler condition $V - m(\mathbf{C}) > \max_{i \in \mathbf{C}} (\gamma_i) f(\mathbf{C})$ identifies slightly fewer cases but is slightly easier to compute. As well as identifying when there is a unique Nash equilibrium, we can use the first parts of the two previous theorems to identify parts of Nash equilibria (when their respective conditions hold).

It is more difficult to characterize Nash equilibria between the two extremes identified above. In general, when $V \geq m(\mathbf{C})$ there are enough funds for each investor who chooses to cash out to receive the full purchase amount, independently of the choices of other investors. Consequently, $\mathrm{cash}(K) = m(K)$ for all $K \subseteq \mathbf{C}$. The main question here, for each investor, is whether they could receive more than their money by choosing to convert. The answer to that question potentially depends on the choices of other investors, since the Cashout Amounts have priority over the funds to be distributed to investors who convert.

Indeed, in some instances of the Liquidity Event game, there is an inherent conflict between the interests of different investors.

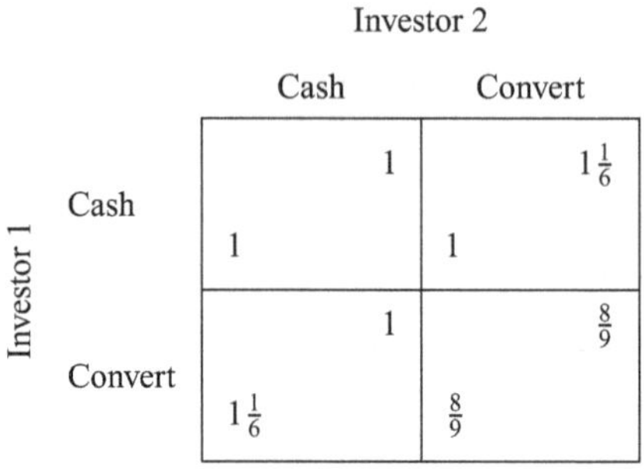

Investor 2

	Cash	Convert
Cash	1 1	$1\frac{1}{6}$ 1
Convert	1 $1\frac{1}{6}$	$\frac{8}{9}$ $\frac{8}{9}$

(Investor 1 labels the rows: Cash, Convert)

Fig. 8.1 A liquidity game with conflicting equilibria

Example 8.1 Consider the symmetric Liquidity Event game, with $\beta = 1$ and $V = 8$, and two investors with $m_i = 1$, $\alpha_i = 1$, $\gamma_i = 3$ for each investor $i = 1, 2$. (Note that the constraint (8.1) is satisfied for this game.) This gives the game depicted in Fig. 8.1.[3]

The game is similar to the Game of Chicken [1] in that there are two Nash equilibria (Cash,Convert) and (Convert,Cash), each with one party gaining a better payoff than the other, but if both play Convert, then both receive a lesser payoff than if they both play Cash. We note that this game has $m_i < \gamma_i \alpha_i$ for all i, so condition (8.3) is not satisfied.

We will soon show that condition (8.3) ensures that conflicts of this kind do not occur, but first we characterize Nash equilibria in terms of inequalities. If K is a Nash equilibrium then for all $i \in K$, we have

$$m_i = U_i(K) \geq U_i(K \setminus \{i\})) = \frac{m_i}{\gamma_i f(\overline{K} \cup \{i\})}(V - m(K \setminus \{i\}))$$

or, equivalently,

$$m(K) - m_i + \gamma_i f(\overline{K}) + \gamma_i \alpha_i \geq V . \tag{8.5}$$

Additionally, for $i \in \overline{K}$, we have

$$U_i(K \cup \{i\}) = m_i \leq \frac{m_i}{\gamma_i f(\overline{K})}(V - m(K)) = U_i(K)$$

which is equivalent to

$$V \geq m(K) + \gamma_i f(\overline{K}) . \tag{8.6}$$

Conversely, if these two conditions are satisfied, then K is a Nash equilibrium.

---

[3] In such diagrams, each box represents the outcome if Investor 1 chooses the action of the label on the left and Investor 2 chooses the action of the label above. Within each box, the number in the south-west corner is the payoff for Investor 1, while the number in the north-east corner is the payoff for Investor 2.

Let $g : \mathcal{P}(\mathbf{C}) \to \mathbb{R}$ be defined by

$$g(K) = \frac{V - m(K)}{f(\overline{K})}$$

for $K \subseteq \mathbf{C}$. Note that for $i \in \overline{K}$ we have $U_i(K) = m_i g(K)/\gamma_i$ when $V \geq m(\mathbf{C})$.

Theorem 8.4 *Assume that $V \geq m(\mathbf{C})$ and condition (8.3) is satisfied. Suppose that there exists a Nash equilibrium, and let K^* be a Nash equilibrium that satisfies $g(K^*) \geq g(K)$ for all Nash equilibria K. Then K^* is an optimum Nash equilibrium with respect to $\preceq$.*

Proof Assume that $V \geq m(\mathbf{C})$ and that there exists a Nash equilibrium. Let $K^* \subseteq \mathbf{C}$ be a Nash equilibrium that satisfies $g(K^*) \geq g(K)$ for all Nash equilibria K. We show that $K \preceq K^*$ for all Nash equilibria K. Let K be any Nash equilibrium. We show that $U_i(K) \leq U_i(K^*)$ for all $i \in \mathbf{C}$. We consider three cases: $i \in K$, $i \in \overline{K^*} \setminus K$ and $i \in K^* \setminus K$.

Since $V \geq m(\mathbf{C})$, we have $U_i(K) \geq m_i$ for all $i \in \mathbf{C}$. It is immediate that for $i \in K$ we have $U_i(K) = m_i \leq U_i(K^*)$.

For i in neither K^* nor K, we have, by definition of K^* and the fact that K is a Nash equilibrium, that

$$U_i(K^*) = m_i g(K^*)/\gamma_i \geq m_i g(K)/\gamma_i = U_i(K) .$$

The remaining case is $i \in K^* \setminus K$. Here, since $i \in K^*$ and K^* is a Nash equilibrium, by (8.5) we have $V - m(K^*) \leq \gamma_i f(\overline{K^*}) - m_i + \gamma_i \alpha_i$. Thus

$$\frac{g(K^*)}{\gamma_i} = \frac{V - m(K^*)}{\gamma_i f(\overline{K^*})} \leq 1 - \frac{(m_i - \gamma_i \alpha_i)}{\gamma_i f(\overline{K^*})} .$$

By condition (8.3), $m_i - \gamma_i \alpha_i \geq 0$, so we have $\frac{g(K^*)}{\gamma_i} \leq 1$. Thus, from the definition of K^* and the fact that K is a Nash equilibrium,

$$U_i(K) = \frac{m_i g(K)}{\gamma_i} \leq \frac{m_i g(K^*)}{\gamma_i} \leq m_i = U_i(K^*) .$$

$\square$

As an immediate corollary, if $V \geq m(\mathbf{C})$, condition (8.3) is satisfied, and there exists a Nash equilibrium, then there exists an optimum Nash equilibrium.

We now establish some results that will lead to an efficient algorithm to detect the existence of a Nash equilibrium.

Theorem 8.5 *Suppose that $V \geq m(\mathbf{C})$ and condition (8.3) is satisfied. If K is a Nash equilibrium, then for all $i \in K$ and $j \in \overline{K}$ we have $\gamma_i \geq \gamma_j$. Moreover, if $\gamma_i = \gamma_j$ then $m_i = \gamma_i \alpha_i$ and $g(K) = \gamma_i$.*

Proof If $i \in K$ then we have $V \leq m(K) + \gamma_i f(\overline{K}) + \gamma_i \alpha_i - m_i$ by (8.5). If $j \in \overline{K}$ then we have $m(K) + \gamma_j f(\overline{K}) \leq V$ by (8.6). Combining the two inequalities, we derive $(\gamma_i - \gamma_j) f(\overline{K}) \geq m_i - \gamma_i \alpha_i$. By condition (8.3), the term on the right hand side of this inequality is non-negative and $f(\overline{K})$ is positive, so it follows that $\gamma_i \geq \gamma_j$. Moreover if $\gamma_i = \gamma_j$ we have $0 \geq m_i - \gamma_i \alpha_i \geq 0$, so $m_i = \gamma_i \alpha_i$ and our inequalities state $m(K) + \gamma_i f(\overline{K}) \leq V \leq m(K) + \gamma_i f(\overline{K})$, so $V = m(K) + \gamma_i f(\overline{K})$. It follows from this that $g(K) = \gamma_i$.   □

We see from this result that if condition (8.3) is satisfied, we can restrict the sets K that need to be considered to identify the Nash equilibria to those sets such that for some $i \in \mathbf{C}$ we have $j \in \overline{K}$ implies $\gamma_j \geq \gamma_i$. This observation will be useful for obtaining efficient algorithms below. Indeed a smaller set of cases suffices. The following result shows that if there is a Nash equilibrium in which some γ_i occurs for both K and $\overline{K}$, then there is an $\approx$-equivalent Nash equilibrium that does not have such a "boundary crossing".

Theorem 8.6 *Assume that $V \geq m(\mathbf{C})$ and condition (8.3) is satisfied. Fix $i \in \mathbf{C}$ and let $E = \{k \in \mathbf{C} \mid \gamma_k = \gamma_i\}$, and $G = \{k \in \mathbf{C} \mid \gamma_k > \gamma_i\}$. Suppose that $Y \subseteq E$. If $Y \neq \emptyset$ and $E \setminus Y \neq \emptyset$ and $G \cup Y$ is a Nash equilibrium, then*

(i) $m_k = \gamma_k \alpha_k$ for all $k \in Y$
(ii) $g(G) = \gamma_i$
(iii) G is a Nash equilibrium

Conversely, if (i)-(iii) then $G \cup Y$ is a Nash equilibrium. Moreover (i) and (ii) imply $g(G) = g(G \cup Y)$ and $G \approx G \cup Y$.

Proof We begin by proving the final (converse) part of the result. We first show that (i) and (ii) imply that $g(G) = g(G \cup Y)$. Note that

$$m(G \cup Y) + \gamma_i f(\overline{G \cup Y})$$

$$= m(G) + m(Y) + \gamma_i f(\overline{G}) - \gamma_i \sum_{k \in Y} \alpha_k$$

$$= m(G) + \gamma_i f(\overline{G}) + \sum_{k \in Y} (m_k - \gamma_k \alpha_k) \qquad \text{(by } \gamma_i = \gamma_k \text{ for } k \in Y)$$

$$= m(G) + \gamma_i f(\overline{G}) \qquad\qquad\qquad\qquad\qquad \text{(by (i))}$$

$$= V \qquad\qquad\qquad\qquad\qquad\qquad\qquad\qquad \text{(by (ii))}$$

It follows that $g(G \cup Y) = \gamma_i = g(G)$.

Next, we show that if $g(G) = g(G \cup Y)$ and (ii) then $G \cup Y \approx G$. For this, we consider three cases $k \in G$, $k \in Y$ and $k \in \overline{G \cup Y}$ and show $U_k(G \cup Y) = U_k(G)$ in each.

- For $k \in G$, we have also $k \in G \cup Y$, so $U_k(G \cup Y) = m_k = U_k(G)$, by definition.
- For $k \in Y$, we have $k \in E$ so $\gamma_k = \gamma_i$. Also, $k \notin G$. Thus

$$
\begin{aligned}
U_k(G \cup Y) &= m_k \\
&= m_k g(G)/\gamma_i \qquad\qquad \text{(by (ii))} \\
&= m_k g(G)/\gamma_k \\
&= U_k(G) \, .
\end{aligned}
$$

- Finally, for $k \in \overline{G \cup Y}$, we have $k \in \overline{G}$, so $U_k(G \cup Y) = m_k g(G \cup Y)/\gamma_k = m_k g(G)/\gamma_k = U_k(G)$.

G is a Nash equilibrium, by (iii), and $G \cup Y \approx G$, so $G \cup Y$ is also a Nash equilibrium, completing the final part of the result.

Suppose now that $Y \subset E$ with $Y \neq \emptyset$ and $E \setminus Y \neq \emptyset$ and $G \cup Y$ is a Nash equilibrium. Claims (i) and (ii) follow by Theorem 8.5. Since $Y \subset E$, we have $\gamma_k = \gamma_i$ for all $k \in Y$. Moreover, since $Y \neq \emptyset$, there in fact exists $k \in Y$, for which also $k \in G \cup Y$. Since $E \setminus Y \neq \emptyset$, there exists $k' \in E \setminus Y$, for which $\gamma_{k'} = \gamma_i = \gamma_k$. Note that $k' \in \overline{G \cup Y}$. Hence, by the second part of Theorem 8.5, with $K = G \cup Y$, we have that $m_k = \gamma_k \alpha_k$ for all $k \in Y$, and $g(G \cup Y) = \gamma_k = \gamma_i$. By definition of g and reasoning similar to that in the first paragraph, we have

$$
V = m(G \cup Y) + \gamma_i f(\overline{G \cup Y}) = m(G) + \gamma_i f(\overline{G}) \, .
$$

It follows that $g(G) = \gamma_i$.

We need to show (iii), that G is a Nash equilibrium. First, we need to show that

$$
m(G) - m_k + \gamma_k f(\overline{G}) + \gamma_k \alpha_k \geq V \tag{8.7}
$$

for all $k \in G$. Let $k \in G$. Then $k \in G \cup Y$ and $\gamma_k > \gamma_i$. Since $G \cup Y$ is a Nash equilibrium and $k \in G \cup Y$, we have

$$
m(G \cup Y) - m_k + \gamma_k f(\overline{G \cup Y}) + \gamma_k \alpha_k \geq V
$$

by (8.5). Equivalently,

$$
(m(G) - m_k + \gamma_k f(\overline{G}) + \gamma_k \alpha_k) + (m(Y) - \gamma_k \sum_{j \in Y} \alpha_j) \geq V
$$

The desired conclusion (8.7) now follows if $m(Y) - \gamma_k \sum_{j \in Y} \alpha_j \leq 0$. But this indeed holds, since we have, for all $j \in Y$, that $m_j = \gamma_j \alpha_j = \gamma_i \alpha_j < \gamma_k \alpha_j$.

Next, we need to show that $V \geq m(G) + \gamma_k f(\overline{G})$ for all $k \in \overline{G}$. Now, $\overline{G} = \overline{G \cup Y} \cup Y$. Since $G \cup Y$ is a Nash equilibrium, we have by (8.6) that $\gamma_k \leq g(G \cup Y)$ for $k \in \overline{G \cup Y}$. Since $g(G \cup Y) = g(G)$, it follows that $\gamma_k \leq g(G)$ for $k \in \overline{G \cup Y}$, which gives the desired inequality in this case. For $k \in Y$, the desired inequality is immediate (and holds with equality) from the fact (ii) that $g(G) = \gamma_i$. This completes the proof that G is a Nash equilibrium.   □

We see from this result that to identify the inequivalent Nash equilibria, it suffices to consider just the linear number of sets $K_\gamma = \{k \in \mathbf{C} \mid \gamma_k \geq \gamma\}$ where $\gamma = \infty$ or $\gamma = \gamma_i$ for some $i \in \mathbf{C}$. Other strategy profiles may be Nash equilibria but, if so, they are $\approx$-equivalent to one of these profiles.

This result directly yields a polynomial time algorithm for deciding the existence of a Nash equilibrium and, in the case that Nash equilibria exist, computing a representative for each $\approx$-equivalence class of the Nash equilibria, as well as an optimum Nash equilibrium. For each $K = K_\gamma$ where $\gamma = \infty$ or $\gamma = \gamma_i$ for $i \in \mathbf{C}$, and for each $k \in \mathbf{C}$, we compute the number $m(K) + \gamma_k f(\overline{K})$ and check whether $m(K) + \gamma_k f(\overline{K}) \geq V - m_k + \gamma_k \alpha_k$ (if $k \in K$) and $m(K) + \gamma_k f(\overline{K}) \leq V$ (if $k \in \overline{K}$). Any instance K satisfying these conditions is a Nash equilibrium, and if none does, then there are no Nash equilibria. The optimum Nash equilibrium can be identified as the one that maximizes $g(K_\gamma)$, by Theorem 8.4.

The time complexity of the algorithm is $O(|\mathbf{C}| \cdot (|\mathbf{C}| + m(b)))$, where b is the bitlength of the numbers defining the problem and $m(b)$ is the cost of multiplying two b-bit numbers.[4] We discuss some special cases below where this computation can be optimized to give a lower complexity.

We will later (see Theorem 8.7) identify a condition, strengthening condition (8.3), under which we have a converse to Theorem 8.6: if G is a Nash equilibrium with $g(G) = \gamma_i$, then for all $Y \subseteq E$, we have that $G \cup Y$ is a Nash equilibrium. However, this converse does not follow just from condition (8.3), as the following example shows. This means that, while we have an algorithm that computes a complete set of representatives of the Nash equilibria, we do not yet have a concrete complete representation of the set of all Nash equilibria, and still need to evaluate each of the (potentially exponentially many) sets $G \cup Y$ individually.

Example 8.2 Consider $\mathbf{C} = \{0, 1, 2\}$, with parameters as given in the following table

i	0	1	2
α_i	1	2	1
γ_i	3	2	2
m_i	$5\frac{1}{2}$	4	2

---

[4] Theoretically, the bound $m(b) = O(b \log(b))$ is known [2], but asymptotically less efficient algorithms may be preferable in practice.

Investor 1 \ Investor 2	Cash	Convert
Cash	10, 16	10, $15\frac{1}{5}$
Convert	$10\frac{5}{6}$, 16	$9\frac{2}{3}$, $18\frac{14}{25}$

Fig. 8.2 A liquidity game with no Nash equilibrium

Let $\beta = 1$ and $V = 13\frac{1}{2}$. Note that here we have $\gamma_1 = \gamma_2 < \gamma_0$, so, with $i = 1$, we have $G = \{0\}$ and $E = \{1, 2\}$. The following can be easily verified using the characterization of Nash equilibria given by conditions (8.5) and (8.6):

- The game satisfies condition (8.1).
- The game satisfies condition (8.3),
- G is Nash equilibrium, and $g(G) = \gamma_1$.
- $G \cup \{2\}$ is a Nash equilibrium. This means with $Y = \{2\} \subseteq E$, we have a "boundary crossing" $2 \in G \cup Y$, $1 \in \overline{G \cup Y}$ with $\gamma_1 = \gamma_2$, as in the assumptions of Theorem 8.6.
- $G \cup \{1\}$ is not a Nash equilibrium, because (8.5) is not satisfied for player 0.

However, while condition (8.3) guarantees that there exists an optimum Nash equilibrium, if any, it is not sufficient to guarantee that there is a Nash equilibrium, as shown by the following example. We identify some sufficient conditions for the existence of Nash equilibria in the following sections.

Example 8.3 Consider the Liquidity Event game, with $\beta = 1$ and $V = 29$, and two investors 1,2 with $m_1 = 10$, $\alpha_1 = 6$, $\gamma_1 = 1$, $m_2 = 16$, $\alpha_2 = 5$, $\gamma_2 = 3$. (Again the constraint (8.1) is satisfied with these parameters.) This gives the game depicted in Fig. 8.2. In this case, there is a cycle of best-reponses 2:Cash $\to$ 1:Convert $\to$ 2:Convert $\to$ 1:Cash $\to$ 2:Cash, where $i : a \to j : b$ denotes that j playing b is a best response to i playing a. Thus, this game has no Nash equilibrium. Note that this game has $m_1 > \gamma_1\alpha_1$ and $m_2 > \gamma_2\alpha_2$, so condition (8.3) is satisfied.

8.5 Special Case: Pre-Money SAFEs

In this section we consider a special case, where $V \geq m(\mathbf{C})$ and $m_i = \gamma_i\alpha_i$ for all $i \in \mathbf{C}$. As discussed in Sects. 8.3.1 and 8.3.2, all the Pre-Money SAFE variants satisfy this condition, as does the Post-Money SAFE with Discount Only.

We show that Nash equilibria are guaranteed to exist in this case. Condition (8.3) is satisfied under this assumption so, by Theorem 8.4, an optimum Nash equilibrium is also guaranteed to exist in this case. We also show that an improvement of the algorithm for computing Nash equilibria is possible.

In this case, the conditions for K to be a Nash equilibrium simplify to the following. For $i \in K$,

$$m(K) + \gamma_i f(\overline{K}) \geq V \tag{8.8}$$

and for $i \in \overline{K}$,

$$V \geq m(K) + \gamma_i f(\overline{K}). \tag{8.9}$$

Equivalently, since $f(K) > 0$, we have that K is a Nash equilibrium when

$$m(K) + \min(\{\gamma_i \mid i \in K\}) f(\overline{K}) \geq V \tag{8.10}$$

and

$$V \geq m(K) + \max(\{\gamma_i \mid i \in \overline{K}\}) f(\overline{K}). \tag{8.11}$$

where we treat the cases of empty sets by $\min(\emptyset) = \infty$ and $\max(\emptyset) = -\infty$ so that, respectively, the constraint (8.10) or (8.11) is trivial when $K = \emptyset$ or $\overline{K} = \emptyset$.

From Theorem 8.5, we have that if K is a Nash equilibrium and $i \in K$ and $j \in \overline{K}$ then $\gamma_i \geq \gamma_j$. Hence $\max(\{\gamma_i \mid i \in \overline{K}\}) \leq \min(\{\gamma_i \mid i \in K\})$, and we have that K is a Nash equilibrium just when V is in the interval

$$[\, m(K) + \max(\{\gamma_i \mid i \in \overline{K}\}) \cdot f(\overline{K}) \,, \, m(K) + \min(\{\gamma_i \mid i \in K\}) \cdot f(\overline{K}) \,] \tag{8.12}$$

In Theorem 8.6 we saw that if K is a Nash equilibrium and there exists $i \in K$ and $j \in \overline{K}$ with $\gamma_i = \gamma_j$, then $G = \{k \in \mathbf{C} \mid \gamma_k \geq \gamma_i\}$ is also a Nash equilibrium, $g(K) = \gamma_i$, and $G \approx K$. The following result shows that in the present special case, we have a converse to this result: for any $Y \subseteq \{k \in \mathbf{C} \mid \gamma_k = \gamma_i\}$, we have that $G \cup Y$ is a Nash equilibrium. This yields a complete characterization of the (pure-strategy) Nash equilibria in this special case. (In general, there may be $G \cup Y$ that are not Nash equilibria, and it is necessary to test each individually.)

Theorem 8.7 *Suppose that* $V \geq m(\mathbf{C})$ *and* $m_k = \gamma_k \alpha_k$ *for all* $k \in \mathbf{C}$. *Fix* $i \in \mathbf{C}$. *Let* $E = \{k \in \mathbf{C} \mid \gamma_k = \gamma_i\}$, *and let* $G = \{k \in \mathbf{C} \mid \gamma_k > \gamma_i\}$. *Suppose* G *is a Nash equilibrium and* $g(G) = \gamma_i$. *Then for all* $Y \subseteq E$, *we have that* $G \cup Y$ *is a Nash equilibrium and* $G \cup Y \approx G$. *Moreover,* $U_k(G \cup Y) = m_k$ *for all* $k \in E$.

Proof Suppose that G is a Nash equilibrium with $g(G) = \gamma_i$ and $Y \subseteq E$. Since $m_k = \gamma_k \alpha_k$ for all $k \in \mathbf{C}$, condition (8.3) is satisfied. We also satisfy conditions (i) and (ii) of Theorem 8.6, so, by the converse part of the theorem, we have that

$G \cup Y$ is a Nash equilibrium, $g(G) = g(G \cup Y)$ and $G \approx G \cup Y$. In particular, we have $G \cup Y \approx G \approx G \cup E$, so for all $k \in E$, since $k \notin G$, we have $U_k(G \cup Y) = U_k(G) = m_k g(G)/\gamma_k = m_k g(G)/\gamma_i = m_k$. □

In this special case, we obtain an improvement of the algorithm for identifying the Nash equilibria up to $\approx$-equivalence. As before, it suffices to consider just the linear number of sets $K_\gamma = \{k \in \mathbf{C} \mid \gamma_k \geq \gamma\}$ where $\gamma = \infty$ or $\gamma = \gamma_i$ for some $i \in \mathbf{C}$. However, the test that we perform for each value of γ can be optimized. Rather than performing a test for each value of γ and each $k \in \mathbf{C}$, we can now just compute the interval (8.12) for each value of γ, and determine whether it contains V. We note that, after sorting the γ_i, each of the terms in the endpoint values of these intervals can be incrementally computed from the corresponding terms in the preceding interval as we scan through $\mathbf{C}$ in order of increasing γ_i, adding or subtracting appropriate values m_k or α_k and adjusting the maxima and minima as we go. As before, an optimum Nash equilibrium can be identified as the one that maximizes $g(K_\gamma)$, by Theorem 8.4.

The complexity of this variant of the algorithm is $O(|\mathbf{C}| \cdot (\log |\mathbf{C}| + m(b)))$, where b is the bitlength of the numbers defining the problem and $m(b)$ is the cost of multiplying two b-bit numbers.

As above, other strategy profiles may be Nash equilibria, but if, so, they are $\approx$-equivalent to one of the strategy profiles considered by the algorithm. However, in this special case, Theorem 8.7 also provides a straightforward way to produce *all* the Nash equilibria from the ones identified by the algorithm. (We are guaranteed that $G \cup Y$ is a Nash equilibrium whenever G is a Nash equilibrium with $g(G) = \gamma_i$, in this case.)

Moreover, the following result shows that, indeed, one of the strategy profiles considered by the algorithm is a Nash equilibrium.

Theorem 8.8 *Suppose that $V \geq m(\mathbf{C})$ and $m_i = \gamma_i \alpha_i$ for all $i \in \mathbf{C}$. Then there exists a Nash equilibrium.*

Proof Consider the sets $K_\gamma = \{k \in \mathbf{C} \mid \gamma \leq \gamma_k\}$, where either $\gamma = \gamma_i$ for some $i \in \mathbf{C}$ or $\gamma = \infty$. Note that $K_\infty = \emptyset$ and for the least γ_i, we have $K_{\gamma_i} = \mathbf{C}$. Also, $\overline{K_\gamma} = \{k \in \mathbf{C} \mid \gamma_k < \gamma\}$.

For $i \in \mathbf{C}$, let γ_i^+ be the least value $\gamma_j > \gamma_i$ for $j \in \mathbf{C}$, else ∞ if there is no such value. Then we have that

$$\min(\{\gamma_k \mid k \in K_{\gamma_i}\}) = \gamma_i = \max(\{\gamma_k \mid \gamma_k < \gamma_i^+, \ k \in \mathbf{C}\}) = \max(\{\gamma_k \mid k \in \overline{K_{\gamma_i^+}}\})$$

Hence

$$m(K_{\gamma_i}) + \min(\{\gamma_k \mid k \in K_{\gamma_i}\}) \cdot f(\overline{K_{\gamma_i}})$$

$$= \sum_{k \in C, \gamma_i \leq \gamma_k} m_k + \gamma_i \cdot (\beta + \sum_{k \in C, \ \gamma_k < \gamma_i} \alpha_k)$$

$$= \sum_{k \in C, \gamma_i^+ \leq \gamma_k} m_k + \sum_{k \in C, \gamma_i = \gamma_k} m_k + \gamma_i \cdot (\beta + \sum_{k \in C, \ \gamma_k < \gamma_i} \alpha_k)$$

$$= \sum_{k \in C, \gamma_i^+ \leq \gamma_k} m_k + \sum_{k \in C, \gamma_i = \gamma_k} \gamma_k \alpha_k + \gamma_i \cdot (\beta + \sum_{k \in C, \ \gamma_k < \gamma_i} \alpha_k)$$

$$= \sum_{k \in C, \gamma_i^+ \leq \gamma_k} m_k + \gamma_i \cdot (\beta + \sum_{k \in C, \gamma_i = \gamma_k} \alpha_k + \sum_{k \in C, \ \gamma_k < \gamma_i} \alpha_k)$$

$$= \sum_{k \in C, \gamma_i^+ \leq \gamma_k} m_k + \gamma_i \cdot (\beta + \sum_{k \in C, \gamma_k < \gamma_i^+} \alpha_k)$$

$$= m(K_{\gamma_i^+}) + \max(\{\gamma_k \mid k \in \overline{K_{\gamma_i^+}}\}) \cdot f(\overline{K_{\gamma_i^+}})$$

That is, the right endpoint of the interval (8.12) for K_{γ_i} is equal to the left endpoint of the interval for $K_{\gamma_i^+}$. Since the leftmost endpoint of these intervals is $-\infty$ and the right endpoint is ∞, one of these intervals contains V, and there must exist a Nash equilibrium. $\qquad\square$

Note also that the (optimum) Nash equilibrium is not necessarily unique, indeed, from Theorem 8.7, we see that there may be an exponential number of Nash equilibria in the worst case. The following example illustrates this phenomenon.

Example 8.4 Consider the Liquidity game with $C = \{1, \ldots, n\}$, $V = n + 1$, $\beta = 1$, and $m_i = 1$, $\gamma_i = 1$ and $\alpha_i = 1$ for each $i \in C$. Plainly $m_i = \gamma_i \alpha_i$ for all $i \in C$ in this case. In this game, all sets $K \subseteq C$ are Nash equilibria, and they all yield the same payouts $U_i(K) = 1$ for all players i. This is obvious for $i \in K$. For $i \notin K$, we have

$$U_i(K) = m_i \frac{(V - m(K))}{\gamma_i(\beta + \sum_{i \in \overline{K}} \alpha_i)}$$

$$= \frac{(n + 1 - |K|)}{(1 + |\overline{K}|)}$$

$$= 1$$

since $|\overline{K}| = |C| - |K| = n - |K|$.

8.6 Special Case: Post-Money SAFEs with Cap

In this section we consider the existence of Nash equilibria in the case where $m(\mathbf{C}) > V$ and $\alpha_i = 0$ for all $i \in \mathbf{C}$. The Post-Money SAFEs with Cap (with or without Discount) are show in Sect. 8.3.3 to fall within this case. We show that Nash Equilibria exist in this case. Condition (8.3) is satisfied in this case so, by Theorem 8.4, one of these equilibria is an optimum.

Theorem 8.9 *Suppose that* $\alpha_i = 0$ *for all* $i \in \mathbf{C}$. *Then there exists a Nash equilibrium.*

Proof In the case where $\alpha_i = 0$ for all $i \in \mathbf{C}$, the conditions for K to be a Nash equilibrium reduce to the following.

$$m(K) + \min(\{\gamma_k \beta - m_k \mid k \in K\}) \geq V \tag{8.13}$$

and

$$V \geq m(K) + \max(\{\gamma_k \beta \mid k \in \overline{K}\}) . \tag{8.14}$$

Suppose that there is no Nash equilibrium. Then for all $K \subseteq \mathbf{C}$, we have $m(K) + \min(\{\gamma_k \beta - m_k \mid k \in K\}) < V$ or $V < m(K) + \max(\{\gamma_k \beta \mid k \in \overline{K}\})$. We derive a contradiction.

Let $\mathbf{C} = \{1, \ldots, n\}$ be sorted so that $\gamma_1 \leq \gamma_2 \leq \ldots \leq \gamma_n$. Define i^* to be the value of $i \in \mathbf{C}$ at which $\gamma_i \beta - m_i$ takes its minimum. Let $\Phi(i)$ be the following proposition:

$$i^* \neq i \text{ and } V < \gamma_i \beta + \sum_{k=i+1}^{n} m_k.$$

In the case $i = n$, we take the summation to be equal to zero. We show that $\Phi(i)$ holds for all $i = 1 \ldots n$, by a reverse induction. Note that this implies that $i^* \notin \{1, \ldots, n\}$, which is a contradiction.

For the base case of $\Phi(n)$, we argue as follows. Note first that, from the fact that $K = \emptyset$ is not a Nash equilibrium, we have that $V < m(\emptyset) + \max(\{\gamma_k \beta \mid k \in \mathbf{C}\}) = \gamma_n \beta$, which is the right hand conjunct of $\Phi(n)$. Secondly, from the fact that $K = \mathbf{C}$ is not a Nash equilibrium, we have that

$$m(\mathbf{C}) + \gamma_{i^*} \beta - m_{i^*} < V . \tag{8.15}$$

If we had $i^* = n$, it would follow that $m(\mathbf{C}) + \gamma_n \beta - m_n < \gamma_n \beta$, which is impossible since $m(\mathbf{C}) - m_n \geq 0$. Thus, $i^* \neq n$, and we conclude that $\Phi(n)$ holds.

For the inductive step, suppose that $\Phi(i), \ldots, \Phi(n)$, where $i > 1$. We show that $\Phi(i-1)$. Consider $K = \{i, \ldots, n\}$. Since this is not a Nash equilibrium, we have that either

$$V < \gamma_{i-1} \beta + \sum_{k=i}^{n} m_k \text{ or } V > \sum_{k=i}^{n} m_k + \min(\{\gamma_k \beta - m_k \mid i \leq k \leq n\}) . \tag{8.16}$$

Suppose first that the second disjunct of (8.16) holds, and let the minimum be attained at value $k^* \in \{i, \ldots, n\}$. Then $V > \gamma_{k^*}\beta - m_{k^*} + \sum_{k=i}^{n} m_k$. However, by the instance $\Phi(k^*)$ of the induction hypothesis, we have that $V < \gamma_{k^*}\beta + \sum_{k=k^*+1}^{n} m_k$. It follows that

$$\gamma_{k^*}\beta - m_{k^*} + \sum_{k=i}^{n} m_k < \gamma_{k^*}\beta + \sum_{k=k^*+1}^{n} m_k .$$

But this yields that

$$\sum_{k=i}^{k^*-1} m_k < 0$$

which is an impossibility, even when the summation on the left is zero. We conclude that the second disjunct cannot hold.

Thus, the left disjunct $V < \gamma_{i-1}\beta + \sum_{k=i}^{n} m_k$ of (8.16) holds, which is the right hand conjunct of $\Phi(i-1)$. If we had $i^* = i - 1$, by (8.15) we would have

$$m(\mathbf{C}) + \gamma_{i-1}\beta - m_{i-1} < V < \gamma_{i-1}\beta + \sum_{k=i}^{n} m_k$$

This yields

$$\sum_{k=1}^{i-2} m_k < 0$$

again an impossibility. Hence $i^* \neq i - 1$, and we have shown $\Phi(i-1)$. □

Since we have that condition (8.3) is satisfied strictly in the case under consideration, by Theorem 8.5, we cannot have a Nash equilibrium with $i \in K$ and $j \in \overline{K}$ such that $\gamma_i = \gamma_j$. The Nash equilibria therefore lie inside the linear number of possible cases $K_\gamma = \{k \in \mathbf{C} \mid \gamma_k \geq \gamma\}$ where $\gamma = \infty$ or $\gamma = \gamma_i$ for some $i \in K$. The algorithm of Sect. 8.4 therefore computes not just a set of representatives, but the set of all Nash equilibria.

However, it remains the case that Nash equilibria are not necessarily unique, as shown by the following example.

Example 8.5 Consider the symmetric two player game with $\alpha_1 = \alpha_2 = 0$, $\beta = 1$, $m_1 = m_2 = 2$, $\gamma_1 = \gamma_2 = 5$ and $V = 6$. The game matrix for this game is shown in Fig. 8.3. There are two Nash equilibria: (Cash,Cash) with payouts $(2, 2)$, and (Convert,Convert) with payouts $(2\frac{2}{5}, 2\frac{2}{5})$, so that we have (Cash,Cash) $\preceq$ (Convert,Convert).

Investor 2

	Cash	Convert
Cash	2 / 2	$1\frac{3}{5}$ / 2
Convert	2 / $1\frac{3}{5}$	$2\frac{2}{5}$ / $2\frac{2}{5}$

(Investor 1 labels rows: Cash, Convert)

Fig. 8.3 A liquidity game with two Nash equilibria

8.7 Mixed SAFE Types

The analysis of the previous sections has assumed that all SAFEs issued by the company are of the same type, that is, all as characterized in Sect. 8.5 or all as characterized in Sect. 8.6. We now present an example that shows that mixing different types of SAFEs can create situations where no pure-strategy Nash equilibria exist. We consider the combination of a Pre-Money SAFE with Cap Only, with principal m_1 and cap c_1, and a Post-Money SAFE with Cap Only, with principal m_2 and a Post-Money Cap c_2.

A first issue is to understand the interaction of Pre-Money SAFEs and Post-Money SAFEs in the calculation of the number of shares issued in conversion. Essentially, this is done by first determining the number of shares to be issued to the Pre-Money SAFE, and then calculating the number of shares that should be issued to the Post-Money SAFE. We follow the notation from Sect. 8.3.3. Let $C = \overline{K}$ be the set of SAFEs that convert. (Note that in this section, payouts $U_i(K)$ are expressed in terms of K, but numbers of shares $s(C)_i$ are expressed in terms of C.)

We first determine the number of shares $s(C)_1$ issued to the Pre-Money SAFE, in the event that this is converted. Recall from Sect. 8.3.1 that the Liquidity Capitalization for the Pre-Money SAFE excludes SAFEs, so it can be treated as a constant, equal to the number of common shares s_{common} in our simple scenario. We obtain that the number of shares issued in conversion to the Pre-Money SAFE is $s(C)_1 = m_1 s_{common}/c_1$, exactly as in Sect. 8.3.1. In the case where only the Pre-Money SAFE converts, the payout to investor 1 is

$$U_1(\{1\}) = \frac{m_1 s_{common}/c_1}{s_{common} + m_1 s_{common}/c_1}(V - m_2)$$

$$= \frac{m_1}{c_1 + m_1}(V - m_2) .$$

The calculations for the Post-Money SAFE follow the description of Sect. 8.3.3. The number of shares issued is

$$s(C)_2 = (m_2/c_2)(s_{common} + \sum_{i \in C} s(C)_i)$$

that is, m_2/c_2 of the total issuance after conversion.

In the case $C = \{2\}$, where only the Post-Money SAFE converts, the payout to player 2 is the same fraction of the remaining value, that is, $U_2(\{1\}) = (m_2/c_2)(V - m_1)$.

In the case where both SAFEs convert, we get that the payout is, similarly, $U_2(\emptyset) = (m_2/c_2)V$. The share issuance in this case is expressed in the equation

$$s(\{1, 2\})_2 = (m_2/c_2)(s_{common} + s(\{1, 2\})_1 + s(\{1, 2\})_2)$$
$$= (m_2/c_2)(s_{common} + (m_1 s_{common}/c_1) + s(\{1, 2\})_2) .$$

Calculating the actual share issuance to investor 2 by solving the equation, we have

$$s(\{1, 2\})_2 = \frac{m_2}{c_2 - m_2} \cdot \frac{m_1 + c_1}{c_1} \cdot s_{common} .$$

It follows that the payout to the Pre-Money SAFE in this case is

$$U_1(\emptyset) = \frac{m_1 s_{common}/c_1}{s_{common} + (m_1 s_{common}/c_1) + \frac{m_2}{c_2-m_2} \cdot \frac{m_1+c_1}{c_1} \cdot s_{common}} \cdot V$$
$$= \frac{m_1}{m_1 + c_1} \cdot \frac{c_2 - m_2}{c_2} \cdot V .$$

Consider now the specific instance with, for the Pre-Money SAFE, $m_1 = 2, c_1 = 7$, for the Post-Money SAFE, $m_2 = 2, c_2 = 4$, and value to be distributed $V = 8.2$. Then the above equations imply that we get the Liquidity game given in Fig. 8.4. (The numbers are approximated to two decimal points.) This has a cycle of best

Fig. 8.4 A liquidity game mixing Pre-Money and Post-Money SAFEs

responses 2:Cash $\to$ 1:Convert $\to$ 2:Convert $\to$ 1:Cash $\to$ 2:Cash. Thus, this game has no pure-strategy Nash equilibrium. Consequently, it also has no optimum.

This means that the combination of Pre-Money and Post-Money SAFEs can potentially create inherent conflicts among SAFE investors at the time of a Liquidity Event. Interestingly, Y Combinator's Post-Money SAFE User Guide [3] recommends against combining Post-Money and Pre-Money SAFEs, or combining SAFEs and convertible notes, the reason given being that this would require a more complex cap-table analysis. The analysis of the previous chapter suggests that, some extra complexity aside, there is not an inherent problem in computing conversion amounts with such combinations. However, mixing of SAFE types can raise difficulties at the time of a Liquidity Event that would require extra-contractual resolution.

8.8 Related Work and Open Problems

We have considered only pure-strategy Nash equilibria in this work, motivated from the perspective that well-designed convertible instruments should guarantee the existence of an optimum pure-strategy equilibrium in order to prevent inherent conflicts between different investors that might require extra-contractual resolution. From this point of view it is a positive result that pure-strategy optima are guaranteed to exist in both Pre-Money and Post-Money SAFEs. However, mixed strategy equilibria may still be worth studying from a theoretical point of view. We have required two different arguments for the two special cases where we have shown that pure-strategy equilibria exist. Example 8.3 shows that not all instances of our game model have a pure-strategy equilibrium. An analysis of mixed strategy Nash equilibria, which always exist, might therefore help to give a unifying treatment.

There exists literature on the computational complexity of computing Nash equilibria [4], generally focused on computing mixed Nash equilibria, a problem that is known to be computationally complex in general. Our polynomial time complexity results do not fit directly into this literature in that we do not represent the game in the input in matrix form: to do so for a Liquidity Event game with n investors would require a representation already of size 2^n. The fact that a problem has high complexity in general also does not inhibit the existence of low complexity instances. Our results show that the particular games we consider have lower complexity than the general problem.

Our analysis in this chapter has assumed that the letter of the contracts will be applied in distributing funds to the investors. Conceivably, such a distribution according to contract may still meet with objections by some party. Depending on the jurisdiction, adjudication of such claims may result in court orders to modify the distribution, on the grounds that some principle of fairness has legal priority over the contractual terms. The possibility of court intervention has not been considered in the present work, but may, for some jurisdictions, be worth considering. The area of bankruptcy theory and "claims problems" [5] provides many different approaches to the "fair" division of competing claims whose total exceeds the amount to be distributed.

Another matter beyond the scope of our work is the question of whether the Liquidity Event should proceed at all. SAFE contracts distinguish between Liquidity Events and Dissolution Events (in which the company is liquidated), and state a simpler distribution rule for Dissolution Events (a simple return of principal), but the Liquidity Event rules could sensibly be used for Dissolution Events also. There is literature [6] on game-theoretic aspects of negotiations between managers of a company facing liquidation and its creditors. The choices in this case concern offers and acceptances of terms for restructuring the obligations of the company in order to maintain it as a going concern, versus liquidation of the company. Game-theoretic reasoning has also been applied in structuring proposals to creditors concerning the legal jurisdiction under which a restructuring is to be conducted [7]. These works are orthogonal to our concerns in this chapter, where it has already been determined that the Liquidity Event game is to be played.

There has also been game-theoretic analysis of convertible notes [8] in which the game is played between the issuing company and an investor, rather than between different investors, as in our analysis. In this work, both the company and the investor have options: the company may choose to convert the note to shares once the share price reaches a certain level, and the investor has an option, after the maturity date, to either recoup principal or convert to shares. The analysis of the game concentrates on the timing of the decision to convert, and is used as a basis for pricing the convertible note. The conversion price for the notes is generally an amount fixed in advance, rather than an amount that depends on the share price of the company, as in SAFE notes. SAFEs also differ in that the underlying shares generally do not trade on public markets, and the condition allowing conversion is an equity round, liquidity event or dissolution, rather than a particular share price being reached.

SAFEs are equity-like instruments, in that they pay no interest and are intended to convert to equity, whereas convertible bonds are more debt-like, in that they pay interest but have an option to convert. However, potentially a similar analysis of the conversion decision in liquidity events may apply to convertible bonds, in which we would interpret m_i as the principal plus remaining interest due, rather than as simply the purchase amount. The details of specific instruments used in practice would need to be investigated to validate this intuition, however.

8.9   Dividends

As noted in Sect. 2.4, a holder of a post-money SAFE will receive dividends whenever common stock holders receive dividends. We discuss dividend distributions in the present chapter since, for SAFE holders, they are determined using the Liquidity Price. (This essentially involves the SAFE holder being treated as if they held the number of shares that they would receive in conversion of the SAFE in a Liquidity event.)

Let D be the dividend amount to be delivered to each share of common stock. Then the Dividend Amount to be delivered to the holder of SAFE i is defined as $D\frac{m_i}{P_{liq}}$, where P_{liq} is the Liquidity Price. Intuitively, this definition is treating the notional ("as converted") number of shares held by the SAFE holder as if these shares had been purchased at the Liquidity Price. This definition is common to all the post-money SAFEs.

The liquidity price is not defined uniformly across the post-money SAFEs. In post-money SAFEs without a Cap, P_{liq} is defined in terms of the fair market value of common stock. It is not clear how the fair market value is to be established when dividends are distributed, since there is no liquidity event to provide a guide. Let p_{fair} be the fair market price per share, however it might be determined. In that case, a SAFE i with Discount would be notionally allocated $s_i = \frac{m_i}{p_{fair}d_i}$ shares, where m_i is the purchase amount and d_i is the discount rate, and receive $D\frac{m_i}{p_{fair}d_i}$ as dividend. A post-money MFN SAFE would be allocated $s_i = \frac{m_i}{p_{fair}}$.

In the SAFE with Cap,[5] the liquidity price is defined by $P_{liq} = c/S_{liq}$, where S_{liq} is the Liquidity Capitalization, which is defined in terms of a Liquidity Event. It is defined as consisting of shares of capital stock, options, and SAFEs and other convertible securities that are not choosing to Cashout. In the case of dividends, there is no actual Liquidity Event and so no possibility of a Cashout. Options, SAFEs, etc. are to be treated on an "as-converted to Common Stock basis." It is not specified how to do this nominal conversion to common stock, but the natural choice is for each instrument to be converted in the way that it specifies in the case of an actual Liquidity Event. Let s_{common} be the number of shares of common stock derived on an as-converted basis from capital stock, options and convertible instruments other than post-money SAFEs with Cap. Then

$$S_{liq} = s_{common} + \sum_i s_i$$

where i ranges over post-money SAFEs with Cap. But we still need to establish s_i.
We have

$$s_i = \frac{m_i}{P_{liq}} = S_{liq}\frac{m_i}{c_i}$$

Notice that we again have a circularity: S_{liq} is defined in terms of s_i and s_i is defined in terms of S_{liq}. We can resolve this circularity in the same way we did for Post-Money SAFEs in the case of Equity Financing events. We have

$$S_{liq} = s_{common} + \sum_i s_i = s_{common} + \sum_i S_{liq}\frac{m_i}{c_i}$$

---

[5] The following also applies to the now-withdrawn post-money SAFE with Cap and Discount.

Let $\xi = \sum_i \frac{m_i}{c_i}$. Solving for S_{liq},

$$S_{liq} = \frac{s_{common}}{1 - \xi}$$

Thus, for SAFE i with Cap we have

$$s_i = S_{liq} \frac{m_i}{c_i} = \frac{s_{common}}{1 - \xi} \frac{m_i}{c_i}$$

In summary, a holder of SAFE i with Discount with discount rate d_i will receive a dividend of $D \frac{m_i}{p_{fair} d_i}$, while a holder of a SAFE i with Cap will receive $D \frac{s_{common}}{1 - \xi} \frac{m_i}{c_i}$. Note that holders of SAFEs with Cap are affected by the fair market value p_{fair}, through the value of the part of s_{common} due to SAFEs with Discount. Like other SAFE holders, a lower value of p_{fair} produces a higher dividend (assuming D is fixed), but its effect is likely overwhelmed by the part of s_{common} due to capital stock and options.

Comparing the two expressions, a SAFE with Cap will receive a better dividend than a SAFE with Discount exactly when

$$p_{fair} s_{common} > \frac{1 - \xi}{d} c$$

This says, very roughly, that, modulo the factor $\frac{1-\xi}{d}$, if the fair market value of the liquidity capitalization is greater than the valuation cap, then the SAFE with Discount will receive the better dividend.

Example 8.6 We adapt Example 1 of Appendix II of the Post-Money Safe User Guide [3]. The first SAFE was purchased for $200,000$ with a valuation cap of $4m$ and a second SAFE was purchased for $800,000$ with a valuation cap of $8m$. The company has $10m$ shares of common stock, no preferred stock, and no other SAFEs, convertible instruments or options. In this case, $s_{common} = 10m$, while, for the SAFEs, $m_1 = 200,000$, $c_1 = 4m$, $m_2 = 800,000$, and $c_2 = 8m$. Suppose the company distributes a dividend of 0.10 per share (so $D = 0.1$) when the fair market price of its shares is 2.

Then $\xi = \frac{200,000}{4m} + \frac{800,000}{8m} = \frac{1}{20} + \frac{1}{10} = 0.15$ and $1 - \xi$ is 0.85. Thus $s_1 = \frac{10m}{0.85} \frac{200,000}{4m} = 588,235$ and $s_2 = \frac{10m}{0.85} \frac{800,000}{8m} = 1,176,471$. Investor 1 receives a dividend of $58,825$ and investor 2 receives a dividend of $117,647.10$. Thus, in addition to $1m$ distributed to holders of common shares, the company distributes $176,472.10$ to SAFE holders.

If, instead, the investors had bought SAFEs with Discount with discount rates of 0.8 (i.e., a discount of 20%) then the dividend for investor 1 is $D \frac{m_i}{p_{fair} d_i} = 0.1 \frac{200,000}{2*0.8} = 12,500$, and the dividend for investor 2 is $0.1 \frac{800,000}{2*0.8} = 50,000$. In this case, the dividend distribution to SAFE holders is $62,500$.

On the other hand, if the fair market price of a share were $0.50 then the investors in SAFEs with Discount would receive dividends of $50,000 and $200,000, respectively, with a total extra distribution of $250,000.

The determination of dividends uses all the machinery designed to handle liquidity events. This is convenient in keeping the SAFE contract short. However, there are some mismatches between the state of affairs in a Liquidity Event and in the issuing of dividends:

- There is no mechanism to establish a fair market price for shares when a dividend is issued, as there might be in a Liquidity Event. This affects the SAFE with Discount and MFN SAFE.
- Liquidity Capitalization includes options:

  > Includes all (i) issued and outstanding Options and (ii) to the extent receiving Proceeds, Promised Options;

  While this makes more sense in a liquidity event, where the options might be exercised, it results in a dilution of the dividend for SAFEs with Cap. The options have not been exercised and, usually, options give no rights to dividends. Similarly, pre-money SAFEs and other convertible securities with no rights to dividends are included, further diluting the dividend. This affects the SAFEs with Cap.

8.10 Conclusion

In summary, we have introduced a general game model for liquidity events in convertible instruments. This model covers all SAFEs. We showed that if the company has issued only one type of SAFE (the types characterized in Sects. 8.5 and 8.6) then the game has optimal Nash equilibria. This shows that, in sharing the value resulting from a liquidity event, the SAFEs determine a mutually agreeable outcome. In the case of Post-Money SAFEs, it shows that the formulation of the payout as the maximum of the Cashout Amount and Conversion amount is sound, since this maximum can be interpreted as the payout received by the SAFE holder in the optimal equilibria. We also outlined polynomial time algorithms to identify all Nash equilibria and, in particular, the optimal ones. However, if the company has issued both Pre- and Post-Money SAFEs with Cap this desirable situation may be lost.

More generally, this analysis shows that the Liquidity Event clauses in these contracts are well-designed, in the sense that they do not create inherent conflicts between the investors, when investments are restricted to a single type of SAFE. However, it also identifies an inability to handle the full range of SAFEs without such conflicts. The identification of optimal Nash equilibria is also helpful in providing an expected outcome: helpful for the company to understand how much money would be needed to pay for Cashouts, and helpful for all to identify the expected number of shares after the liquidity event (significant in a direct listing or an IPO).

References

1. Rapoport A, Chammah AM (1966) The game of chicken. Am Behav Sci 10(3):10–28
2. Harvey D, van der Hoeven J (2021) Integer multiplication in time $O(n \log n)$. Ann Math
3. Y Combinator (2018) Post money safe user guide. https://www.ycombinator.com/docs/Post%20Money%20Safe%20User%20Guide.pdf. Accessed Dec 2024
4. Nisan N, Roughgarden T, Tardos E, Vazirani V (eds) (2007) Algorithmic game theory. Cambridge University Press
5. Thomson W (2003) Axiomatic and game-theoretic analysis of bankruptcy and taxation problems: a survey. Math Soc Sci 45(3):249–297
6. Schwartz A (1993) Bankruptcy workouts and debt contracts. J Law & Econ 36(1):595–632
7. Turner D (2016) The restructuring game, how game theory is being applied in complex restructuring scenarios. https://www.pwc.co.uk/business-recovery/assets/the-restructuring-game.pdf. Accessed Oct 2021
8. Brennan MJ, Schwarz ES (1980) Analyzing convertible bonds. J Financ Quant Anal 15(4):907–929

Chapter 9
Dissolution Events

As discussed in Chap. 2, all SAFEs—whether Pre-Money or Post-Money—entitle their holder, in the case of a Dissolution Event, to the Purchase Amount (also called the Cash-Out Amount), or a *pro rata* amount in the case that there are insufficient funds.

The two kinds of SAFEs differ in the specificity in which they order the different claims on the Proceeds. The Liquidation Priority clause in the Post-Money SAFEs can be seen as an updating of the Dissolution Event clause in the Pre-Money SAFEs. It clarifies the treatment of convertible notes and makes priorities among different classes of claimants more explicit. It also makes one change: the priority of Preferred stock holders is improved from being junior to SAFE holders, to par. This is understandable in the context of the increasing use of lightweight funding (such as SAFEs) where the Preferred stock holders might be former SAFE holders whose SAFEs have converted in a previous equity round.

The effects of a Dissolution Event are relatively easily formulated for each kind of SAFE. As above, m_i denotes the purchase amount of SAFE i.

For the Pre-Money SAFEs, let V be the value of the assets legally available for distribution. Then, if $\sum_i m_i \leq V$, then the holder of each SAFE i will be returned their purchase amount m_i. The remainder is distributed to capital stock holders. If $\sum_i m_i > V$, then each SAFE holder receives

$$V \frac{m_i}{\sum_i m_i}$$

There is nothing left for capital stock holders.

For the Post-Money SAFEs, we take V to be the remainder of the Proceeds after all senior claims (as defined in the Liquidity Priority clause) have been satisfied. Recall that all Preferred Stock is on par with SAFEs. The amount due to Preferred stock

R. van der Meyden and M. J. Maher, *Simple Agreements for Future Equity (SAFE)*, Blockchain Technologies, https://doi.org/10.1007/978-981-96-3920-5_9

holders is determined by the details of the Preferred shares, which may vary.[1] Thus we will address Preferred stock holders in aggregate. Let V_P be the total amount due to Preferred stock holders. Then if $V_P + \sum_i m_i \leq V$, then each SAFE holder will be returned their purchase amount m_i. Preferred stock holders will receive all they are due, too. The remainder is distributed to common stock holders. If $V_P + \sum_i m_i \geq V$, then each SAFE holder receives

$$V \frac{m_i}{V_P + \sum_i m_i}$$

In this case, common stock holders receive nothing.

Clearly, if there are a significant number of Preferred shares, then the returns to a Post-Money SAFE holder in a dissolution are substantially reduced in comparison to a Pre-Money SAFE, assuming the same remainder of Proceeds V. Given the posited binary attitude of investors [1, 2] and the presumably small value of V for a failing business, this might be seen as a distinction without a difference.

In the event that there are both Pre-Money and Post-Money SAFEs issued by the company, a potentially awkward situation arises when there are Preferred shares: these two methods of dividing V are in conflict. The Pre-Money SAFE requires that V "will be distributed with equal priority and *pro rata* among the Dissolving Investors," where the Dissolving Investors are all the SAFE holders (see clause 1(c) of the Pre-Money SAFE) and not Preferred stock holders. The Post-Money SAFE requires that the investor's rights are "[o]n par with payments for other Safes and/or Preferred Stock" (clause 1(d)(ii) of the Post-Money SAFE). These requirements are incompatible.

It can reasonably be argued that this can be resolved by recognizing that Post-Money SAFEs are not SAFEs in the sense of the Pre-Money SAFE. Indeed, Post-Money SAFEs did not exist at the time the Pre-Money SAFE was formulated, so "all other Safes" might be interpreted as: all other Pre-Money SAFEs. Further, the Post-Money SAFE makes the statement that "this Safe is intended to operate like standard non-participating Preferred Stock" and the Pre-Money SAFE is clear that it is to have priority over Preferred stock. This all suggests that Pre-Money SAFEs would have priority over Post-Money SAFEs in this case. Of course, this argument is weaker if the Pre-Money SAFE were executed after 2018, when the Post-Money SAFE was introduced.

However, when there are both Pre-Money and Post-Money SAFEs but no Preferred stock, the situation is unclear. The argument above might continue to hold, and give Pre-Money SAFEs priority. Alternatively, since both kinds of SAFE require all SAFEs to be on par, there is an arguable case that Pre-Money and Post-Money SAFEs should receive equal priority. We consider this to be a point with respect to which SAFEs require clarification.

---

[1] Even though Safe Preferred shares are essentially identical to Standard Preferred shares, there might be other Preferred shares.

References

1. Green JM, Coyle JF (2016) Crowdfunding and the not-so-safe Safe. Va Law Rev Online 102:168–182
2. Coyle JF, Green JM (2015) Contractual innovation in venture capital. Hast Law J 66(1):133–183

Chapter 10
Towards a Better SAFE

The previous chapters have identified a number of weaknesses in the financial structure of SAFE contracts. We have argued that, while post-money SAFEs address some issues with pre-money SAFEs, and better enable the company to finance its early activity by issuing multiple SAFEs, there remain some deficiencies. In particular, Chaps. 7 and 8 identify weaknesses arising from the fact that SAFEs create obligations on the company that are sometimes, in total, beyond the capacity of the company to meet. This creates situations of inherent conflict among the interests of different SAFE investors. In the present chapter, we present a proposal for improved SAFE contract terms that avoids such conflicts. (We state the proposal in general terms, rather than attempt to formulate precise legal language, since this would likely be jurisdiction-dependent.) But we begin by collecting several weaknesses that we noticed.

10.1 Weaknesses in SAFEs

There is no doubt that the post-money SAFEs are an improvement on the pre-money SAFEs, particularly concerning the SAFEs with Cap. The proliferation of conversion methods for equity financing has become unnecessary. However, there are several weaknesses that we noticed, and we list here. To say they are problems is not to say that they cannot be dealt with in practice, with good will among the parties. However, when there is not good will, they are potential pitfalls.

- A company with a mixture of pre- and post-money SAFEs faces several problems on termination. In equity financing, determining the share allocation is difficult; in a liquidity event, SAFE holders can be pitted against each other; in a dissolution, the relative priority of pre- and post-money SAFEs is unclear. Y Combinator advises against mixing SAFE types, but we say more strongly: don't do it.

R. van der Meyden and M. J. Maher, *Simple Agreements for Future Equity (SAFE)*, Blockchain Technologies, https://doi.org/10.1007/978-981-96-3920-5_10

- Upon a liquidity event, the share of value going to a SAFE with Discount is determined by the fair market price p_{fair}. This value is required to be "due and payable ... immediately prior to, or concurrent with" the liquidity event. If there is a change of control, then generally there is a transaction that led to the change of control, and which can be used as the basis for identifying a fair market price. If there is an IPO, then the underwriting due diligence leading to the IPO offering price provides a solid basis for an estimated market price.
  However, if there is a direct listing, there is no basis on which to determine a market price before the direct listing; it is only *after* the direct listing that there is a market. Even then, the shares may be very thinly traded. As a result, in this case, the liquidity price is not well-defined.
- When dividends are distributed, there is not a liquidity event on which to base a fair market price. Thus, again, the liquidity price is not well-defined for SAFEs with Discount, and so the dividend distributed to such SAFEs can be somewhat arbitrary.
- When calculating dividends for SAFEs with Cap, the liquidity capitalization includes many "shares" that do not attract dividends. (See Sect. 8.9.) Consequently, the dividends provided to such SAFEs are reduced.
- It is possible for the outcome of an equity financing to be a change of control (for example, the new investor buys more than half the company, or a converted SAFE receives more than half the shares). In this case, the event can be considered an Equity Financing or a Liquidity Event (or even both). Perhaps the ambiguity is deliberate, to keep the contract simple and leave resolution of this issue to be based on the particular circumstances, but it might be worthwhile to clarify in the SAFE that only one of these two clauses can be executed.
- For a SAFE i with Cap, if $m_i \geq c_i$, then the Discounted Valuation method fails to allocate shares. But such a SAFE is non-sensical and there seems no legitimate use for it. It might be better, at least in theory, to require $m_i < c_i$ in the SAFE.
- More generally, when raising funding using SAFEs with Cap, it is essential to ensure that $\xi < 1$ (that is, $\sum_i \frac{m_i}{c_i} < 1$), so as not to commit more than 100% of the company in the case that the company is successful. This is indirectly required by the company representation (b) that the

  > execution, delivery and performance by the Company of this Safe is within the power of the Company

  but a clearer and more direct statement in the SAFE or accompanying documentation is advisable.
- The Discounted Valuation method is, in general, unable to calculate an allocation of shares to post-money SAFEs with Cap Only, for some reasonable company valuations. (See Sect. 7.5.) Our proposal to address this is in the following section.

Of course, it is easier to identify problems than it is to design a solution that avoids the problems while achieving design goals (such as simplicity in the case of the SAFE). However, we do have a proposal that may be a step towards such a solution for some of these problems.

10.2 Resolving Conflicting Entitlements

Equation (7.17) provides a particularly simple understanding of the promise made to the SAFE investor in scenarios where the multiple post-money SAFEs are issued, and converted using the Discounted Valuation method. In cases where $v_{pre} > c_i$, the Post-Money SAFE with Cap Only promises the SAFE investor shares valued at a proportion m_i/c_i of the inherent pre-money valuation v_{pre}. That is, in this case this SAFE promises the investor a share m_i/c_i of the company immediately before being diluted in the equity round. In addition to this, there is downside protection: if $v_{pre} \leq c_i$, then the SAFE investor is promised shares of value equal to the money m_i they invested.

We have shown in Sect. 7.5 that there are situations where these promises cannot be met for all investors, even when the condition $\sum_i m_i < v_{pre}$ for the applicability of the Discounted Valuation method to pre-money SAFEs holds. There are various approaches that one might take to address this deficiency of the Post-Money SAFE contract in such situations. One is to take all promises at face value, so that the value of the promises exceeds the amount of value v_{pre} that is to be distributed. In this view, the situation is akin to that studied in bankruptcy theory [1], where we have we have a set of creditor entitlements or claims $E_1, \ldots, E_n$ whose sum exceeds the value V of the estate to be distributed. The question raised by this is how to determine a "fair" distribution to the creditors. The literature provides a large number of alternative approaches, with some dating back as early as the Babylonian Talmud [2]. A diversity of justifications for the alternatives can be found, drawing on axiomatic characterizations of fairness, as well as game-theoretic methods. In principle, any of the approaches from the literature could be adopted, with the SAFE revised to explicitly state the fair distribution method applied in a situation of conflicting entitlements.

The situation is actually richer than that studied in bankruptcy theory, since, in addition to the entitlements of the investors on the basis of promises made in their investment contracts, there is an additional factor that could be taken into consideration in determining a distribution, namely, the money m_i invested. This suggests that there is an even richer space of justifiable options available than in bankruptcy theory. For example, rather than base the distribution solely on the entitlements E_i, one could do a distribution based instead just on the individual investments m_i. Other schemes that take both the E_i and the m_i into account can also be envisaged.

Pro-rating claims is one general approach for determining distribution values. More precisely, suppose that an amount V is available for distribution. Consider a set I of investors, such that each has a Claim Amount E_i in the event causing a distribution to be made. (These claim amounts may be calculated from the specific details of the event). Having determined the claim amounts, we obtain a total claim $E = \sum_{i \in I} E_i$. We can now determine a *Distribution Amount* v_i for each investor $i \in I$ as follows: if $V \geq E$, then $v_i = E_i$ for each investor i. If $E > V$, then v_i is the pro-rated claim for investor i:

$$v_i = \frac{E_i}{E} \cdot V$$

We call this scheme *Pro-rating by Claim Amount*.

Alternately, in case each investor i is associated with an original investment amount m_i, we could define

$$v_i = \frac{m_i}{\sum_{i \in I} m_i} \cdot V$$

in the case $E > V$. We call this scheme *Pro-rating by Investment Amount*.

In general, the question of which distribution approach should be used to resolve conflicting claims is ultimately one of values. Work in bankruptcy theory [1, 2] has sought to justify distribution schemes from more basic principles, but the selection of principles is also ultimately a question of values, so this just pushes the problem back, rather than solve it.

However, in the specific case of SAFE contracts, we have a further design consideration: conservatism. We would like to return at least the investment amount whenever possible. Pro-rating by Claim Amount does not guarantee this, in general, while Pro-rating by Investment Amount does. If $V \geq \sum_{i \in I} m_i$, then it follows that

$$v_i = \frac{m_i}{\sum_{i \in I} m_i} \cdot V \geq \frac{m_i}{\sum_{i \in I} m_i} \cdot \sum_{i \in I} m_i = m_i$$

so the investor gets at least their money back. Thus, while Pro-rating by Claim Amount may be a reasonable approach in general, Pro-rating by Investment Amount seems a more reasonable approach to use in SAFEs.

Whichever distribution scheme is used for investment contracts (including SAFEs), if it is written into these contracts, the company should take care to issue contracts to its investors in such a way as to ensure that the scheme is the same in all instruments at the given rank in the priority order. The contracts issued should also be consistent in defining the priority order.

We now develop a way that SAFEs could be stated, based on a Pro-rating by Investment Amount (PIA) approach to resolving conflicting entitlements/claims, in order to avoid the problems we have identified with existing SAFEs. For definiteness, we will refer to this notional SAFE as the Pro-rating SAFE, or PSAFE. However, there is no such document, we are simply addressing the mechanics of how Events are resolved.

Rather than directly describing a share issuance, like pre-money SAFEs, or stating a formula for share issuance that sometimes does not have solutions, like post-money SAFEs, we propose that PSAFEs be formulated using the following:

1. For each type of event, the PSAFE states a claim to an amount of *value* in the company.
2. The PSAFE states that in situations where the total claims exceed the total amount of value available for distribution, we use PIA to determine a *distribution amount*.

A distribution of cash and/or shares, depending on the type of event, will be made, determined by the distribution amount.

We note that when PSAFEs are used in combination with other instruments that state rules inconsistent with PIA, we have a contradiction, so the company needs to take care to avoid this, for example, by issuing PSAFEs only. (Instruments that do not state a rule for resolving excess claims are, arguably, consistent.)

As an example, we give a concrete instantiation for PSAFEs designed on the model of the Post-Money SAFE with Cap Only. We describe the application of the idea for each event type.

In the case of Equity Financing events, we derive a number of shares to be issued to SAFE investors, determined from an *inherent* pre-money (Cap Table view) valuation V of the company at the time of the equity round. Here I is the set of SAFE investors, and other investors holding instruments that convert in the equity round. (We also assume that investment amounts m_i and claim amounts E_i are well-defined for these other investors.)

As with the post-money SAFEs, PSAFEs are issued shares prior to the equity round investor's money being added, and that the value of these shares is determined with respect to the pre-money valuation, so that share issuance to the equity round investors is conservative.

In particular, for PSAFE investor $i \in I$, the PSAFE defines the Claim Amount for Equity Financing events similarly to the Post-Money SAFE. PSAFE investor i's claim is $E_i = m_i$ if $V \leq c_i$ and $E_i = (m_i/c_i)V$ otherwise. This definition ensures that always $E_i \geq m_i$. Thus, in the case $V > E$, we have $v_i = E_i > m_i$, so this instrument is conservative for the investor in this case. Since we use PIA, we also have $v_i \geq m_i$ if $E > V \geq \sum_{i \in I} m_i$. Thus, whenever $V \geq \sum_{i \in I} m_i$, this instrument is conservative for the PSAFE investor.

These distribution amounts can then be used to issue shares. In the case of an Equity Financing event, the PSAFE requires that after conversion to shares, but before shares are issued to new investors, investor $i \in I$ holds a proportion v_i/V of the company. Note that, since the value V represents the inherent (Cap Table view) valuation of the company, which is unaffected by an issuance of shares for no new money, the shares of proportion v_i/V represent value equal to v_i.

In case $V < E$, this scheme yields the result that investors in the set I own 100% of the company, so existing shareholders in the company—in particular, the founders and other common stock holders—are left with nothing. This could be handled by the board deciding on a founder incentivization package after the equity round, or aborting the equity round and negotiating a bespoke structure with the affected parties.

We can apply the same general approach to Liquidity events. Again, each PSAFE i should provide a claim amount E_i. We give the definitions of this for our example below, after discussing some issues specific to Liquidity events. Note that existing Preferred shares can be included in this scheme, in case, as in the post-money SAFEs, it is desired to rank PSAFEs and Preferred shares at the same level. To do so, we simply require that Preferred shares are associated with an investment amount m_i

state a claim amount E_i. It is common for Preferred shares to state the investor's rights, in liquidity events, as a multiple of the original investment amount, so this assumption is reasonable.

For Liquidity events, a number of issues arise. The Liquidity Valuation V here represents the valuation of the company at the time of the Liquidity event (again, determined using the Cap Table view), derived as either a fair market value determined in good faith by management, or a valuation derived from the details of an acquisition or IPO underwriting.

To handle the priority of Cashouts over conversions, we need to distinguish the available cash V_{cash} from the total available value V. We have a set $K \subseteq I$ of PSAFE investors choosing Cashout. In this case, we apply Pro-rating by Investment Amount twice, first for the Cashout investors, and then for remaining unsatisfied claims at the same priority as the remaining PSAFEs.

Given the values E_i for $i \in K$, we apply the PIA, using valuation V_{cash}. Provided the total amount of cash V_{cash} available is greater than $\sum_{i \in K} E_i$, we have $v_i = E_i$, and otherwise a pro-rated amount of V_{cash}.

Having completed the distribution of cash to investors choosing Cashout, we then proceed to the second stage. In general, claims will be updated after the first stage as follows. For investors i in $I \setminus K$, the claims remain unchanged $E_i' = E_i$. For investors i in K, their claim is reduced by the value they received in the first stage: $E_i' = E_i - v_i$. The distribution is based on the valuation $V' = V - \sum_{i \in K} v_i$. We again apply the PIA to give a distribution of the remaining value.

This describes the issuance determination in Liquidity events in general. For the specific SAFE modeled on the Post-Money SAFE, we have $E_i = m_i$ for a PSAFE investor i choosing to Cashout, and $E_i = (m_i/c_i)V'$ for a PSAFE investor choosing to Convert.

It would be desirable to reconsider the game-theoretic analysis of liquidity events of Chap. 8 for these new definitions. We leave this for future work.

Dissolution events are simpler. Here V is the total value available to be distributed in the dissolution, after all creditors at higher priority than SAFE investors have been paid (since all value is to be paid out as cash). Again we apply Pro-rating by Investment Amount for PSAFEs and all other investors at the same priority. For the specific PSAFE modeled on the Post-Money SAFE, we have $E_i = m_i$.

References

1. Thomson W (2003) Axiomatic and game-theoretic analysis of bankruptcy and taxation problems: a survey. Math Soc Sci 45(3):249–297
2. Aumann RJ, Maschler M (1985) Game theoretic analysis of a bankruptcy problem from the Talmud. J Econ Theory 36:195–213

Chapter 11
Summary of the Analysis

In Parts II and III of this book, we have analyzed the technical terms of SAFE contracts with respect to some specific scenarios, with the aim of developing an understanding of how well these technical terms address a number of requirements one might expect such a contract to have, and how use of these contracts might affect the decision-making and exercise of rights by the participants.

Part II considered scenarios in which the company has issued a single SAFE, and focuses on issues dealing with the conversion of the SAFE in an Equity Financing event. Part III extended this analysis to situations where the company has issued multiple SAFEs, and additionally deals with Liquidity and Dissolution events.

The present chapter provides a summary of these parts, giving a high level overview of the issues for readers whose primary interest is in the topic we take up in Part IV: the implementation of SAFE contracts by smart contracts on a blockchain. In particular, this chapter highlights a number of key points from Parts II and III that are pertinent to the design and implementation of SAFEs as smart contracts. We assume that the reader is familiar with the content of Part I, in which the motivation and main terms of SAFE contracts are introduced.

11.1 Summary

As discussed in Part I, SAFEs are a contract between a company and an investor in which the investor provides money and the company promises to deliver shares, according to a specified formula, when the company receives a formal round of equity funding. Alternately, in a liquidity event (say, the company is acquired or makes an IPO), the investor also receives a share of the proceeds (or their money back). SAFEs also provide for a recovery of funds in case the company is dissolved.

SAFEs have been designed with the intention to be a simple and layperson friendly method for financing startups. Indeed, at roughly six pages in length, they are much shorter than the documents usually required for an equity round. From a transaction point of view, SAFEs are rather simple. An investor pays money and, later, the company assigns shares to the SAFE holder (or, sometimes, pays money). This simplicity and the brevity of the contract make SAFEs ideal for a case study of automating these contracts on a blockchain.

Initially, four different types of SAFE were introduced (Pre-Money SAFEs, in the terminology of this book) differing primarily in the formulas used to describe the conversion the SAFE to shares or other proceeds in Equity Financing, Liquidity, and Dissolution events. However, since that time, a new form of these contracts (Post-Money SAFEs) has been introduced, designed to address a changed pattern of usage of SAFE contracts, in which companies issue multiple SAFE contracts before conducting an equity financing round.

Our analysis of SAFEs has been based on two main concerns: the payoff to the participants in terms of the monetary value and proportional shareholding delivered as a function of uncertain variables, and the consequences of the players independently seeking to optimize this value.

One objective of investors is to avoid losses. We have used the term *conservative* (for an investor, possibly given some assumptions about the uncertain variables) to describe a process that ensures that the investor does not not suffer an immediate paper loss.

Intuitively, SAFE contracts have been designed using a conditional formula to determine the conversion of the SAFE to shares or monetary value, in an attempt to prevent losses for the SAFE investor. If the company has performed well by the time of the SAFE is converted, the SAFE aims to give the investor a share in the upside but, if not, the SAFE is designed to return value to the SAFE investor equal to their original investment amount, if possible.

When the usual understanding of the mechanics of equity rounds is applied to equity rounds in which SAFE contracts convert, there is a tension between the interests of the SAFE investor and the equity round investor, resulting from the issuance of new shares to the SAFE investor at the time of the equity round. On the usual understanding, the price of new shares issued in an equity round is determined from the valuation of the company just prior to the equity round (the pre-money valuation) by

$$\text{price} = \frac{\text{pre-money valuation}}{\text{pre-money capitalization}} \tag{11.1}$$

where "pre-money capitalization" is the number of shares issued by the company prior to the equity round. Moreover, in a typical equity round, we also have

$$\text{price} = \frac{\text{post-money valuation}}{\text{post-money capitalization}}$$

where

$$\text{post-money valuation} = \text{pre-money valuation} + \tag{11.2}$$
$$\text{new investor's money}$$

and

$$\text{post-money capitalization} = \text{pre-money capitalization} + \\ \text{new investor's shares}\ .$$

These equations have the intuitively correct consequence that, in the course of the equity round, the new investor exchanges their money for shares of equal value.

However, when a consequence of the equity round is the forced issuance of shares to the SAFE investors in conversion of their SAFEs, we in fact have

$$\text{post-money capitalization} = \text{pre-money capitalization} + \\ \text{new investor's shares} + \\ \text{new SAFE shares}$$

with the consequence that the new investor is immediately diluted as a consequence of the round, and receives shares of lesser value than their investment. On this understanding, the equity round is not conservative for the new investor. Indeed, in spite of the intentions of the design of the SAFE contract, it is not guaranteed to be conservative for the SAFE investor either.

Due mainly to attempts to moderate this dilution, several different methods arose in practice to calculate the number of shares due to convertible instruments such as SAFEs when there is an equity financing. The different methods are permissible for use with SAFEs because the notion of pre-money valuation (of the company) is not defined in the contract.

Some of the confusion can be clarified by recognizing that there are two distinct ways to account for convertible instruments such as SAFEs:

The Liability view   in which a SAFE is considered a liability that lowers the value of the company
The Cap Table view   in which a SAFE is considered as representing virtual shares on the company's capitalization table.

These views are explained in Chap. 3. The different conversion methods can then be better understood and justified once their implicit view of the SAFE is recognized. Some methods, however, are ad hoc, without a clear justification. Chapter 4 gives a detailed explanation for these issues, based on a simple scenario in which the company has issued a single SAFE (a Pre-Money SAFE with Cap Only).

Our approach is to represent background domain knowledge as equations among variables arising in an equity financing. (See Fig. 3.1 for the complete set of equations.) The accounting views can also be represented as equations. Finally, the Equity Financing clause of a SAFE, defining the number of shares to be provided to the SAFE holder, can be represented as conditional equations. Many of the methods mentioned above, when applied to SAFEs, involve circular definitions. Using algebraic manipulations afforded by the equational framework, for each method we are able to obtain closed-form expressions defining the value of each SAFE, and the number and proportion of shares going to the SAFE holder, the company, and the equity investor.

It is important for these three parties to understand the relative merits, for them, of the different conversion methods. In Chap. 6, we provide this comparison and consider how the three parties might react to this information, and "play the game" of negotiating their contract parameters and the design of the equity round. Our analysis there shows that, for the single SAFE scenario, when the participants use an accounting view that is appropriate to the conversion method used, for most of the conversion methods, their negotiations result in identical consequences for the value they will receive in an equity round, which is, moreover, guaranteed to be conservative when the company has not lost value after the SAFE investment. Moreover, for a given single Pre-Money SAFE with Cap Only, there is an equivalent Post-Money SAFE with Cap under these "rational" interpretations. (We also show that when the participants may fail to match the conversion method to the appropriate accounting method, the outcomes of the game are much less predictable, and depend on the relative negotiating powers and legal stances of the players.)

The upshot of this analysis for the single Pre-Money SAFE scenario is that one the methods of calculating the parameters of an equity round, the "Discounted Valuation" method, canonically represents the method used by rational players. We therefore focused on this method in Part III when taking up issues of equity round conversion when multiple SAFE contracts have been issued.

In many ways, the Post-Money SAFEs are an improvement on the Pre-Money SAFEs. They provide both the investor and the company a simple conceptual view of what shareholding will be delivered to the investor in the context of an equity round: in all cases, at least a predetermined proportion of the company, immediately before dilution by the new investor's money while remaining conservative, whenever possible. This outcome depends on the valuation of the company at the time of the equity round, but is independent of the other SAFEs the company has issued. By contrast, in scenarios involving multiple Pre-Money SAFEs, the proportional shareholding received depends not just on the valuation of the company at the time of the equity round, but also on what other SAFEs have been issued by the company.

One salient point concerning Post-Money SAFEs is of relevance to their implementation as smart contracts—instead of providing an explicit formula enabling direct computation of the number of shares to be issued in the equity round, they state a *constraint* on the share issuance, requiring that some simultaneous equations be solved to derive the issuance. Our analysis shows that this approach to specifying the conversion consequences of the SAFE has one disadvantage—for Post-Money SAFEs, when converted using the rational Discounted Valuation method, there are situations in which there are no solutions to these simultaneous equations, whereas Pre-Money SAFEs would have been convertible using the Discounted Valuation method. In effect, Post-Money SAFEs create obligations on the company that sometimes cannot be met. In Chap. 10, we have proposed new SAFE conversion terms that remain simple to understand, while guaranteeing convertibility in all cases where Pre-Money SAFEs would have been convertible.

In addition to Equity Financing events, SAFEs may terminate as a result of a Liquidity event (like an IPO or an acquisition of the company), and through the dissolution of the company. We also analyzed each of these events in Part III.

In a liquidity event, investors holding Pre-Money SAFEs have the choice to Cashout, and get their money back, or Convert to shares and receive a share of proceeds depending on the resulting proportional shareholding. The investors choosing Cashout are paid first. As a result, in scenarios where multiple SAFEs have been issued, liquidity events create a game-theoretic situation, in which the choices of other SAFE investors can potentially affect the payout. Our analysis shows that when only Pre-Money SAFEs have been issued, there is (up to equivalence of payouts) a unique optimal solution to the game. Thus, there is no inherent competition among SAFE holders, and no need for strategizing in making the choice of Cashout or Convert.

In a liquidity event, the holder of a Post-Money SAFE has no choice; they receive the "maximum of the two options." Prima facie, in multi-SAFE scenarios, it is not obvious that there is such a maximum, because of interactions between the payouts of the different SAFEs. Our analysis shows that if only Post-Money SAFEs with Cap have been issued, there is again a unique optimal payout.

In a dissolution, all debts must be paid before SAFE and stock holders can share the remaining value of the company. There is a hierarchy of claims on this value, with holders of preferred stock having priority over holders of common stock.

Our analysis shows that a number of problems can arise when mixing Pre-Money and Post-Money SAFEs. In liquidity events, the resulting game may have no optimum, and SAFE holders then need to take the possible choices of other SAFEs into account, leaving the outcome of the game hard to predict. In dissolution events, the Pre-Money SAFEs require that payment to SAFEs has a higher priority than stockholders, whether preferred or common stock. On the other hand, Post-Money SAFEs require the same priority as other SAFEs and preferred stock. These two requirements are incompatible. Thus the mixing of Pre- and Post-Money SAFEs can result in a situation where it might not be possible for the company to satisfy its obligations to all SAFE holders.[1]

11.2 Challenges in Implementing SAFEs

There are many challenges that arise when trying to implement SAFEs as smart contracts. Most of these will be described and addressed in Part IV. Here we focus on the challenges that can be identified as a result of our analyses.

One important issue, that is already evident from Part I, is the variety and changing nature of the SAFE. The change from Pre-Money to Post-Money SAFEs altered most of the "mechanics" of the SAFE, even if the broad ideas behind it were preserved. Furthermore, since that time there have been new versions of the Post-Money SAFEs that have made further, but smaller, changes. We have even proposed changes ourselves! (Chap. 10) Moreover, SAFE investors are likely to seek additional variations

---

[1] This same problem can arise with a liquidity event when the company does not have enough funds to cover all Cashout investors.

in response to individual circumstances. It is not unusual that contracts evolve over time, in response to experience and a changing market, so these changes are likely to continue. An implementation needs to be able to accommodate such changes as gracefully as possible.

Arising from these changes is the problem of dealing with legacy fundraising instruments. This presents itself in the form of companies using a mixture of instruments in their fundraising (without an intervening equity round). Companies may have used convertible notes, and then moved to SAFEs, but we focus on the change from Pre-Money to Post-Money SAFEs. The different mechanisms of these two kinds of SAFEs may require different implementations but it is, nevertheless, important to abstract and represent the commonalities, to make the implementation adaptable to further changes.

Our analyses have mostly focused on increasing understanding of events in a SAFE. With this improved understanding, we have identified specific problems ensuing from mixing SAFE types. Liquidity events can lead to strategizing among SAFE holders. Dissolution and liquidity events can expose inconsistencies in liquidation priorities, leaving it *prima facie* impossible to satisfy the terms of all SAFEs. Computing share allocations to SAFEs in an Equity Financing becomes more complicated. Furthermore, even without mixing types, post-money SAFEs may be unconvertible at some valuations. Any implementation will need to support resolution of these problems.

Part IV
SAFE Smart (Legal) Contracts

Chapter 12
Blockchain and Smart Contracts

In this chapter, we give a brief introduction to blockchain and smart contracts to set context for the smart contract representation of SAFE contracts that is the main topic of this part of the book.

To a large extent, many of the details of the design of the distributed computation platforms called blockchains are orthogonal to our main concern of smart contract representation of SAFE contracts. However, a basic understanding of the way that these platforms distribute trust will help to understand the motivation for using this class of computation platforms for representing financial instruments such as SAFEs.

12.1 Blockchain

Blockchain, also known as Distributed Ledger Technology, refers to computational systems that allow communities of independent actors to maintain agreement on an evolving computational state, which typically represents, among other things, the state of ownership of certain value bearing assets. Actors in the system interact via computer software-hardware systems called *nodes*, which communicate over the internet in order to process requests submitted to the system by the actors.

Certain of these nodes may have the specific role of maintaining the state of agreement and checking that other nodes are operating according to the rules of the platform. These nodes are called *validators*. Not all participants are required to operate validators: other roles in these systems include *archival nodes*, which maintain the history of the computational state as a service to the community, and *thin clients*, which track only the parts of the state of the system that are directly relevant to their interests, and submit requests, typically called *transactions*, that cause changes to the state of the computation being maintained by the system.

© The Author(s), under exclusive license to Springer Nature Singapore Pte Ltd. 2025
R. van der Meyden and M. J. Maher, *Simple Agreements for Future Equity (SAFE)*,
Blockchain Technologies, https://doi.org/10.1007/978-981-96-3920-5_12

A common type of thin client is wallet software, which stores the encryption keys controlling the cryptocurrency belonging to a user, and tracks payments sent or received by the user.

The term "blockchain" derives from the data structure used to represent the history of transactions in Bitcoin [1], but it has come to be used to refer to a broader class of distributed computation platforms that have arisen since Bitcoin demonstrated the feasibility of this type of platform for the application of digital cash.

Distributed Ledger Technology, in the form of Fault-Tolerant Distributed Systems [2, 3] long predates the emergence of Bitcoin in 2009, but these technological precedents all required that the set of validator nodes be limited in size, relatively static, and have their members subject to approval. The key innovation of Bitcoin was to show how a fault-tolerant ledger could be securely maintained by an *open* and *variable* group of participating validators, with any node free to join or leave the network at any time.

Blockchain systems are fault-tolerant in the sense of operating correctly even when some of the participating validators are faulty. In order to achieve this, the nodes operate a *consensus protocol*, which is software that prescribes the pattern of messages to be sent between the participants in order to maintain their agreement on the state of the computation.

There exist many different designs for these protocols; a deep discussion of which is not pertinent to our concerns in this book. The interested reader is referred to introductory texts on distributed algorithms [3], and recent surveys of consensus protocols that have emerged since Bitcoin [4]. The main points of concern to us are a number of dimensions of difference between the available blockchain platforms.

One of the key dimensions of difference lies in the type of fault against which the protocol is resilient, the precise specification of the way in which the protocol protects against these faults, and the assumptions about the distribution of faulty nodes that need to be satisfied in order for the specification to be satisfied. Fault types against which these protocols may provide protection include:

- **Crash failures**: In these failures, a faulty node behaves correctly until some time at which it crashes, and stops operating. At the time of the crash, the node sends an arbitrary subset of the messages it was supposed to send.
- **Omission failures**: In this case, a faulty node may misbehave by failing to send and/or receive an arbitrary subset of the messages that it is supposed to send.
- **Byzantine failures**: In this fault model, a faulty node may deviate in arbitrary ways from the behavior specified by the protocol. This encompasses both random errors and malicious behavior by a party that is attempting to gain advantage by subverting the specified behavior of the system. Moreover, malicious faulty nodes may collude in the way that they deviate from the protocol, coordinating their attempted attack on the protocol.

The greatest challenge for the design of a protocol that defends against these failure types is posed by Byzantine failures. Many protocol designs are nevertheless able to defend against such failures, and blockchain protocols in open networks typically aim to be resilient against such failures. However, this comes at a cost in the amount of communication and computation required, so some protocols protecting only against the weaker omission and crash failure types are also in use, particularly in closed, enterprise settings.

The requirements for a consensus protocol state that, subject to some assumptions about the distribution of faulty nodes, the nonfaulty validator nodes are successful in maintaining agreement about the evolving computational state resulting from processing the transaction requests submitted by nodes participating in the network. This requirement encompasses both *safety* and *liveness* properties. The safety properties state that, at each time, the nonfaulty nodes have formed a mutual agreement upon a computation state that is a valid result of processing some set of valid transaction requests. The liveness properties state that this agreed state continues to evolve in response to the submitted transactions. There are tensions between these requirements and a further requirement that the protocol be resilient to a breakdown of network connectedness, and protocols differ with respect to whether they sacrifice safety or liveness in this situation.

The sense in which the specification is satisfied also differs between protocols. A foundational result in distributed computing due to Fisher, Lynch, and Patterson [5] states that it is impossible to satisfy a formal specification of consensus under the assumptions that (1) communication is *asynchronous* in the sense that nodes do not have a common clock, and messages may take an arbitrary amount of time to be delivered, (2) the only failures are crash failures, and at most one node in the network may fail, and (3) the consensus protocol is *deterministic* in that it prescribes for each node a unique next action at each stage of the computation. Since crash failures are the weakest type of failures, different protocols respond to this theoretical impossibility by modifying one of the assumptions (1) or (3). Modifying (1) are protocols that work only in *synchronous* or *weakly synchronous* networks, where a global clock and an upper bound on message delivery time (known or not) can be assumed. Modifying (2) are protocols that operate probabilistically rather than deterministically, or satisfy the specification only in a probabilistic sense.

Bitcoin is among the protocols that operates according to a probabilistic protocol. It does so via a "proof of work" mechanism, in which validators compete for the right to contribute the next entry (in the form of a bounded size "block" of transactions) into the agreed history of transactions, by solving a computationally difficult cryptographic puzzle whose solutions are (pseudo-)randomly distributed. The effect of this is that, at each stage, the "winning" validator is randomly selected according to a distribution that weights validators by the amount of computational resources that they contribute to the search for a solution. Random selection of validators ensures that no single validator is able to control the selection of transactions, ensuring that transactions are not *censorable*, in the sense that a valid transaction cannot be prevented from entering the history, so long as sufficient validators exist that are prepared to include this transaction.

A further aspect of Bitcoin's protocol, the "longest chain" rule, moreover deals with situations where multiple conflicting proposed histories exist. Through this rule, the majority of validators (by computational resources contributed) are able to ensure agreement on their preferred history of transactions, even in the face of simultaneous puzzle solutions being found or a malicious minority attempting to create an alternate history that is to their benefit.

The Proof of Work mechanism has been much criticized for the large amounts of energy consumption it requires for its computations, and there has therefore been a considerable amount of effort to develop alternate consensus protocols. In enterprise settings, where closed networks with a limited number of validators are preferred, and weaker failure models than Byzantine failures may sometimes be assumed, consensus protocols in the style of the pre-Bitcoin protocols are often used, which do not require the large energy consumption of Bitcoin. The Linux Foundation's Hyperledger blockchain standardization project [6] includes a number of protocols of this kind.

For open networks, the approach now preferred in many blockchain platforms, among them Ethereum [7], Cardano [8], and Algorand [9] is a "Proof of Stake" rule, which uses a source of randomness weighted not by computational power but an amount of cryptocurrency held by the validator. In some of these protocols, this cryptocurrency must moreover be "staked", that is, posted as collateral at risk in the case malicious behavior by the validator is detected. Ethereum and Cardano draw their randomness by combining randomness proposed by the validators; Algorand instead uses cryptographic techniques to generate pseudo-randomness. The benefit of the Proof of Stake approach is that it avoids the need for large amounts of energy consumption in order to operate the network.

Common to all blockchain platforms is their use of Public Key cryptography in the form of cryptographic signatures to provide authentication and integrity to transactions. When submitting a transaction for processing, the user signs the transaction data using their private key. Validators use a corresponding public key to verify the correctness of this signature, and ensure that the transaction has been requested by an appropriately authorized party.

In open cryptocurrency blockchains, these public keys are generally not associated through a Public Key Infrastructure to user identities, so that digital assets whose ownership is represented on the blockchain are held anonymously, with the user identified only through their public key. This still allows user behavioral profiles to be inferred from open blockchain data when users reuse public keys, or transact with entities such as cryptocurrency exchanges whose public keys are public knowledge. A class of "Privacy-coin" blockchain-based cryptocurrencies therefore also exists, which use advanced cryptographic techniques to allow users to obscure their transactions in order to protect against such inferences from transaction data.

Open blockchains, following the precedent set by Bitcoin, typically represent ownership of a cryptocurrency associated with the platform. This cryptocurrency plays several roles in the network. In blockchains such as Bitcoin, intended primarily as peer-to-peer payment systems, this cryptocurrency is the main motivation for use of the platform. However, the cryptocurrency typically also plays a secondary role of incentivizing the validators, who are rewarded in this currency, by way of transaction fees and "block-rewards" for the work they do in running the consensus protocol. Requiring a fee payment to be associated with a transaction moreover helps to prevent adversarial attacks in the form of "network spam"—large amounts of transaction traffic intended to cause congestion and interfere with user activity.

By contrast, closed enterprise blockchain networks are typically not associated with a novel cryptocurrency, relying instead on fiat currency holdings in traditional financial institutions when transacting value. Such systems have one fundamental difference with cryptocurrency systems.

The base truth about ownership of a cryptocurrency is held on the blockchain itself, secured through a cryptographically signed record maintained by the agreement of the distributed network of validators. An exchange of ownership of a cryptocurrency holding requires that the owner sign a transaction using their private cryptographic key, of which they are the unique holder, and for the distributed network of validators to agree upon the validity of the signed transaction. So long as the user ensures their unique ownership of their private key, this places the user in complete control of their holding: it cannot be transferred without the user's approval. On the other hand, a fiat currency holding in a traditional financial institution is controlled by that institution, requiring merely a change in that institution's database to be lost or transferred. This means that fiat currency holdings are subject to a greater amount of trust than holdings of cryptocurrency.

Although they have been touted as being "trustless", even cryptocurrency systems are not completely free of trust assumptions: they rely upon trust that the hardware and software protect a user's private keys, that the blockchain software is free from bugs and security vulnerabilities, and that the network of validators does not include more than a threshold of malicious validators.

12.2   Smart Contracts

Beyond showing the viability of a cryptocurrency managed by a distributed network, one of the innovations of Bitcoin was to allow users to specify rules for the transfer of their cryptocurrency holdings, in form of "smart contract" programs to be executed by the blockchain validators when checking the validity of a transaction. The language in which such rules are expressed is a simple bespoke programming language called Bitcoin Script.

Coinage of the term "smart contract" has been attributed to Nick Szabo, a computer scientist with interests in economics and law, in a blog in 1997 [10]. The fundamental idea is that multi-party agreements concerning future transactions be secured not through the use of contracts, and enforced by a legal system, but instead expressed as computer programs that are executed by a distributed network of transaction validators.

For example, Bitcoin enables joint control of a cryptocurrency holding to be enforced through code that specifies a "multi-signature policy" such as the requirement that a transaction be signed by two parties before it is considered valid.

A slightly more complex policy, also implementable in Bitcoin Script, is a rule that says that any two out of three signatures suffice for approval of the transfer.

This allows the enforcement of escrow arrangements, where two transacting parties are either able to agree upon a transfer, both signing the transaction, or one of the parties, together with a trusted escrow agent who rules in favor of that party, both sign in order to transfer the funds. Note that this programmatic way of implementing escrow is significantly more secure than traditional approaches where the funds are held by the escrow agent, and at risk of misappropriation by the agent. Placed under the control of a 2/3-signature smart contract, the escrow agent does not have sole power to transfer the funds, and would need to collude with one of the parties to do so, making it significantly less likely that the funds will be transferred for the sole benefit of the escrow agent.

Each holding of an amount of Bitcoin is associated with a Bitcoin Script program that takes as input the details of a transaction attempting to transfer this holding. If the execution of the program on this input evaluates to "True", the transaction is considered to be valid and the holding is transferred once the transaction is included in the agreed blockchain history.

Bitcoin Script has limited expressiveness and, in practice, validators on the Bitcoin blockchain accept only a small number of the programs that could potentially be written in this language. However, the demonstration that simple forms of agreement between mutually distrustful parties could be automatically enforced without reliance on slow and expensive legal processes has motivated the development of blockchain platforms that provide a stronger level of support for the implementation of smart contracts.

12.3 Ethereum

First among the smart contract motivated blockchain platforms, but soon followed by many others, was Ethereum. Proposed in 2013 by Vitalik Buterin, Ethereum was implemented through a company formed in 2014, which financed its development effort through a crowdfunding of Bitcoin contributions, with a promise that the investment be would be converted to a holding of the Ether, the native cryptocurrency of the Ethereum platform to be developed. Once the platform commenced operation, further issuance of Ethers began to be made in the form of rewards to the network validators, similarly to the issuance policy in Bitcoin.[1]

One of the limitations of Bitcoin Script is that it does not permit programs containing looping constructs (instructions such as *"while condition do action"*). This means that many functions cannot be implemented. Ethereum, by contrast, aims to provide a smart contract programming language that is "Turing Complete", meaning that it can express any program that can be expressed in any other programming

---

[1] The legal status of the Ethereum crowdfunding has been controversial, and subject to investigation by the US Securities and Exchange Commission as an illegal securities offering, though there have been statements by SEC officers that operation of the Ethereum platform may now be "sufficiently decentralised" that Ether no longer constitutes a security [11].

language. (In practice, there is a bound on the running time of programs in Ethereum transactions, so that, in fact, not all computations are possible, even if they can be expressed.)

The state of the Ethereum ledger represents cryptocurrency holdings by an amount of value associated with a user's public key, called an *address*. Additionally, some addresses on the ledger are associated with smart contracts. In Ethereum, a smart contract is similar to an object in an object-oriented programming language: it has a state that records the values of a set of variables in the smart contract, but it is also associated with a set of functions. These functions are comprised of program code that may update the value of the smart contract's variables, and may also make calls on other smart contracts. Each function call is executed effectively atomically, that is, it either executes to completion without interruption, or, in case an error condition arises during the computation, it *aborts*, and the state of the object is *reverted* to its state when the function was called.[2]

A smart contract may also hold an amount of cryptocurrency, and its code may perform the operation of transferring some of that value to another address. This means that the value in a smart contract is controlled by the functions in the smart contract. The only way that the value in a smart contract can be transferred is by transactions that make function calls on the smart contract. The smart contract code can thereby enforce conditions on such transfers, and enable a transfer only when its conditions have been met.

When validating an Ethereum transaction, the validators run the code in any function call in the transaction, in order to update the state of the smart contract. Ethereum's code is deterministic, so that all validators running the transaction will come to the same conclusion about how the state of the Ethereum ledger should be transformed as a result. Since running code incurs a cost for the validators, they are compensated by a fee amount that is determined by metering the running of the code—each computation step is associated with a number of units of "gas", and the total amount of gas expended in a transaction's computation, multiplied by a per unit "gas price" (denominated in *Wei*, the smallest fraction of an Ether) is paid to the validator.[3] (This establishes a market in computation, where validators will give preference to running code from transactions offering the higher gas price.) Each transaction sent to a smart contract may also transfer an amount of cryptocurrrency to the smart contract.

---

[2] However, it is possible for another function call to the object to be caused to execute when a function makes a call to one of its own functions, or calls a function on another object which then calls back to the original object. Calls to other objects therefore require careful programming to avoid security vulnerabilities.

[3] This amount must be offered by the user when submitting the transaction; when the gas runs out, computation aborts.

12.4   Ethereum's Programming Environment

Mutiple smart contract languages have been developed for a variety of blockchain platforms, but due to the market share of Ethereum leading to a greater number of developers being familiar with the Ethereum smart contract programming framework, competitive chains are tending to also support this framework.

Programming directly at the level of the primitive instructions provided by computer hardware requires thinking in terms of bits and bytes (0's and 1's, and finite sequences of these values). Programming at this level is done primarily by developers of computer operating systems and programming languages, whose goal is to develop abstractions for use by the general programmer.

Most application programming is done using high-level languages such as Java or Python, which abstract away from the specifics of the machine on which the code will run, enabling the programmer to develop code that will run on computers with differing architectures and instruction sets.

The programming environment on the Ethereum blockchain is based on a number of levels of abstraction. At the lowest level, the Ethereum Virtual Machine (EVM) provides an instruction set for changing the state of various types of memory, including a temporary memory that exists only during the course of a computation, as well as longer term *storage* that represents states resulting from complete computations in response to transactions. The EVM is itself an abstraction from the instruction set of specific computer hardware, designed to facilitate portability of the code to multiple different machines. Accordingly, Ethereum code at this level is called *bytecode*, following terminology from other platforms such as the Java Virtual Machine (JVM). One of the main differences between EVM and JVM-like environments is that the definition of the EVM includes metering of the cost of each basic instruction in terms of an amount of "gas", used to assign a cost to an overall computation requested in a transaction.

Most programming of smart contracts in Ethereum is done not at the bytecode level, but in a higher level language called Solidity, that is compiled (translated) to bytecode. Designed to be accessible to JavaScript programmers, Solidity is a form of object-oriented programming language with a number of novel aspects related to the EVM model of memory and the value-transferring aspect of Ethereum transactions.

For a full exposition of the Solidity programming language, the reader is referred to its documentation [12] or introductory texts [13]. We give just a minimal introduction sufficient to help the non-technical reader to understand some of the code snippets in the following chapters.

Figure 12.1 presents an example of a Solidity program, a simplification of a contract [14] for the ERC-20 token standard, as implemented by Consensys, a software development company founded by Jospeh Lubin, one of the founders of Ethereum itself. The code defines a `contract class`, a notion that is similar to that of a *class* of *objects* in an object-oriented programming language. Contract classes describe the overall structure of smart contract *instances*. For any class, there may be many instances active on the blockchain, each with its own distinct state at any moment of time.

```
contract EIP20 {
    uint256 constant private MAX_UINT256 = 2**256 - 1;
    mapping (address => uint256) public balances;
    mapping (address => mapping (address => uint256)) public allowed;

    constructor (uint256 _initialAmount)    {
        balances[msg.sender] = _initialAmount;
            // Give the creator all initial tokens
        totalSupply = _initialAmount;
            // Update total supply
    }

    function transfer(address _to, uint256 _value)
      public returns (bool success) {
        require(balances[msg.sender] >= _value);
        balances[msg.sender] -= _value;
        balances[_to] += _value;
        emit Transfer(msg.sender, _to, _value);
        return true;
    }

    function transferFrom(address _from, address _to, uint256 _value)
            public returns (bool success) {
        uint256 allowanceAmount = allowed[_from][msg.sender];
        require(balances[_from] >= _value && allowanceAmount >= _value);
        balances[_to] += _value;
        balances[_from] -= _value;
        if (allowanceAmount < MAX_UINT256) {
            allowed[_from][msg.sender] -= _value;
        }
        emit Transfer(_from, _to, _value);
        return true;
    }

    function balanceOf(address _owner)
      public view returns (uint256 balance) {
        return balances[_owner];
    }

    function approve(address _spender, uint256 _value)
      public returns (bool success) {
        allowed[msg.sender][_spender] = _value;
        emit Approval(msg.sender, _spender, _value);
        return true;
    }

    function allowance(address _owner, address _spender)
      public view returns (uint256 remaining) {
        return allowed[_owner][_spender];
    }
}
```

Fig. 12.1 An example of a Solidity smart contract

Smart contract instances are comprised of a number of variables that store the state of the instance. In the example, the variables `balances` and `allowed` are declared to be part of this state. There is also a declaration of a constant `MAX_UINT256`, and the remainder of the code defines several *functions* that operate on the variables. (Conceptually, these are the same for every instance of the smart contract.)

The variable `balances` is declared to have type

```
mapping (address => uint256).
```

This means that values of this variable are mappings that take as input an `address` and yield an output of type `uint256`. The type `address` is the basic representation of identity in Ethereum: it is the public key associated to a user's private key, and smart contract instances are also identified by their address. The type `uint256` represents unsigned integers with 256 bits in their binary representation. The `balances` variable is used to store the number of tokens that users hold. (The value is zero when a user does not hold any tokens.)

When a variable is declared using the keyword `public` the variable may be read from outside of the code of this smart contract; a "getter" function is automatically constructed that may be called on the smart contract to read the current value of the variable. (In case of variables like `balances` of type `mapping`, rather than returning the whole mapping, this getter has a parameter for the key at which the value of the mapping should be returned.) By contrast, the constant `MAX_UINT256` is declared to be `private`, so that it may be used only in the scope of the smart contract. The variable `allowed` is declared as a mapping from two addresses to a number. This mapping is used to represent delegations, whereby one address allows another to transfer its tokens.

An instance of a smart contract is created by calling its `constructor` with values for its arguments (just an integer `_initialAmount` in the example) that are used to initialize the smart contract's variables. In the case of this contract, `_initialAmount` tokens are created ("minted" in the terminology of the field), and the creator of the smart contract is set as the owner of all these tokens. The expression `msg.sender` refers to the address of the user submitting (and signing) the transaction that calls the function in which the expression occurs. Thus, the constructor sets the balance of the creator of the smart contract instance to be the total initial amount of the token supply.

The remainder of the code defines the functions associated to the smart contract type. Each has arguments and a return value. For example, the arguments of the function `transfer` are an address `_to` and a number `_value`, and this function returns a Boolean value that can be referred to in the code as `success` (though this is not needed in this code). The keyword `public` states that this function may be called from outside of the code of this smart contract. The keyword `view`, used in defining functions `balanceOf` and `allowance`, expresses that the function will not alter the values of any variable in the contract.

As already noted, functions, in Solidity, execute *atomically*. They either return successfully, or *abort* due to an error condition arising during the computation. If a function call aborts, it *reverts*, so that the state of the system is reset to what it was at the start of the function's computation. One of the instructions that may abort in the example is the statement

```
require(balances[msg.sender] >= _value)
```

A `require` statements tests a condition, and aborts if it is false. This capability is commonly used to ensure that pre-conditions for the safety of an operation have been met. In this example, the `require` statement ensures that a user does not transfer more than their balance.

Finally, `emit` statements do not affect the state of the computation, but merely post an entry in the form of an "event" into a log recorded as part of the history of the ledger. This can be read by users of the blockchain to determine that certain functions have been executed.

As can be seen from the above example, Solidity is a type of imperative object-oriented programming language. There are many other constructs in the language. We will explain these as we encounter them in the following chapters.

12.5 Smart Contract Applications

There were some attempts to exploit the capabilities of Bitcoin Script to implement smart contracts, and this remains a topic of interest in the Bitcoin ecosystem, but the availability of more expressive smart contract platforms like Ethereum has enabled significantly more projects. Applications that have been developed include

- **Fungible tokens**: A fungible asset is one whose value derives from its type, but not its identity, so that it has equivalent value to any asset of the same type. Common stock in an enterprise and paper money bills are examples: owners generally care about the number they own, but not about their identity (e.g., serial numbers). The data contained in a smart contract can record ownership of some type of fungible asset by associating a number of items held with the cryptographic address of an owner; the functions of the smart contract can enforce that only the owner is able to make a transfer of ownership, or delegate the right to make such a transfer. On Ethereum, the ERC-20 interface for fungible token smart contracts is a standard designed to simplify the development of exchanges for trading of different types of fungible assets. (An example of a smart contract complying with this standard was discussed in the previous section.)

  From 2016-2018, following the approach used by the Ethereum project itself, the ERC-20 standard was used by many projects that raised crowdfunding for blockchain and blockchain application development activities by *Initial Coin Offerings* (ICO), in which investors transferred Ethers to a smart contract that represented rights of investors with respect to the outcomes of the development effort. (The use of the word "coin" in ICO derives from the fact that, in the case of a project developing a novel cryptocurrency, these rights would be, as with the

Ethereum project itself, an entitlement to some amount of the novel cryptocurrency once the software was launched.) In many cases, these projects were not associated with a legal entity, and many proved to be fraudulent [15]. While it was viewed by its proponents as democratizing startup finance, this form of fundraising was highly controversial, appearing to violate securities laws concerning disclosures and investor sophistication or net worth, and regulator actions against a number of projects have led projects to subsequently take more care to be legally compliant. The term *Security Token Offering* has since been used for cryptocurrency-based fundraising for legal entities, with investor rights represented in smart contracts.

- **Non-fungible tokens**: by contrast with fungible assets, the identity of a *non-fungible* asset is a significant factor in its value, and an asset of some type may consequently be valued quite differently from another asset of the same type. Works of art and real estate are examples of this type of asset. Smart contracts have also been used to represent ownership of various types of non-fungible assets. The Ethereum ERC-721 Non Fungible Token (NFT) standard [16] describes an interface for such smart contracts.

  An early example was CryptoKitties, [17], digital images of cats, each distinct, usable in a software application that enabled 'breeding" of cats to generate new instances with attributes "genetically" derived from those of their parents. Trading of CryptoKitties was a popular speculative game for a period. Representation of ownership of digital artworks, and rights with associated communities, using NFT smart contracts became popular in 2020-2021, with a speculative bubble in which some works were sold for very high valuations [18]. Online exchanges [19] have been developed to support the trading of these assets.

- **Atomic Swap**: One of the benefits obtainable from smart contracts is the disintermediation of trusted third parties that are commonly used to secure financial transactions, enabling these to be replaced by the execution of code and secured instead by the decentralized network of blockchain validators. This can result in significant efficiencies, a decrease in costs associated with management of collateral, and an increase in security. The multi-signature escrow application mentioned above is one example of this.

  A related application is a smart contract for an *atomic swap* of digital assets, which allows two parties to exchange digitally represented assets without risk of losses resulting from either party, or a trusted intermediary, failing to fulfill their obligations in the exchange. This solves the problem known in financial markets as "delivery versus payment." We describe this application, which is used as a component in our smart contracts for SAFEs, in greater detail in Sect. 15.1.1. The capacity for blockchain platforms to secure trades using atomic swaps, as well as to reduce costs from reconciliation of records held at independent organizations, has motivated the development of blockchain-based platforms for Stock Exchanges. For example, the Australian Stock Exchange has had a project aiming to replace its clearance and settlement system using a blockchain system [20].

- **Decentralized Autonomous Organizations**: The smart contracts that represent investor rights can be given extra functionality beyond merely representing a shareholding. For example, processes for shareholder voting can be implemented in the smart contract, and payments of returns to investors can similarly be automated. The resulting "on-chain" governance structure of a project can thereby be secured by the smart contract code preventing deviations from a pre-defined set of rules, and the blockchain history providing a transparent and available history of events. The term "Decentralized Autonomous Organization" [21] has been used to describe this approach to securing organizational governance.

- **Decentralized Finance**: This class of applications uses smart contracts to encode the functionality of financial services such as *decentralized exchanges* [22], which allow trading in digitally represented assets without the risks associated with the use of human operated and centralized exchange services, which have had instances of misappropriation of client funds [23]. Other examples include *lending protocols* which allow users to borrow cryptocurrency by posting digital assets as collateral [24], with the smart contract automatically liquidating this collateral when the borrower fails to meet programmatically specified obligations.
- **Provenance records**: A non-financial, but nevertheless significant application of blockchain and smart contracts, has been to provide secure provenance records. This class of applications includes systems that provide integrity to the tracking of the movement and transfer of ownership of products, so that in the event of product recalls, a history is available that allows the source to be identified, e.g., of contamination of a agricultural products. For consumers, the availability of a trustworthy record of provenance allows for purchase decisions based on environmental and safety concerns.
- **Multi-party Commercial Processes**: Finally, blockchain and smart contracts can be used to provide a common source of truth for multiple organizations concerning the state of cross-organizational processes. Applications in this class include supply chain management, trade finance, and international shipping [25]. Processes that were previously managed using paper documents such as bills of lading can be digitized, provided with integrity through digital signatures, and the blockchain used to ensure a secure record of events.

As is apparent from this list of applications, a common type of application for smart contracts is the representation, in digital form, of some type of asset, and the automated enforcement of the rights associated to the asset. Our discussion, in the following chapters, of the smart contract implementation of SAFE contracts is in this vein.

References

1. Nakamoto S (2008) Bitcoin: a peer-to-peer electronic cash system. https://bitcoin.org/bitcoin.pdf. Accessed Dec 2024
2. Pease M, Shostak R, Lamport L (1980) Reaching agreement in the presence of faults. J ACM 27(2):228–234
3. Lynch N (1996) Distributed algorithms. MIT Press
4. Davidson M (2023) State machine replication and consensus with Byzantine adversaries. Technical Report IR 8460, NIST. https://csrc.nist.gov/pubs/ir/8460/ipd. Accessed Dec 204

5. Fischer MJ, Lynch NA, Paterson MS (1985) Impossibility of distributed consensus with one faulty processor. J ACM 32(2):374–382
6. Hyperledger. https://www.hyperledger.org. Accessed Dec 2024
7. Ethereum. http://ethereum.org. Accessed Dec 2024
8. Cardano. https://cardano.org. Accessed Dec 2024
9. Algorand. https://algorand.co. Accessed Dec 2024
10. Szabo N (1997) Formalizing and securing relationships on public networks. First Monday 2(9)
11. Hinman W (2018) Digital asset transactions: when Howey met Gary (Plastic): remarks at the Yahoo finance all markets summit: Crypto. https://www.sec.gov/news/speech/speech-hinman-061418. Accessed Jun 2024
12. Solidity a statically-typed curly-braces programming language designed for developing smart contracts that run on Ethereum. https://soliditylang.org. Accessed Jun 2024
13. Antonopoulos A, Wood G (2019) Mastering Ethereum: building smart contracts and dapps. O'Reily
14. Tokens. https://github.com/Consensys/Tokens. Accessed Jun 2024
15. Zetzsche DA, Buckley RP, Arner DW, Föhr L (2019) The ICO gold rush: It's a scam, it's a bubble, it's a super challenge for regulators. Harv Int Law J 60(2)
16. Smith C et al ERC-721 nonfungible token standard. https://ethereum.org/en/developers/docs/standards/tokens/erc-721/. Accessed Sept 2024
17. Cryptokitties. https://www.cryptokitties.co. Accessed Nov 2024
18. Farago J (2021) Beeple has won. here's what we've lost. New York Times. https://www.nytimes.com/2021/03/12/arts/design/beeple-nonfungible-nft-review.html. Accessed Jun 2024
19. Opensea NFT exchange. https://opensea.io. Accessed Jun 2024
20. ASX (2021) CHESS and CHESS replacement. https://www2.asx.com.au/markets/clearing-and-settlement-services/chess-replacement. Accessed Dec 2024
21. Hassan S, De Filippi P (2021) Decentralized autonomous organization. Internet Policy Rev 10(2). https://doi.org/10.14763/2021.2.1556.
22. Uniswap. https://smarter.uniswap.org. Accessed Dec 2024
23. Young SD, Keoun B (2022) The epic collapse of Sam Bankman-Fried's FTX exchange: a crypto markets timeline. Coindesk. https://www.coindesk.com/markets/2022/11/12/the-epic-collapse-of-sam-bankman-frieds-ftx-exchange-a-crypto-markets-timeline. Accessed Dec 2024
24. Aave. https://app.aave.com/. Accessed Dec 2024
25. World Trade Organisation (2020) Blockchain and DLT in trade: where do we stand? https://www.wto.org/english/res_e/publications_e/blockchainanddlt_e.htm. Accessed Dec 2024

Chapter 13
Difficulties in Formalizing SAFE Contracts

Legal contracts are intended for human-operated processes such as negotiation, execution (i.e., signature), performance, compliance, arbitration, and litigation, so are necessarily expressed in natural language. A consequence of this is that legal contracts are often subject to indeterminacy, which presents difficulties for their formal representation as computer code in smart contracts. One might expect that contracts in the narrow, technical domain of corporate finance are more amenable to formalization. In fact, even in this technical domain, natural language documents may suffer from indeterminacy. In the present chapter, we identify a number of specific ways that this difficulty arises for SAFE contracts.

We begin with a brief review of terminology used in the legal literature to describe forms of indeterminacy in legal texts, and some of ways that the law may deal with the difficulties this causes. We then identify some particular occurrences of indeterminacy in SAFEs. We then discuss other difficulties that arise when formulating SAFEs as smart contracts: incompleteness of the contract, the openness of the context with which the contract interacts, the dependency of its interpretation on the legal context, and differences between the time at which an action occurs and the time at which contractual provisions relating to the action can be enforced. Our focus in this chapter is on identifying issues that need to be dealt with: subsequent chapters describe strategies for dealing with these issues, the actual formalization we have developed that applies these strategies, and its evaluation.

13.1 Indeterminacy in Legal Text

There is a variety of forms of indeterminacy in legal text, described using different overlapping terminologies (which themselves can suffer from vagueness).

© The Author(s), under exclusive license to Springer Nature Singapore Pte Ltd. 2025   173
R. van der Meyden and M. J. Maher, *Simple Agreements for Future Equity (SAFE)*,
Blockchain Technologies, https://doi.org/10.1007/978-981-96-3920-5_13

The term *open texture* was introduced by Waismann [1] to refer to the problem that an empirical concept, however carefully defined, might have uncertain applicability in unforeseen circumstances. The Oxford Dictionary of Philosophy [2] gives an illuminating example. In general, the term *mother* is quite clear. Nevertheless, advances in reproductive technology have led to the situation where past uses of the term have not distinguished between "the mother that produces the ovum, the mother that carries the foetus to term, and the mother that rears the baby." Questions of applicability of concepts in unforeseen situations can be an issue even in the rigorous setting of mathematics: Lakatos [3] showed that, historically, the concept of *polyhedron* has repeatedly had to be adapted to cope with unforeseen examples and counterexamples. A more apposite example of Waissman's notion of open texture arises in the white paper associated to The DAO [4] (a crowdfunding smart contract) which essentially says that terms are specified by specific smart contract on "the Ethereum blockchain." An attack on this smart contract led to a bifurcation ("hard fork") of the Ethereum blockchain into two distinct blockchains: the Ethereum blockchain and the Ethereum Classic blockchain. As Grimmelmann [5] points out, previously this reference was clear but, after the hard fork, both blockchains might now be referred to by this text.

Hart [6, 7] reused Waismann's term, in application to Law, but seemingly expanded it to include vagueness, which Waismann had distinguished from his notion of open texture. Schauer [8] views Waismann's open texture as *potential* vagueness that circumstances might bring to light, whereas Hart's notion also encompasses *actual* vagueness. Actual vagueness can be deliberate and obvious: requiring actions be taken "in a timely manner," or taking "all reasonable precautions." Such uses allow the requirements to be viewed in the context of actual events, rather than attempting to anticipate all eventualities. Vagueness can also arise simply out of the normal use of words. A famous example is a rule that no vehicles are allowed in a public park [7, 9, 10]. The rule seems clear, but what about a truck on a plinth as a war memorial? Or a wheelchair? Or skis? Thus, a consequence of the existence of open texture and vagueness in legal text is that adjudicators can be called upon to apply judgment in its interpretation (judicial discretion) [11]. As a result of rulings by the adjudicators in specific cases, the understanding of the concepts evolves over time.

Works vary in how they interpret "ambiguity" in discussing indeterminacy. We prefer to restrict that term to situations where the use of words or phrases with double meanings, the structure of the text, or inconsistency in the text, lead to indeterminacy, as distinct from vagueness of language. For example, a military regulation requiring a commander to "sanction unauthorized actions performed while out of communication with central command," is ambiguous, based on the dual meanings of "sanction". Of course, the context of the surrounding regulation is likely to disambiguate the meaning of this text. On the other hand, a check with £245 written in numbers and "two hundred pounds" in words[1] is ambiguous due to inconsistency, but with no chance of disambiguation on the basis of the document alone. A further

---

[1] See *Saunderson v Piper*, Bingham's Reports of New Cases in Common Pleas and Other Courts, 1839.

distinction is made between *patent ambiguity*, which is apparent from the legal text alone, and *latent ambiguity*, where the text is clear but the particular context makes the meaning unclear (for example, reference to "my nephew, John" when there are two nephews named John) [12].

In many jurisdictions of contract law, the doctrine of *contra proferentem* attempts to resolve such situations when the context and ordinary, natural meaning of the words are not sufficient. It is described variously as preferring the interpretation that works against the drafter of the contract, or the party with greater bargaining power, or the party seeking to rely upon the ambiguous text [13, 14].

There is a further notion of *defeasibility* in the law: that apparent straightforward application of the plain language of a law can fail due to specific circumstances. A classic example is the 1889 case of a man who killed his grandfather and then claimed to inherit the estate[2]. Despite clarity of the will, and the law, that the grandson should inherit the estate, the court ruled against the grandson, based on the (higher) principle that "no man should profit from his own wrong."[3] This aspect of legal reasoning is distinct from issues of open texture, vagueness, and ambiguity.

13.2   Indeterminacy in SAFEs

An example of the phenomenon of indeterminacy in the case of SAFE contracts is the term "Equity Financing" used in the key clause of the SAFE contract. The Pre-Money SAFE contracts define this term this as follows:

> **"Equity Financing"** means a bona fide transaction or series of transactions with the principal purpose of raising capital, pursuant to which the Company issues and sells Preferred Stock at a fixed pre-money valuation.

This definition contains terms that raise multiple issues of interpretation.

- What does it mean for a transaction to be "bona fide"? This is a legal term denoting that the transaction is in "good faith", and not fraudulent. This is a prime example of an "open-textured" concept, that is hard to give an a priori formalization.
- How do we determine whether a transaction was done "with the principal purpose of raising capital"? The relevant facts for such a determination are likely to be the actions of human agents (their spoken assertions and written communications, and other real-world actions) that may be complex to interpret. An example of a transaction not done for purposes of raising capital is an issuance of stock in exchange for services.
- The definition explicitly allows that the Equity Financing event consists of a "series of transactions." This raises the question of how we know which transactions are part of the event.

---

[2] *Riggs v. Palmer* in the New York Court of Appeals, 1889.

[3] A more recent occurrence of a similar situation is the question of whether a presidential pardon could, in effect, pardon the pardoner.

With respect to the question of how to identify the transactions in an Equity Financing event, that the sale of Preferred shares should be at a "fixed pre-money valuation" in the Pre-Money SAFE suggests that two sales of Preferred Shares, at prices that imply different pre-money valuations, are not both part of the same Equity Financing event. Alternately, one could take the interpretation that each transaction in the series should be at a "fixed pre-money valuation." If we adopt this alternative reading then sales of preferred shares at different valuations could be considered part of the same Equity Financing. (Post-Money SAFEs explicitly allow for this in their Equity Financing clause.) Thus we see that there is lexical ambiguity in the definition of Equity Financing. This is distinct from the open texture of the phrases used in this definition, because it turns on whether the fixed valuation applies to each transaction individually or to the series of transactions as a whole.

Another ambiguity arises from an overlap in the definitions of Equity Financing and Liquidity Event. If an equity financing results in a shareholder becoming a majority owner, then a change of control and, hence, a liquidity event has occurred. It seems obvious (although not specified in any way) that the SAFE holder should not get the benefit of both the Equity Financing and Liquidity Event clauses, but it is unclear which clause should have priority.

Patent ambiguities like these can usually be addressed, once identified, by a slight clarification of the contract (including any formalization). However, latent ambiguities are more difficult, since they may become apparent long after the formalization. For example, the reference to "all other Safes" in clause 1(c) of pre-money SAFEs became ambiguous only after the introduction of post-money SAFEs, which led to confusion in the priority of the two kinds of SAFEs in the event of a dissolution, as discussed in Chap. 9.

Both open texture, in Waismann's sense, and latent ambiguity can arise from unforeseen circumstances. One source of unforeseen circumstances—which, nevertheless, is foreseeable—is *aliasing*: the situation where two (or more) nominally different roles (presumed independent) are taken by the same entity.[4] If, in an equity financing, the founders are also the equity investor, then they can manipulate the valuation to their own advantage, and to the detriment of SAFE investors. An artificial increase in the valuation will reduce the number of shares obtained by SAFE investors, leaving the existing shareholders with a greater share of the company, and the company in essentially the same financial state. (See Example 6.1.) Corporate law has provisions against self-dealing, which might apply in this situation. In a SAFE contract, this behavior is also guarded against by the restriction to "bona fide ... transactions with the principal purpose of raising capital." However, the question of whether this concept applies in a particular case may require resolution in court.

The apparent difficulty for formalization of open texture concepts is that it is unclear what primitive concepts should be included, that potentially the volume and complexity of relevant detail could be very large, and that evolving case law may both broaden the scope of relevant detail and force changes to whatever formalization is adopted.

---

[4] The examples of defeasibility in the previous section arise from aliasing.

13.3   Incompleteness

One way that legal contracts can be indeterminate is through *incompleteness*, the failure to define terms or stipulate the consequences of all possible contingencies. In this respect, they have a structure that differs significantly from code. Whereas imperative code (provided it always terminates) has a logical flow that provides an output for every input, natural language contracts are expressed in a collection of declarative sentences, which state conclusions about certain situations. The coverage of all possible "input situations" may be incomplete. Moreover, some terms may be left undefined. Empirical studies show that business contracts are often incomplete [15]. This observation is the starting point for an entire area of economics [16].

Incompleteness is to be expected in the SAFEs, where design considerations of simplicity and layperson-friendliness outweighed completeness [17]. An example that arises in the case of the Pre-Money SAFE contract with Cap Only is that the text makes use of the terms "pre-money valuation" (the new investor's valuation of the company before an equity round) and "price" (of shares paid by an investor in an equity round), but does not state how these concepts are related.

These are terms of art in the domain of venture finance and, in the context of equity rounds not involving convertible instruments, are related by the equations discussed in Chap. 3 and listed in Fig. 3.1, where p_{new} denotes the price and v_{pre} the pre-money valuation. However, these equations are inconsistent in the context of converting SAFEs (Sect. 3.2), with different consistent subsets. Thus, the SAFEs are incomplete, and trying to remedy the incompleteness with domain knowledge leads to different understandings of the contract. As a result, as we saw in Chap. 4, multiple different methods could be used to convert SAFEs to shares during equity financing.

We might take the view that the incompleteness identified above constitutes a deficiency in the natural language contracts that should be corrected when developing corresponding smart contracts. However, as discussed in Sect. 6.3, there is evidence that the incompleteness is deliberate.

The SAFE holder is potentially the loser from the negotiations between the founders and the equity investor. It is not clear whether the they have any negotiation rights in this matter: this question is another point of incompleteness in the contract.

For another example, the Equity Financing clause in the Post-Money SAFE leaves a small lacuna: in the event that cases (1) and (2) of this clause produce the same number of shares, it is not specified what result is entailed. (See discussion in Sect. 5.1.) Assuming "greater" is interpreted as greater or equal, it still is not clear which type of shares—Safe Preferred or Standard Preferred—should be issued to the SAFE investor: the text is then inconsistent, although the difference may be immaterial if, as seems to be the intention, the only difference between Safe and Standard Preferred shares is the price paid. This is a latent ambiguity, although to some it might be patent. (This issue is precisely the sort of "edge case" that can be detected as the benefit of a careful formalization and verification.)

Although this seems too minor to reach court, under *contra proferentem* (on the "against drafter" interpretation), Y Combinator, as SAFE investor, would receive Standard Preferred shares,[5] since they drafted the contract; otherwise, (on the "against party with greater bargaining power" interpretation), the SAFE investor might expect Safe Preferred shares since the company, as party to the transaction that sets the valuation, has the power to influence that valuation. However, if the company issues Standard shares to the SAFE investor, who then sues for Safe shares, it is the SAFE investor who is relying on the ambiguous text and so might lose under *contra proferentem* (on the "against relying party" interpretation). This variance of conclusions depending on the interpretation of the legal principle itself adds to the indefiniteness of the contract!

For our present purposes, the main point is that SAFE contracts, as they are treated in practice, may not be deterministic. This is particularly true with respect to the relationship between pre-money valuation and price, and consequently may also leave open precisely how to calculate the number of SAFE shares to be issued, if only a pre-money valuation is given.

13.4 Open Structure

Another of the difficulties that needs to be addressed in formalizing the SAFE contract is the *open structure* of the context in which it applies. That is, there are many—and changing—elements of the environment in which it operates, that may not even be mentioned in the SAFE. A key aspect of this context is that the Company has a *capitalization table*, listing shareholders and the number and types of shares each holds. A SAFE itself has a particular structure, but it is not self-contained. Instead, it creates new rights and obligations with respect to the Company's future issuance of shares and, thus, the future state of the capitalization table. To properly formally represent a SAFE, we need to model the way that it depends upon and affects its environment.

One approach that the smart contract programmer might take to accommodate this is to include, in a smart contract representing the company's financial structure, method calls that correspond to the issuance of a SAFE note. However, this is insufficient: the Company may issue a SAFE note to an investor, but it could equally enter into any number of other contracts, each with a completely different structure. Already, SAFE notes come in multiple forms: the four original Pre-Money SAFEs, and the four new Post-Money versions. Other firms have developed their own variants on the same theme. Further variations may arise from a negotiation between the Company and the early investors. Some of the work in this book points to additional variations. The term "other convertible securities" used in the definition of Company Capitalization encompasses an unbounded variety of contract forms. Given

---

[5] Under the presumption that Safe Preferred shares are preferable to Standard Preferred shares.

this infinite variety of potential contracts, it is not feasible for each possibility to be anticipated and enabled by including a set of functions specific to the structure and logic of the contract.

A related point is that the offer of the new investor in the term sheet that triggers the price calculation may be stated in a number of different ways, e.g., "$5M for a 30% share of the company post-money" or "$5M at a pre-money valuation of $10M."

Another type of openness in what we need to deal with is the governance structure of the company, which evolves over time. In particular, we expect that one of the consequences of an equity round will be that a new governance structure is instituted—it is typical for the lead investors to take board positions. However, the details of this structure will be one of the points of negotiation in the equity round, and cannot be predicted ahead of time. A smart contract representation of the company therefore needs to allow flexibility in the representation of the governance rules.

Liquidity events may also have many forms that are difficult to predict in advance. Terms of an acquisition will be set by an acquirer, and may involve offers mixing cash and shares, constraints on existing employees, and post-acquisition financial structure of the company. Corporate restructurings may have any number of forms, including spinouts, splits into multiple entities and change of domicile. If the company itself exists after the liquidity event, it is likely to have a changed governance structure.

13.5   Legal Context

Another consideration which might be considered an issue of open texture, incompleteness, or open structure is the legal context in which the contract is agreed upon. The text of a conventional contract does not stand alone: it inhabits a world of laws (statutes and common law) and enforcement mechanisms. This context can affect both the text of a contract and the way it is interpreted. The presence of laws can lead to the text of a contract failing to address situations that are already covered by those laws, while the enforcement of the contract by a court can apply terms that are not in the text of the contract (known as implied terms).

Dependency on the legal context can be explicit in the contract. For example, the SAFE contract defines the notions of "change of control" (of the company) and "accredited investor" by reference to the definition in statutes (specifically, the Securities Exchange Act and the Securities Act). The meaning of these definitions might change as a result of new precedents or by legislative amendments to the statutes. Thus there is dynamism in the context on which the interpretation of contract depends. The SAFE also refers to actions to allow "such Change of Control to qualify as a tax-free reorganization for U.S. federal income tax purposes." In this case, the validity of the actions rely on perceptions of policy and procedures of the Internal Revenue Service.

However, implicit dependencies on the legal context are more widespread. Particularly in the case of a SAFE, where simplicity is a central feature of the contract, the contract does not specifically address possible contingencies (for example, when the equity investor is also a founder). Furthermore, common law will imply terms to a contract that are not present in the text [18]. For example, the implied covenant of good faith and fair dealing[6] implies that the company issuing the SAFE must conduct its business in a reasonable manner: the company may not simply issue a SAFE, pay the consequent income as bonuses to the founders, and declare bankruptcy. Consumer and financial statutes might also imply terms in a SAFE. In a legal environment without these implied terms, the contract would need additional language to provide suitable protection against such behavior. Courts can also designate parts of a contract as "invalid, illegal or unenforceable," as acknowledged in clause 5(e) of the SAFEs.

Consequently, the effect of a SAFE is not defined only by the text of the contract, but also by the whole legal environment for which it was developed. A smart contract implementation that addresses only the actions and limits expressed in the text of the SAFE will fail to give full effect to the intent of the SAFE, unless it (and the parties to the contract) operates under the laws, regulations, and enforcement mechanisms that the SAFE was designed for.

These are issues that cannot be addressed by formalization of the text of the contract alone. They motivate the use of smart contracts in tandem with—and not instead of—a conventional contract. The effect of nuncospectivity is another motivation.

13.6 Nuncospectivity

Conventional contracts are generally enforced retrospectively—after events related to the contract have occurred (or have failed to occur). Occasionally, when awarding injunctions, for example, enforcement is addressed prospectively, before the events occur.

However, smart contracts can provide enforcement only (or mainly) at the time of the event—a perspective we might call *nuncospect*.[7] To see the problems this presents, we return to the definition of Equity Financing discussed in Sect. 13.2.

The Equity Financing definition refers to a "series of transactions." At the time of the first transaction (or, indeed, any other transaction), it may not be possible to know whether there will be a later transaction that should be interpreted as part of this series. With hindsight, it is possible to identify transactions that have similarities in time of execution or other characteristics that suggest that the transactions should be considered as comprising a series of transactions, rather than as individual transactions. Similarly, the "bona fides" and "the principal purpose" of a transaction can become clearer in retrospect, when there are other events to create a context. Such events can appear irrelevant to the smart contract, for example, transfer of money

---

[6] Kirke La Shelle Co. v. Paul Armstrong Co. in the New York Court of Appeals, 1933.

[7] From the Latin *nunc* meaning now, and *spectare* meaning to look.

between two accounts not appearing in the contract. Thus the nuncospective view of smart contracts limits the extent that the terms of a SAFE can be detected or enforced by smart contract alone.

In the immediately following chapters, we will focus on what smart contracts can enforce nuncospectively. We return to the issue of retrospective interpretation in Sect. 16.3.

References

1. Waismann F (1968) Verifiability. In: Harré R (ed) How I see philosophy. Palgrave Macmillan, London
2. Blackburn S (1996) The Oxford dictionary of philosophy. Oxford University Press
3. Lakatos I (1976) Proofs and refutations. Cambridge University Press, Cambridge, UK
4. Decentralized Autonomous Organization (DAO) Framework. https://github.com/blockchainsllc/DAO. Accessed Nov 2024
5. Grimmelmann J (2019) All smart contracts are ambiguous. J Law Innovat 2:1–22
6. Hart HLA (1961) The Concept of Law. Clarendon Law Series, Clarendon Law
7. Hart HLA (1958) Positivism and the separation of law and morals. Harvard Law Rev 71(4):593–629
8. Schauer F (2013) On the open texture of law. Grazer Philosophische Studien 87(1):197–215
9. Fuller LL (1958) Positivism and fidelity to law: a reply to Professor Hart. Harv Law Rev 71(4):630–672
10. Schauer F (2008) A critical guide to vehicles in the park. NY Univ Law Rev 83(4):1109–1134
11. Reyes Molina SA (2020) Judicial discretion as a result of systemic indeterminacy. Can J Law Jurisprudence 33(2):369–395
12. Anon (1911) Ambiguity. In: *Encyclopedia Britannica*. Encyclopedia Britannica, Inc
13. American Law Institute (1981) Restatement (second) of contracts. American Law Institute
14. Alden S, Ottaway A, Tetstall J (2012) Drafting contracts: guidance on managing ambiguity. https://www.mondaq.com/australia/contracts-and-commercial-law/163072/drafting-contracts-guidance-on-managing-ambiguity. Accessed Dec 2024
15. Macaulay S (1963) Non-contractual relations in business: a preliminary study. Am Sociol Rev 28(1):55–67
16. Hart O (2017) Incomplete contracts and control. Am Econ Rev 107(7):1731–1752
17. Coyle JF, Green JM (2015) Contractual innovation in venture capital. Hastings Law J 66(1):133–183
18. Knowler J, Rickett C (2011) Implied terms in Australian contract law: a reappraisal after University of Western Australia v Gray. Monash Univ Law Rev 37(2):145–161

Chapter 14
Strategies for Formalization

While the difficulties discussed in Chap. 13 present challenges to the formalization of SAFE contracts as smart contracts, they do not prevent smart contracts providing useful levels of security for this application. In this chapter, we describe a number of strategies addressing these difficulties that we have applied to develop a smart contract representation of SAFE contracts, and the scenario in which they operate.

Underlying several of the difficulties are forms of indefiniteness in the text defining SAFE contracts. One can take a number of attitudes to this issue. One is that indefiniteness in the text is a weakness in the design of the contract, and insist that only well-designed contracts that are free of indefiniteness should be represented as smart contracts. This leads one to first increase the precision of the contract to eliminate ambiguity and incompleteness. Alternately, one can take the attitude that the contract's indefiniteness is intentional, to allow the parties to the contract room for negotiation, or to accommodate unforeseen circumstances.

Depending on the attitude taken, the blockchain can be used to give integrity to the cap table data and assure terms of the SAFEs to various degrees. At one end of the spectrum, we have automated enforcement of highly precise terms. At the other, the blockchain enforces the use of processes whereby human agents resolve the indefiniteness. As multiple human agents are typically involved in a contract and its operation, the design of such processes requires design decisions to be made about the powers that each of these agents should have with respect to the resolution of indefiniteness.

In what follows, we develop an approach that encompasses both these attitudes to the question of how far to formalize the terms of a natural language SAFE contract, and makes it possible to cover a range of points on this spectrum, depending on how much formalization is undertaken.

R. van der Meyden and M. J. Maher, *Simple Agreements for Future Equity (SAFE)*, Blockchain Technologies, https://doi.org/10.1007/978-981-96-3920-5_14

14.1 Increasing Precision

Our analysis of SAFE contracts has identified a number of cases where the text of the contracts is ambiguous. Particularly where the ambiguity is unexpected or has a minor impact on the legal meaning, it may make sense to remove the ambiguity in the formal representation.

Increasing precision in a contracts terms when developing a smart contract implementation may alter or refine the literal meaning of the source contracts. However, so long as the parties to the contract are aware of the modifications, understand them, and can agree to them, this could be argued to be a benefit rather than a problem.

One example where we may wish to increase precision is with respect to the question of whether the SAFE holder has any rights to influence the terms of an equity round, on which the SAFE is silent. The contract could be modified to grant such rights in various forms, for example, a right to approve or a right to arbitration in case of an objection. Even if neither of these options is programmed into the smart contract, then *de facto*, the smart contract is more precise in that it definitively answers the question in the negative. (We discuss this further in Sect. 14.3.)

Another example relates to the meaning of "the greater of" a number of Standard Preferred Stock and a number of Safe Preferred Stock, issued in the context of an Equity Financing event for a Post-Money SAFE. This raises the question of which type of shares should be issued in case of equality. (See Sect. 13.3.) Particularly as the two types of stock operate identically in this case, it makes sense to resolve the ambiguity in one direction when expressing the issuance in code. This is an increase of precision that both parties are likely to find acceptable.

Similarly, in a situation where an equity financing causes a Change of Control (and hence a Liquidity Event), we might choose to treat the situation only as an Equity Financing. This increase in precision might be agreed to by the parties beforehand, since the situation is unlikely, but if the increase of precision is unnoticed until the event occurs, then the disadvantaged party could object.

14.2 Limitation of Fact Patterns

When developing a smart contract representation, even where the attitude adopted is that open texture and incompleteness in a natural language contract should be replaced by precision, it is not necessarily required to resolve all possible questions of interpretation that could arise. The issues of open texture discussed in Sect. 13.1 bear on an "interpretative perspective," in which, given a set of facts, we seek to determine whether a concept applies. However, smart contracts are not necessarily required to deal with arbitrary fact patterns. This is because smart contracts themselves may impose a more restricted model of the factual ground for concepts than one might find in the real world. The code can limit the possible fact patterns that can be constructed, and the developer has the option to design the code so as to avoid fact patterns that raise difficult questions of interpretation. It suffices that the parties to the smart contract are prepared to accept its limited flexibility, and that they are in agreement that the code correctly captures the concepts of concern within this limited range.

As one concrete example of this, consider the question of which transactions, in particular, which issuances of Preferred shares, should be considered to be part of an Equity Financing[1]. If we design the smart contracts so that arbitrary patterns of share issuance are supported, then we need to address this question of interpretation. However, by appropriate choice of operations, and constraints on when these operations are enabled, we can ensure that this question does not arise.

Suppose that in addition to the basic operation of issuing shares, we design the system so that there is a compound operation of conducting an equity round, with the effect of issuing shares, among other things. Further, suppose that whenever a SAFE contract has been issued but remains unconverted, we constrain the smart contract for the company so that when shares are issued outside of the context of this equity round operation, the type of the shares may not be Preferred shares. That is, the company may issue Common stock, but not Preferred stock, except by use of the equity round operation. Then, in the fact patterns that arise from the smart contract, every issuance of Preferred shares, that is of relevance to the terms of the SAFE, is necessarily part of an equity round. We therefore avoid the difficulty of interpreting which issuances of Preferred shares should count as part of an Equity Financing: they are explicitly marked as being so through having been generated by the equity round operation.

14.3 Use of a Human Oracle

An external source of information, used by a smart contract in its decision-making, is known in the area as an *oracle*. Oracles may be machines or human. An example of a machine oracle is a data feed from a sensor, that it is used by a smart contract to determine payouts. One application of this that has been mooted is insurance payments or discounts relating to international shipments based on temperature measurements from sensors in the shipping containers [1]. Another common example is the price feeds from centralized exchanges used by decentralized finance smart contracts to make decisions concerning exchange rates or collateral liquidation [2]. Alternately, a smart contract may rely on input from humans to make decisions. An example of this is the input provided by human arbitrators to resolve disputes between parties interacting via the blockchain [3]. In either case, the input from the oracle should be authenticated as coming from a trusted source. This is typically done by having the source cryptographically sign the data or transaction that they are providing.

As already noted, when dealing with indefinite terms and incompleteness in a contract, there is a spectrum of approaches that can be used to determine whether a term applies in a given situation.

---

[1] Recall that Equity Financing is defined as a sale of Preferred shares.

At one end of the spectrum is full automation. This requires that the conditions under which the term applies be exactly expressed in code, and that any ambiguities or incompleteness in decision-making be resolved. Limitation of fact patterns may help with this, by ruling difficult cases out of scope. Nevertheless, in any formalization, the possibility that code will deviate from a legal system's interpretation of the contract in unforeseen situations remains. (We discuss how such situations can be handled in Sect. 17.6.)

At the other end of the spectrum, we can leave the decision about whether a particular condition applies to be made by human agents. An issue when using human oracles is *which* human agents should have the power to make this decision. Alternatives include:

- One party to the contract is trusted to assert satisfaction of the condition.
- All parties to the contract need to agree on the satisfaction of the condition.
- A trusted third party asserts that the condition holds.
- All parties need to agree, but in case of disagreement, a third party arbitrates the dispute.

In practice, the choice of approach may depend on which parties to the contract are affected by the particular condition, as well as the degree of trust between the parties. In the worst case, there is always ex post recourse to the courts (we discuss this in Sect. 17.5).

One of the instances in SAFE contracts where trust is required in a human to provide input is the invocation of code representing equity financings, liquidity events, and dissolutions. SAFE contracts do not grant the SAFE holder the right to influence the timing or terms of funding decisions, IPOs, sale of shares or assets, etc., although they do determine when these actions are construed as one of the events in the SAFE contract, and constrain the consequences of those actions in the performance of the event. Thus, SAFE holders might not be aware of impending events. Hence, it is the company management that is best placed to identify that one of the events (as defined in the SAFE) is occurring, and it is therefore natural that the ability to invoke the code corresponding to one of these events, and state its broad parameters, reside in the hands of company management.

However, this does place a certain amount of control over the interpretation of indefinite terms in the hands of company management. There are cases, such as an equity financing that causes a change of control, that, arguably, could be interpreted as either an Equity Financing event or a Liquidity event occur. Our implementation trusts the CEO/Board to make the determination of which case applies.

Another case of power resting with company management is with respect to the amount of cash available for distribution to SAFE holders in a liquidity event that is intended as a tax-free reorganization. The SAFE explicitly gives the company the right to reduce the amount of cash available for distribution in this case. Our implementation expands this right of company management to all cases of liquidity event. This is an example where we both rely upon a human oracle and increase the precision of the original source contracts.

14.4 Propose-and-Verify Implementations

A further strategy that we apply in our smart contract representation of SAFEs is to use what we call a *propose-and-verify* implementation pattern. Rather than have the smart contract code *calculate* the values of certain variables needed to perform events, we have a human oracle *propose* values of these variables, and then use the smart contract code to *verify* that the values proposed are consistent with the formalization of the legal contract. This approach has a number of different benefits. As described in the following subsections, it facilitates dealing with incompletely stated relationships between variables, the representation of declaratively stated relationships, and enables us to move complex computations off-chain.

14.4.1 Dealing with Incompletely Stated Relationships

As noted in Sect. 13.3, one of the forms of incompleteness in SAFE contracts, concerns the relationship between pre-money valuation and equity round price. One full-automation approach to dealing with the incompleteness would select a specific formula for computing price from pre-money valuation and hard-coding this formula into the smart contract. This would be an increase in the precision of the contract, and require that the SAFE investor and the company agree on the conversion approach at the time the SAFE contract is issued.

An alternative, which preserves the incompleteness in the SAFE, is to admit that SAFE investors have negotiation rights in the equity round. This leads to implementations in which the company submits on-chain a proposal for the structure of the equity round, and both SAFE investors and equity round investors are given the opportunity, on-chain, to provide their consent to the proposal, to enable it to proceed. Alternately, by withholding consent, an investor can block the proposal from progressing, incentivizing the parties to negotiate a revised proposal. In this approach, the proposed issuance of shares to a SAFE investor could be any number, open to negotiation by the human agents, so it is not necessary for the full details of the Equity Financing clause to be represented in smart contract code: this can be left in natural language form.

Advantages of this approach are that it leaves open-textured concepts to be dealt with by human intelligence, and gives significant flexibility with respect to choice of resolution of the incompleteness in the contracts. Moreover, the chain-code is easier to implement and verify, since it is required only to code the proposal and acceptance process, rather than the potentially intricate structure of the SAFE contracts.

On the other hand, this approach provides no assurance of the correctness of calculations used to determine the share issuance in an equity round, so there is a higher risk of errors. It is also the case that a SAFE investor is given the power to hold up an equity round simply by withholding their consent, which is against the spirit of "automatic conversion" in the SAFE contracts, and the fact that these contracts do not explicitly grant the SAFE investors powers to influence the form of the equity round.[2]

14.4.2 Representation of Declaratively Expressed Relationships

On a full-automation approach to formalizing the Equity Financing and Liquidity Event clauses, one approach would be to have the number of shares to be issued in conversion of the SAFE be computed by a smart contract representing the SAFE, given appropriate inputs. For example, in the case of the Pre-Money SAFE, given the pre-money valuation and price of the equity round, the number of shares to which the SAFE is converted is expressed in the Equity Financing clause as a simple rule that can be easily coded.

In the case of the Equity Financing and Liquidity Event clauses of Post-Money SAFEs, things are less straightforward, however: here these clauses state a constraint in the form of an equation on the shares to be issued in conversion. The combined SAFEs therefore generate a set of simultaneous equations that need to be solved in order to determine the correct issuance to each SAFE investor. Conditions for solvability of such simultaneous equations may be complex, particularly if the company has issued multiple convertible instruments of different types. For constraints written in sufficiently expressive forms, the solution problem is, indeed, likely to be intractable, making the code for computing the SAFE issuance overly complex for a resource (e.g., gas cost) constrained blockchain setting. In full generality, with convertible instrument terms written in arbitrary code, the solution problem will be undecidable, and there may be no program that always computes a unique solution, even if one exists. Even where the problem is decidable, formally verifying the correctness of code that generates and solves equation systems may be quite complex.

An alternative to computing the correct issuance on-chain is to require that the company propose a concrete number of shares to be issued to each convertible instrument investor as well as each other investor, and to have the smart contract code representing these investors *verify* that the proposal complies with the terms of the convertible instruments. For contracts like the Post-Money SAFEs, which state these terms as equations, this means that the code need only verify that the proposed issuance solves the equations, rather than compute a solution. This is both computationally more tractable, and easier to code. Moreover, it means that the code of the smart contract more closely resembles the natural language form of the legal SAFE, which simplifies the job of assuring the SAFE investor that the smart contracts will perform as expected. Formal verification of code that merely verifies correctness of

---

[2] More complex versions of this approach might address this problem by allowing an equity round to proceed with only the consenting SAFEs being converted, leaving the others to be resolved through off-chain legal processes. But this creates additional problems since the resulting cap table state will be "incomplete", which complicates performance of subsequent actions.

the number of shares issued, rather than computes it, is likely to be significantly simpler.

A further advantage of an approach based on verification rather than computation is that it is potentially more tolerant of indeterminacy in the original natural language SAFE contract, should that be considered desirable. A computation approach would require that indeterminacy be eliminated in order to yield a deterministic result. On the other hand, if we treat the SAFE as stating a set of constraints on the equity round, we can tolerate the existence of multiple solutions.

14.4.3 Price Computations

As noted in Sect. 4.3, the fact that a company has issued SAFE contracts means that a sophisticated equity round investor will want a reduced share price in order to ensure that the value of their shares is not immediately diluted by the SAFE issuance. In general, price is a matter for negotiation between the company (which would prefer a higher valuation and price) and the investor (who would prefer the opposite), but Eq. (3.4) provides a rational basis on which a price can be computed from a valuation. Since terms for SAFE share issuance in this equation are given by potentially complex conditions, which may circularly depend on the share price and/or valuation, we again have a problem in the form of a set of equations to be solved.

It would be desirable to provide computational support for this problem. However, both the requirement to be able to determine the price by negotiation and the same computational complexity considerations as in Sect. 14.4.2 imply that it is preferable to leave this support to off-chain computations. When there are multiple SAFEs with different conversion conditions, and we use the "Propose-and-Verify" approach, it would similarly be desirable to have off-chain computational support for constructing equity round proposals that will be accepted by the on-chain verification code. As our focus in this work is on smart contract aspects of systems supporting SAFE contracts, we do not attempt to address this problem.

14.5 Programmable Open Structure

We have noted a number of aspects of the scenario in which a company uses SAFEs that have a very broad and unpredictable set of structures. The company's governance structure, and the rules concerning distributions to shareholders are examples of such open structure.

To deal with this issue when developing a smart contract representation of a company that uses SAFEs, we can take advantage of higher order expressive power provided by programming languages. "Higher order", here, refers to the fact that we can abstract and encapsulate data structures, programs and behaviors, and provide the encapsulated artifacts as inputs to be used in computations.

As a simple example of this type of abstraction, suppose we have a list of integers (type Int), and we would like to write code that combines the numbers in the list, using operations to be defined later. We make two assumptions about the unknown operations: they have the types

```
function empty() returns Int
function f(Int,Int) returns Int
```

That is, function empty, given no inputs, returns an integer, and function f, given two integers, returns an integer. Given a list a of integers (typed a : Int[]), we can write the following pseudo-code for combining the elements of the list using these (not yet implemented) abstract functions:

```
result = empty();
for i == 1 to length(a) do { result := f(result,a[i]) ; }
return result;
```

If the list has length n then, starting from the value empty(), this code computes first f(a[1],empty()), then f(a[2],f(a[1],empty())), etc, and finaly returns the value f(a[n],......f(a[2],f(a[1],empty()))). More concretely, if we take empty() = 0 and define f(a,b) = a+b, then the code computes a[n]+a[n-1]+...+a[1]+0, that is, the *sum* of the numbers in the list. Alternately, if we we take empty() = 1 and define f(a,b) = a*b (the product), then the code computes a[n]*a[n-1]*...*a[1]*1, that is, the *product* of all the numbers in the list. For another example, take empty() = 0 and define f(a,b) = 1+b. Then the code computes 1+1+...+1+0, that is, it returns the *length* of the list a.

The point of this is that by abstracting the functions empty and f, and thinking of them also as *inputs* of the computation, we see that the same code can compute many different useful functionalities. This type of abstraction is the essence of higher order computation.

Solidity provides a number of forms of higher order expressiveness. One is the ability to encapsulate multiple functions, specified only by their types, in a class of smart contracts. For example, we could define an abstract class IntFunctions of smart contracts, whose specific refinements contain concrete implementations of a function empty and a function f with types as above. The code above can then take an instance of this abstract class as input, which provides it with specific instances of these functions to use in the computation.

Below, we use this encapsulation capability to represent various open structure functionality of SAFE contract scenarios in a separate smart contract, an instance of which is provided at "runtime", that is, during the performance of the SAFE contracts. Particular applications of this are to represent the highly variable and unpredictable governance structure of the company after conversion of the SAFE contracts in an Equity Financing or Liquidity event, and the potentially complex structure of distributions to shareholders in liquidity events. So long as the structure and processes to be represented can be expressed in code (for which we can use the strategies of the previous sections of this chapter), we can encapsulate the unpredictable complexities in an object or smart contract that can be provided as an input at runtime. We refer to Sects. 15.1.2, 15.2.4 and 15.2.5 for the details.

14.6 Previous Work

The problem of formalizing open-textured concepts (in Hart's sense) was addressed by Bench-Capon and Sergot in [4]. They critically examine several possibilities. One approach, which they called *approximation*, is simply to replace a vague concept by a similar sharp concept. In the context of a legal decision support system, they proposed what is essentially an incremental form of approximation, involving the inductive generation of rules both for and against an object being an instance of the concept, allowing a user to make an informed choice as to whether the object is an instance of the concept. As the authors say,

> Resolution of open texture is at bottom a matter of choice, and it is the duty of the user to make that choice

We will argue in Sect. 15.2.2, that the spectrum of possibilities discussed by Bench-Capon and Sergot arises when implementing SAFEs as smart contracts. Full automation would correspond to approximation. Alternately, in the spirit of the above quote, use of a human oracle (Sect. 14.3) can be combined with the propose-and-verify approach (Sect. 14.4), by allowing a human to provide input to set the conditions under which a proposal will be verified.

As an aside, we can see analogous considerations concerning automated decision-making. Whatever concept is being adjudicated by automated decisions, there can be cases demonstrating the open texture of the concept. Moreover, when the decision concept concerns people, the "foundational indeterminacy of human identity" [5], appears analogous to Waismann's form of open texture when compared to a sharp concept defined by a profile. Article 22 of the European General Data Processing Regulation (GDPR), expresses a right not to be subject to automated decision-making. We can see this as a continuation of the sentiment of [4] above. Almada [6] proposes building contestability into all stages of an automated decision-making system as an improvement over *post hoc* appeals against decisions. A combination of human input and the propose-and-verify approach has similarities to contestability.

Other approaches proposed for dealing with open-textured concepts have included "case-based" techniques [7], including machine learning techniques such as clustering and comparing a current case to "similar" cases (e.g. by computing a nearest neighbor to already decided cases according to some metric) [8], use of artificial neural networks [9], and proposals based on prototypes and deformations, subject to a coherence criterion [10]. These works address open texture in Hart's sense, but not in Waismann's original sense. They attempt to learn distinctions based on past cases, but the possibility of unexpectedness that is central to Waismann's sense is, by its very nature, not addressable by this approach. The context for work that addresses issues of open texture includes legislation [11] and case law [12]. These works do not address contract formalization. A specific class of contract (like SAFEs) with specific open-textured terms might not generate a sufficient number of cases, as compared to statutes, to make the above approaches applicable.

More recently,[3] the "Rules as Code" movement has worked on systems providing support for drafting legislation in several countries [14]. The legislation is represented in a form that can then be interrogated for inconsistencies and unintended consequences. This is a tool for analysis of draft legislation, but also potentially a supplement to the legislation that can provide a basis for tools to support navigation of the legislation and other services.

Of the literature that addresses contract formalization [15], relatively little appears to deal with indeterminacy. Grosof and Poon [16] and Reeves et al. [17] address infrastructure issues around computerized contracts, but they presume a kind of defeasible logic to express the contract.[4] More recently, Catala [19], following Lawsky [20], is based on default logic. Peyton Jones et al. [21] designed a functional domain-specific language (DSL) for describing technical aspects of a class of financial contracts, and Stipula [22] is a DSL based on concurrency theory and deontic logics. Kowalski and Datoo [23] use a controlled natural language to represent the structure of legal clauses, and standardize the legal wording of such clauses. Other attempts to develop controlled natural languages for contract representation are Lexon [24] and L4 [25]. None of these works attempt to handle the various forms of indeterminacy in written contracts. One exception is the work of Daskalopulu [26], which applies logical techniques, including Bench-Capon and Sergot's [4] proposal to express open texture concepts using rules to express conditions under which a case can or cannot be included within the scope of a concept.

It is interesting to note that the TAXMAN work of McCarty [27], applying AI to the interaction of corporate reorganization and tax law, deals with an issue similar to ours with *bona fide* equity rounds: how to interpret a sequence of corporate transactions. McCarty notes that in some cases, the law treats the consequence of a sequence of events differently from the set of consequences of the events individually. Tax law has hinged on questions relating to the *intent* underlying a sequence of corporate transactions, on the basis of which legislation may make allowances with respect to tax consequences that would usually apply. We will argue that smart contracts do not need to represent such reasoning on-chain; it suffices that they support amendments arising from legal rulings in which such reasoning has been applied (See Sect. 17.6.)

---

[3] Although building on a long history. See [11] and even [13].

[4] Defeasible logic has also been proposed to express regulations [18].

14.7   Combination with a Legal Contract

While the strategies discussed in the previous sections may address a significant number of the difficulties raised in Chap. 13, they may leave a residue of risks against which the smart contract formalization does not provide protection.

Hardest to address using a formalization strategy are the difficulties arising from legal context, discussed in Sect. 13.5, and of nuncospectivity, discussed in Sect. 13.6. Using human input, as proposed in Sect. 14.3, leaves the risk that a person delegated to provide this input, will make self-interested or corrupt decisions concerning the input they provide. Such input should be provided "in good faith," a notion that is inherently difficult to formalize, and which needs to be left to human judgment.

To address such residual risks, we believe that the smart contract needs to be combined with a legal contract, making a hybrid form that is being called a "smart legal contract" [28]. We expand on this topic in Chap. 17. As in the methodology of Ricardian contracts [29], discussed in Sect. 17.4, this agreement to accept the code's interpretation of the legal concepts may itself be signed by the parties as a legal contract.

We present, in the following chapter, how the formalization strategies of the present chapter can be applied to develop a smart contract representation of SAFE contracts and context in which they operate.

References

1. Bocek T, Rodrigues BB, Strasser T, Stiller B (2017) Blockchains everywhere - a use-case of blockchains in the pharma supply-chain. In: 2017 IFIP/IEEE symposium on integrated network and service management (IM), Lisbon, Portugal, May 8–12, 2017. IEEE, pp 772–777
2. Werner S, Perez D, Gudgeon L, Klages-Mundt A, Harz D, Knottenbelt W (2022) Sok: decentralized finance (DEFI). In: Herlihy M, Narula N (eds) Proceedings of the 4th ACM conference on advances in financial technologies, AFT 2022, Cambridge, MA, USA, Sep 19-21, 2022. ACM, pp 30–46
3. Kleros (2020) Dispute revolution: the Kleros handbook of decentralized justice. Kleros, augmented edition
4. Bench-Capon T, Sergot M (1988) Towards a rule-based representation of open texture in law. In: Walter C (ed) Computer Power and Legal Language. Quorum Books, New York, pp 39–61
5. Hildebrandt M (2019) Privacy as protection of the incomputable self: from agnostic to agonistic machine learning. Theor Inquiries Law 20:121–83
6. Almada M (2019) Human intervention in automated decision-making: toward the construction of contestable systems. In: Proceedings of the seventeenth international conference on artificial intelligence and law, ICAIL. ACM, pp 2–11
7. Ashley KD (1992) Case-based reasoning and its implications for legal expert systems. Artif Intell Law 1:113–208
8. Popple JD (1993) SHYSTER: a pragmatic legal expert system,. PhD thesis, Australian National University
9. Bench-Capon TJM (1993) Neural networks and open texture. In: Oskamp A, Ashley KD (eds) Proceedings of the fourth international conference on artificial intelligence and law, ICAIL '93. ACM, pp 292–297

10. McCarty LT, Sridharan NS (1981) The representation of an evolving system of legal concepts: II. prototypes and deformations. In: Proceedings of the 7th international joint conference on artificial intelligence, IJCAI, pp 246–253
11. Sergot MJ, Sadri F, Kowalski RA, Kriwaczek F, Hammond P, Cory HT (1986) The British Nationality Act as a logic program. Commun ACM 29(5):370–386
12. Sanders KE (1991) Representing and reasoning about open-textured predicates. In: Proceedings of the third international conference on artificial intelligence and law, pp 137–144
13. Artosi A, Pieri B, Sartor G (eds) (2013) Leibniz: logico-philosophical puzzles in the law. Springer
14. Waddington M (2021) Rules as code. Law in Context 37(1):179–186
15. Kimbrough SO, Wu DJ (eds) (2005) Formal modelling in electronic commerce. Springer
16. Grosof BN, Poon TC (2004) Sweetdeal: representing agent contracts with exceptions using semantic web rules, ontologies, and process descriptions. Int J Electron Commer 8(4):61–97
17. Reeves DM, Wellman MP, Grosof BN (2002) Automated negotiation from declarative contract descriptions. Comput Intell 18(4):482–500
18. Antoniou G, Billington D, Maher MJ (1999) On the analysis of regulations using defeasible rules. In: 32nd annual Hawaii international conference on system sciences (HICSS-32). IEEE Computer Society
19. Merigoux D, Chataing N, Protzenko J (2021) Catala: a programming language for the law. Proc ACM Program Lang 5(ICFP):1–29
20. Lawsky SB (2017) A logic for statutes. Florida Tax Rev 21(1):60–80
21. Peyton Jones S, Eber JM, Seward J (2000) Composing contracts: an adventure in financial engineering, functional pearl. In: Odersky M, Wadler P (eds) Proceedings of the fifth ACM SIGPLAN international conference on functional programming (ICFP '00). ACM, pp 280–292
22. Crafa S, Laneve C, Sartor G (2022) Stipula: a domain specific language for legal contracts. In: Proceedings programming languages and the law. ACM. https://popl22.sigplan.org/home/prolala-2022#program
23. Kowalski R, Datoo A (2022) Logical English meets legal English for swaps and derivatives. Artif Intell Law 30(2):163–197
24. Lexon (2024) Plain text programming. Online, https://lexon.org. Accessed Nov 2024
25. Watt SJ, Goodenough O, Wong MW (2023) Deontics and time in contracts: an executable semantics for the L4 DSL. InL Sileno G, Spanakis J, van Dijck G (eds) Legal knowledge and information systems - JURIX 2023: the thirty-sixth annual conference, Maastricht, The Netherlands, 18-20 Dec 2023, Frontiers in artificial intelligence and applications, vol 379. IOS Press, pp 119–124
26. Daskalopulu A-P (1999) Logic-based tools for the analysis and representation of legal contracts. PhD thesis, Imperial College London
27. McCarty LT (1976) Reflections on TAXMAN: an experiment in artificial intelligence and legal reasoning. Harvard Law Rev
28. Allen JG, Hunn P (eds) (2022) Smart legal contracts: computable law in theory and practice. Oxford University Press, Oxford
29. Grigg I (2004) The Ricardian contract. In: Proceedings of the First IEEE international workshop on electronic contracting. IEEE, pp 25–31

Chapter 15
Architecture for SAFE Smart Contracts

We describe the architecture we use as the basis for a Solidity implementation. The main features are a decomposition of behaviors into finite state machines and the use of pre-existing design patterns, including the atomic swap and controller patterns. Snippets of code are used to illustrate our approach. A Solidity implementation of the architecture and a number of SAFE contracts is available at https://github.com/RonVanderMeyden/SAFESmartContracts.git. The present chapter is a revision and expansion of a brief abstract presented in [1].

15.1 Design Patterns

A number of general design patterns prove to be useful in the development of smart contract representations of SAFE contracts. In this section, we give a general introduction to these patterns.

15.1.1 Atomic Swap

Suppose that two parties A and B have agreed to exchange digital assets a and b, respectively. If the transfers are made sequentially (e.g., first A transfers a to B and B transfers b to A), each party faces the risk that the other party will not abide by their end of the bargain (e.g., B could receive a but then refuse to transfer b to A, leaving A with a loss). Smart contracts provide a method to overcome this difficulty, in the form of *atomic swap* transactions [2, 7].

Atomic swaps can involve multiple blockchains, but for our purposes, in this chapter, swaps on a single chain will suffice. These can be effected by a smart

© The Author(s), under exclusive license to Springer Nature Singapore Pte Ltd. 2025   195
R. van der Meyden and M. J. Maher, *Simple Agreements for Future Equity (SAFE)*,
Blockchain Technologies, https://doi.org/10.1007/978-981-96-3920-5_15

contract that allows the following operations, in the case of a swap of asset a held by party A for an asset b held by party B:

SwapContract(A, a, B, b):

    Deposit(A, a): A transfers a to this contract
    Withdraw(A, a): A recovers a from this contract
    Deposit(B, b): B transfers b to this contract
    Withdraw(B, b): B recovers b from this contract
    Swap: if this contract holds both a and b, then
                  transfer b to A and transfer a to B

We omit detailed code, giving just the interface of the smart contract and a sketch of its functionality. Each line describes an operation performable in the contract. There are operations *Deposit* whereby the parties can transfer their assets to the control of the smart contract, so that the assets come under the smart contract's control. Once both assets have been transferred, the operation *Swap* transfers each asset back to the opposite party. In case the other party fails to perform their *Deposit* action in a timely fashion, the action *Withdraw* can be used to recover an asset that has already been deposited, so that the swap then fails. The effect is that the swap happens atomically, either taking full effect, or leaving both parties with effective control over their original asset (i.e., either holding it, or able to retrieve it from the swap contract).[1]

Usually such swaps are asset for asset (e.g., one form of cryptocurrency for another), but richer forms of swap smart contract can be constructed (e.g., a multi-party contribution of cryptocurrency or approval in exchange for a change of state in a smart contract.) We use such a richer form of swap below to secure aspects of equity rounds.

15.1.2 Controller

Code expressed in an object-oriented programming language consists of objects with a number of variables, and operations that act on these variables. Frequently, these operations need to be subjected to access controls, so that they may be performed only by agents authorized according to some policy.

A common example of this is a situation where a unique agent may perform some operations. In Solidity code, this is captured by the *owner* pattern (see, e.g., [3]). To distinguish this kind of ownership for purposes of access control from the notion of ownership of shares in a company, we will henceforth use the term *controller* rather than "owner" for the former. The pattern is implemented by including in the smart contract a variable `controller` of type `address` that stores the address of the controller. Each operation that may be performed only by the controller then

---

[1] See [7] for an analysis of the powers of the agents in such contracts in terms of strategy logic operators.

contains a check that the address from which the operation is called is the address of the controller. In Solidity, this may be done concisely by first defining a modifier as follows:

```
modifier OnlyController {
    require( msg.sender == controller,
            "Only controller can call this function."
    );
    _;
}
```

This modifier can then be applied in the declaration of each operation that may be performed only by the controller, with the effect that the check is performed at the start of the operation. For example,

```
function set_controller (address newcontroller)
    public OnlyController {
        controller = newcontroller ;
    }
```

restricts the function `set_controller` that changes the controller, so that it may be performed only by the current controller.

Typically, the value of the `controller` variable is the address of the public key of a human agent. However, we note that once a smart contract has been created with the controller pattern, it is possible to impose richer forms of access control by taking the value of `controller` to be the address of a *controller smart contract* that encodes the logic of the richer form of access control. For example, to impose a 2/3 majority vote as the access control rule on the operation f uniquely performable on a smart contract C by its controller, we can create a new smart contract V that first checks that a majority of the three voters agree to perform operation f on C, and if so, makes the call $C.f$ as agent V. Once we set C.`controller` to be equal to the address of V, the smart contract V will be the only agent that is able to perform $C.f$, and it will only do so in case of a 2/3 majority approval.

Similarly, suppose we wish to restrict the powers of a (human or smart contract) controller γ of a smart contract C, so that γ is permitted to perform certain operations f on C if and only if property ϕ_f holds of C. To do this, we may create a controller smart contract R that also contains an operation named f for each operation f on C to be so constrained. The code for $R.f$ first checks that condition ϕ_f holds of C (this requires calls to *view* functions of C, which do not change the state of C) and then calls $C.f$ in case the condition ϕ_f is satisfied, or aborts otherwise. The contract R is defined to have its own controller. We then arrange to have γ set the controller of C to be the address of R, and set the controller of R to be γ. This has the effect that still only γ may perform operations on C, but it needs to do so via the contract R, and its powers to do so are restricted by condition ϕ_f for each operation f.

In settings where changes of access control policy are necessary, it is desirable to include the function `set_controller` among the functions so controlled. (The function to change the controller of R needs to be given a different name to avoid a clash with the mirrored `set_controller` that calls C.`set_controller`.)

We use this *constrained controller* pattern in the smart contract code below.

15.1.3   Payout

An entity such as a company will frequently have *obligations* to pay certain amounts to other parties. Such obligations can be met by transferring the required amounts. However, in the context of smart contracts on the Ethereum platform, direct transfers have some disadvantages. When the recipient is a smart contract (e.g., representing a "wallet" of the intended human recipient), the transfer causes code to run in the recipient smart contract. The motivation for this is that the recipient may need to perform book-keeping operations as a result of the transfer. The disadvantages for the sender in this is that, first, they are charged "gas costs" to pay for the computation of the recipient, and second, that it exposes them to a risk of attack by the recipient. The recipient may perform not just their own computation, but also make a function call on the sender, that is applied to a state of the sender in which it may have only partially completed the computation triggering the transfer. Unless the sender has been carefully programmed to be secure and correct in the face of such "reentrant calls", this may constitute an exploitable security vulnerability. Indeed, vulnerabilities of kind have in fact been exploited: the infamous hack of The DAO smart contract [8], used a "reentrancy attack" to steal a large amount of cryptocurrency from a crowdfunding smart contract.

One of the defenses recommended to protect a smart contract against reentrancy attacks is for the smart contract not to make a transfer to another smart contract, but to simply mark the funds as withdrawable and to require the intended recipient to perform a withdrawal operation. This has the advantage that the recipient will pay the gas costs associated to the transfer (including its own book-keeping costs), and that the transfer will be effected when the state of the sending smart contract is not that of a partially complete computation.

However, this defense also has some disadvantages. It leaves the funds to be transferred in the sending smart contract, co-mingled with other funds. From the point of view of the recipient, this means that the funds remain vulnerable to use by the sending smart contract. Even when the sending smart contract has been programmed to preserve the funds for the recipient, the analysis of the smart contract code to establish that this property holds may be complex, particularly when the sending smart contract consists of a large body of code. From a legal perspective, this complexity may impede the sender from establishing that it has met its payment obligation.

These problems can be addressed by use of a simple Payout smart contract, with the following functionality:

Payout:

   add_payment(*m*,*A*,*R*): add amount *m* to the account of *A*, with reason *R*

   transfer(*A*, *m*, *B*): *A* transfers an amount *m* from this contract to *B*.

By performing *add_payment*(*m*,*A*,*R*), (in which the reason *R* is a string explaining the payment) the sending smart contract transfers the amount *m* to the Payout smart contract, in such a way that only the intended recipient *A* will be able withdraw this amount, transferring it to some other party using the operation *transfer*(*A*, *m*, *B*). Since this is the only functionality of the Payout contract, its code is short and simple, and it can be easily verified that only *A* will be able to withdraw the funds intended for it. In particular, these funds are no longer under the control of the sending smart contract.

In effect, through the use of such a Payout contract, the sender can definitively establish that it has met its payment obligation by transferring funds from its own control to the control of the recipient. We believe that this structure will meet legal standards of the payment obligation having been met, so long as the intended recipient has assented to use of the blockchain platform and payment in cryptocurrency. Below, we use a Payout contract to secure payment obligations of a company in Liquidity and Dissolution events.

15.2   Architecture and Implementation

Factors, noted in Chap. 13, that influence the design of our smart contract solution for SAFE contracts, are the *open structure* of the situation we are modeling, and the issue that Post-Money SAFE contracts are stated as a *constraint* on the promised shares rather by a method to compute their number. To address that there are multiple forms of SAFE contract, we have abstracted these to an abstract contract (instantiated by subclassing) that places constraints on the Company and the way it manages an equity round. This abstraction also enables us to handle a variety of policies concerning the rights that the SAFE holder has with respect to approval of the equity round, by encoding these rights directly into the code of the smart contract representing the SAFE. To handle the openness and dynamic nature of the governance structures of the company over time, we use the constrained controller pattern. Effectively, these aspects of the implementation apply higher order programming capabilities to represent open structure (see Sect. 14.5).

To address the fact that Post-Money SAFEs state but do not solve a constraint system on the equity round and liquidity event, rather than have the smart contract *calculate* the number of shares to be issued to the SAFE investor, we have the company *propose* the structure of an equity round, and have this proposal *verified* for correctness by the smart contract (see Sect. 14.4). This also enables the SAFE

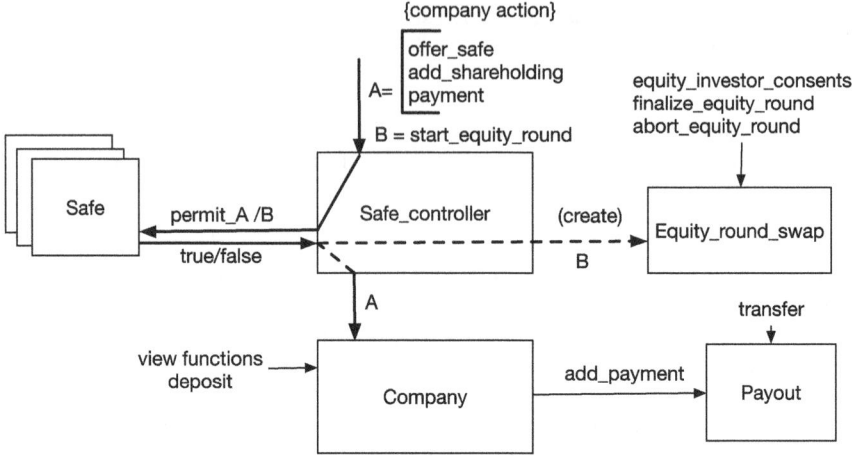

Fig. 15.1 Architecture with Safe_controller

contract abstraction to have instances that are both nondeterministic (a way to handle incompleteness in the legal text) and arbitrarily expressive in its constraints without requiring costly on-chain constraint solving. Executing the offer-acceptance process, an equity round and a liquidity event of a SAFE are "long-running" and composed of multiple transactions. We treat them as *compound* events that are serialized with other events using a finite state-machine.

The architecture of our solution is dynamic, with its principal configurations shown in Figs. 15.1 and 15.3. Figure 15.4 depicts the finite state-machine view of some of the components, that underpins some key correctness properties.

In overview, a Company smart contract represents the financial structure of a company, in particular, its share structure and financial contracts issued. We use smart contracts of type Safe to represent the constraints that the SAFE contracts issued by a company place on the actions A that the company may perform. Actual enforcement of these constraints is achieved by setting the controller of the Company smart contract to be a Safe_controller smart contract. The Safe_controller mediates all access to the Company actions, while SAFE contracts are in force, as depicted in Fig. 15.1. To provide assurance to the investors that the equity round will be conducted fairly, the atomic swap pattern is used in the form of a Equity_round_swap smart contract, with the left diagram in Fig. 15.3 indicating the form of the architecture, while an equity round is active. If the equity round fails, the architecture reverts to that of Fig. 15.1. To handle the post-round authorization regime, the design allows a new controller to be installed after the equity round completes.

Each of these smart contracts is discussed in greater detail in the remainder of this section. The more detailed discussion in the following subsections is based on a Solidity code implementation.

15.2.1   Company Financial Structure

We first lay out the context in which SAFE instruments operate: the financial structure of a company. From a financial point of view, the structure of a company consists of information including (1) its accounting journals (henceforth, we use the term "books"), and (2) its capitalization table (or "cap table"). The books record information such as assets, liabilities and transactions. The cap table records the state of shareholder equity in the company, i.e., the number of shares owned by each shareholder.

Instruments such as convertible bonds have a mixed nature: they may have properties similar to a loan (which would be a liability, hence recorded in the books), but may also constitute an obligation to issue shares (which impacts the shareholder structure, so should be recorded in the cap table). Part of the design intent of the SAFE, which involves no interest payments, and therefore lacks loan-like features, was that it should be represented in the cap table. On the other hand, other parties, e.g., tax officers, or investors valuing the company, may take a different view. Accounting standards contain detailed rules governing whether, depending on its precise construction, an instrument should be represented in the books or in the cap table for purposes of compliance with reporting obligations [9]. In some cases, the instrument needs to be decomposed into parts represented in the books and parts represented on the cap table. We take no stance on this issue, and simply represent the convertible instruments independently of both the books and the cap table, leaving accounting status as an independent consideration.

Since our primary focus is the representation of SAFE contracts and their effects on the cap table over time, we take a highly simplified view of the company books: we assume that the only assets or liabilities recorded on-chain are some of the company's holdings of the cryptocurrency native to the blockchain. (The code has no dependency on the books, but we explicitly model payments to and from the company, including payments of principal for SAFEs and shares purchased.)

Company transactions such as issuance of SAFE instruments and equity rounds involve the transfer of money to the company. Even in a possible future where it is common for investors to hold cryptocurrency that they wish to invest, it may well remain the case that most investments are made from a larger base of fiat currency holdings. Moreover, a company that receives cryptocurrency investments is likely to need to transfer the money received to parties who prefer payment in fiat. A full treatment of SAFE instruments registered and secured on blockchain should therefore accommodate fiat payments in some way. We do not attempt to cover this problem in the present work, and assume that all money transfers are made in the native cryptocurrency of the blockchain. This assumption will enable us to determine the full extent of the security benefits that could be obtained, in principle, from a blockchain and smart contract treatment of SAFE instruments. Inclusion of fiat payments, except in the form of fiat represented on-chain by a "stablecoin" or Central Bank Digital Currency, is likely to weaken the security guarantees that can be provided. We leave the question of what would be lost in adding a coverage of fiat payments for treatment elsewhere.

Based on the above assumptions, our on-chain representation of the state of a company will be a smart contract of class `Company`, that

- may hold cryptocurrency, representing some of the assets of the company, and
- represents the cap table of the company, in the form of a record of shares held by each investor in the company,
- represents the convertible instruments issued by the company (assumed to be SAFE contracts only).

Concrete instantiations may refine this abstract representations in various ways. In the remainder of this section we develop an abstract description of this smart contract and discuss various design questions that arise.

With respect to the record of shareholdings in the company, a first issue is to represent possible types of shares in the company. The text of the SAFE contract mentions several share types: Common Stock (held by the founders), Preferred Shares (typically held by equity round investors), and SAFE Preferred Shares (issued to the SAFE investors in conversion of the SAFE). Each share type corresponds to a bundle of rights relating to company governance, voting, dividends, participation in future equity rounds, corporate transactions such as sale of the company, and preference order and/or amounts for distribution of money in the event of a company liquidation. After an equity round, founder and employee stock is generally subject to vesting schedules, so effectively differs from other Common Stock. In general, it is open to a company to define other share types.

The most general treatment of the record of shareholdings should therefore allow for flexibility in the definition of the type representing shares. Technically, this could be done by representing shares as an abstract object interface that can be subclassed to define different share types. Since our main focus is on SAFE contracts and their effects on payments and the *number* of shares issued, we do not attempt to develop such a detailed representation in the present work. Instead, we assume that the three share types mentioned in the SAFE contract are the only possible types of shares. This allows us to represent the general idea of a share as an enumerated type[2]

```
enum ShareType {Common,
                Preferred,
                Safe_Preferred}
```

An individual shareholding can therefore be represented as a tuple

---

[2] One disadvantage of this representation is that the differences in the rights associated with each of these share types are not directly bundled with their definition. Instead, the code of company operations will need to include cases for each of the share types. We leave the development of alternative representations that address this issue for treatment elsewhere, but note that an architectural structure for a solution can be seen from our presentation of SAFE smart contracts below, using a controller to restrict the company to comply with the terms of the SAFE contracts it has issued.

```
struct Shareholding {address   payable owner;
                     ShareType  class;
                     uint number; }
```

consisting of the owner of the shares, the share type and the number of shares held. (We treat the shareholder identity very simplistically as a pseudonymous cryptographic address. For legal purposes, further identifying information would very likely need to be represented, raising privacy questions, particularly in the setting of public blockchains. We discuss privacy issues in Chap. 18.) The keyword `payable` used in this definition marks an address as one to which payments can be sent. Functions can similarly be declared payable if payments can be attached when they are called. Non-payability is used as a defensive measure, to prevent unintended transfers.

The record of shares held in the company can now be represented as an array of such shareholding entries, represented in the smart contract by a variable `shares` declared as[3]

```
Shareholding[] shares;
```

Similarly, the record of SAFEs issued by the company can be represented by an array of entries:

```
Safe[] safes_issued;
```

where a SAFE is represented by type `Safe`.

The company's books are represented just as a value of cryptocurrency holdings. We can generally rely upon the blockchain platform to maintain the value of cryptocurrency held in the `Company` smart contract. In the setting of Solidity, this can be accessed inside the code for the smart contract via the construct `address(this).balance`.

It should be possible to update these variables by means of various operations. In order to focus on the main issue in this work—cash payouts and conversion of SAFE contracts to shares—we assume SAFEs and shares are not transferable, and that the company may not cancel shares or restructure the cap table by means of stock splits or merges. Initially we consider a minimal set of operations on `Company`:

- `add_shareholding` to issue new shares,
- `add_SAFE` to issue a new SAFE
- `payment`, which transfers some of company's cryptocurrency holdings, to pay for goods and services
- `deposit`, used by other parties to transfer cryptocurrency to the company

---

[3] Additional use of a mapping in Solidity would allow for more efficient lookup, but since our main concern in this work is to understand the logic of the situation and the ways that smart contracts support enforcement of security properties, we do not attempt to develop code that is optimal on measures such as time, space, or gas usage.

However, these operations are not unrestricted: only a duly authorized officer of the company should be able to issue new shares or SAFEs, and there are circumstances in which the company will be constrained in performing such actions. For example, equity round term sheets (preliminary contracts setting out the expected conditions of an equity round as a basis for more detailed negotiations) may constrain the company from changing its capital structure until closing or abandonment of the equity round. After an equity round, the governance structure of the company is likely to have changed, for example, by instituting a board with membership from the new investors, and issuance of new stock will typically require approval of the board and/or a majority vote of shareholders. Further, as discussed in Sects. 13.2 and 14.2, to avoid questions of interpretation such as whether an issuance of Preferred shares implies the occurrence of an Equity Financing event, we would like to limit the possible fact patterns so that Preferred shares can only be issued in the context of an explicit Equity Financing event. Although not in the scope of our limited model, restrictions on rights to transfer stock may also apply, since transfers among investors may change voting power. Share transfers may also be subject to rules imposed by the jurisdiction of incorporation [10].

To capture such constraints, we therefore require a means of access control on the performance of actions on the cap table. This can be done in an abstract way using the Controller pattern from Sect. 15.1.2. We take the original controller to be the agent (address) creating the `Company` smart contract, and include a function `set_controller` whereby the controller (and only the controller) may transfer control to another agent (possibly a smart contract). After an equity round, the new controller could be the (address of the) director of the board, or a smart contract that implements the desired governance structure and rules. For technical reasons, an additional controller value is used in our code at intermediate stages of an equity round. (See Sect. 15.2.4.)

A listing of the functions in `Company`, and other smart contract classes used in the implementation, is presented in Fig. 15.2.

15.2.2   Safe Contract Representation

We next present a representation of the terms of a SAFE contract using a smart contract. The representation captures constraints that a SAFE places on the company's actions.

SAFEs come in multiple forms: Pre-Money SAFEs and Post-Money SAFEs, each of which may include a cap and/or a discount. We develop a representation that can accommodate all of these types of SAFE. Our approach allows a company to issue a set of SAFE contracts of variant types. We believe the general approach can be extended to handle other types of convertible instrument, including interest bearing convertible bonds, KISS (Keep It Simple Securities) [4], etc.

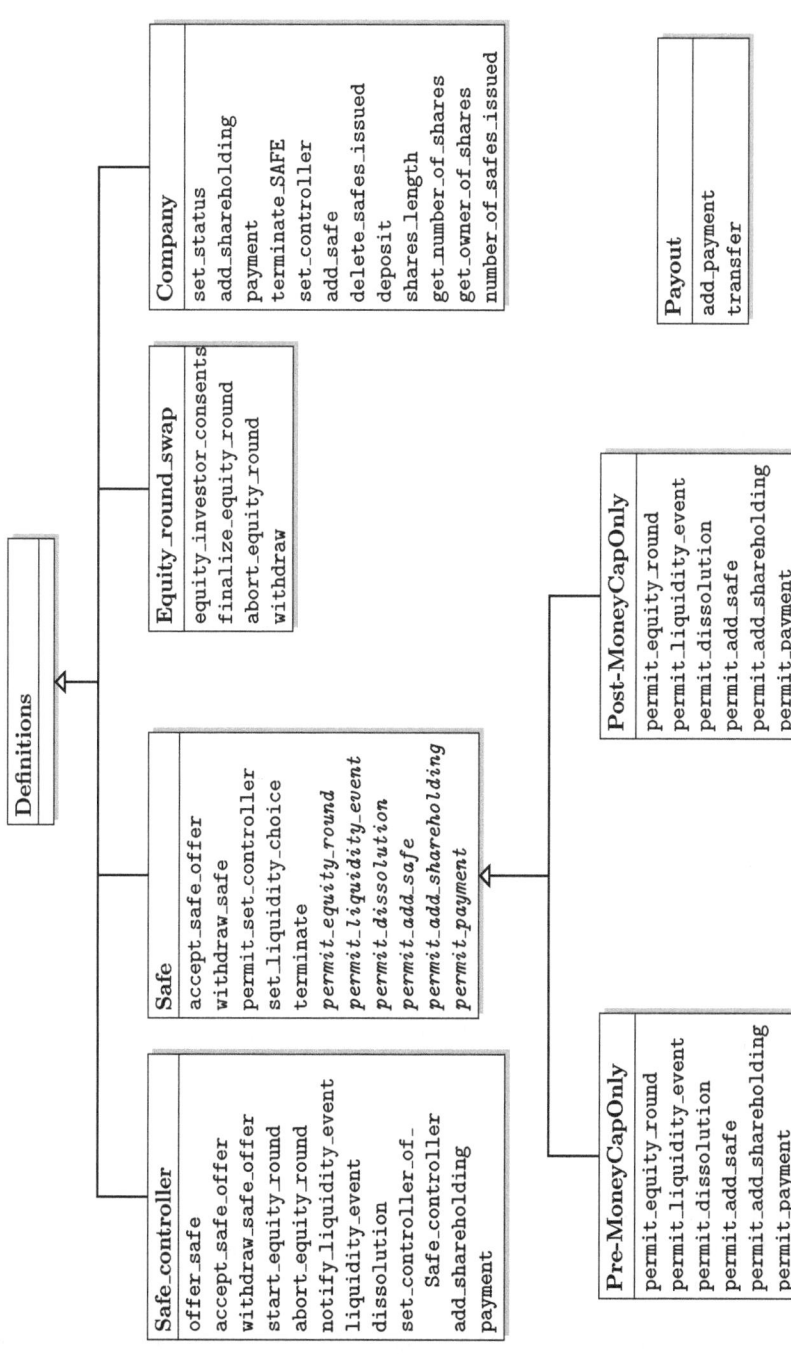

Fig. 15.2 Class diagram for the architecture and implementation

We achieve generality by defining an abstract smart contract class `Safe`, which may be subclassed to instantiate specific types of SAFE. The main variables of this abstract class are

```
uint public principal;
address payable public owner;
Company public company;
uint public deadline;
Safe_status public status = Safe_status.Offered ;
Termination_reason termination_reason = Termination_reason.None
```

representing, respectively, the amount paid for the SAFE (in Wei), the party holding the SAFE, the smart contract for the company being invested in, a deadline by which the offer of the SAFE needs to be accepted by the investor, the status of the SAFE, and the reason (if any) for termination of the SAFE. The status is given by a value from the enumerated type

```
enum Safe_status   {Offered,
                    Active,
                    Withdrawn,
                    Terminated}
```

A `Safe` is offered to the investor in state `Offered` and becomes `Active` once accepted with payment of principal. A deadline is given for acceptance. (The offer and acceptance process is described below in Sect. 15.2.3.) On termination, the type of event triggering termination of the SAFE is recorded with a value from the enumerated type

```
enum Termination_reason {None,
                    Equity_round,
                    Liquidity_event,
                    Dissolution}
```

Specific subclasses of `Safe`, for example `Pre-MoneyCapOnly`, add variables such as

```
uint public cap;
```

to represent the valuation cap used in converting this particular type of SAFE to shares in the context of an equity round. The set of such parameters of a SAFE vary among the different types of SAFE, so each subclass defines its own constructor to allow the values of these variables to be set at the time of creating a new `Safe` instance.

Once a company has issued a SAFE, it needs to comply with restrictions imposed by the SAFE. In the `Safe` smart contract itself, we capture the logic of these constraints, but do not actually enforce them. Enforcement will be done using the controller pattern, by setting a controller for the company that queries the Safe contracts to determine which actions on the company are consistent with all the SAFE contracts that the company has issued. We discuss the Safe_controller contract used for this purpose in more detail in Sect. 15.2.3. For now, it suffices to note that the Safe smart contract has a variable declared as

```
Safe_controller public safe_controller;
```

that is used to record the Safe_controller that is used to enforce the SAFE constraints. Functions accept_safe_offer and withdraw_safe_offer on the Safe are called via the safe_controller to update the status of the Safe, as described below.

The logic of constraints that the SAFE places on the company is captured by functions in the Safe smart contract named permit_A, where A is an action (function) on the Company. The parameters of permit_A and A are the same. The function permit_A may query the state of the contracts at company and safe_controller and returns a boolean value, which expresses whether the company is permitted to perform action A in that state. These functions will in fact be called only when no equity round is active, and no SAFE is presently on offer, because of the serialization policy described below, so we focus on that case.

These functions are virtual in Safe and are given concrete definitions in the subclasses for specific types of SAFE, such as Post-MoneyCapOnly. For example, in our implementations of specific SAFEs, function

```
permit_add_shareholding(
     address payable, // owner of shares
     ShareType,        // type of shares
     uint              // number of shares
) public virtual view returns(bool)
```

returns true when it is consistent with the constraints placed on the company by the SAFE for the company to issue the given number of shares of the given class to an investor.

The Solidity keyword virtual indicates that this declaration is describing only the interface of the function, and that actual code for the function needs to be provided in the concrete subclasses of the abstract class Safe. The permit_A are the only virtual functions of the class Safe. Parameter variable names have been elided and replaced by comments (which follow the "//") in the declaration because the function is virtual and therefore lacks a body that references the variables.

When no equity round is active and the share class is Common, this returns true, but it returns false if the share class is Preferred or Safe_Preferred. (We apply here the simplifying assumption of Sect. 14.2, requiring that all issuances of Preferred or Safe_Preferred shares must be part of a properly constructed equity round.)

This leaves the question of what this function should return in case there is an active equity round. An apparent difficulty is that whereas the share issuance associated to the equity round should be permitted, all other requests to issue shares should be rejected, as part of the *stand-still* provisions for the equity round. (These provisions aim to keep the financial state of the company stable during the round in order to assure the equity investors that the financial position of the company will not change materially while the round is in progress. Were this to occur, it would impact the value of the shares they are acquiring.) A further reason to reject such extra requests to add shares during the course of the equity round is that, conceptually, they should be sequenced after the equity round has completed, at which point the company would typically be under a different control regime, which might not approve the transaction.

Prima facie, this requires us to be able to distinguish requests to add shares that are part of the equity round from others. However, we can avoid doing so by taking the following view of the events in which shares are added in execution of an equity round: rather than being treated as independent events requiring individual approval, these are part of a compound "equity round" event, which is authorized as a whole. (See the discussion of `permit_equity_round` below.) Thus, `add_shareholding` events that are part of the equity round are not referred to the `Safe` for approval via a call to `permit_add_shareholding`. This understanding makes it correct for `permit_add_shareholding` to simply return `false` while an equity round is active.

Similarly, function

```
permit_add_SAFE(uint,              // principal
                uint,              // cap
                address payable ) // owner
```

returns `true` if the state of the company is such that a new SAFE may be issued. In general, a SAFE does not prohibit addition of other SAFEs when an equity round is not active.

In the case of function

```
function permit_payment(address,        //recipient
                        uint,           //amount to pay
                        string calldata)  //reason
```

the result is `true`, since a SAFE generally does not constrain the company's rights to spend its funds.

Since `set_controller` is one of the operations of `Company`, we similarly should have an operation

```
permit_set_controller(address payable)   // new controller
```

In general, a SAFE does not constrain changes of the agent (e.g., CEO or a delegate) who may perform company actions.[4] However, while a `Safe` is active, the controller of the `Company` smart contract should be a `Safe_controller`, in order to automatically enforce the rights of the SAFE investors. The agent who may perform company actions is therefore not the controller of smart contract `Company`, but of the smart contract `Safe_controller`. The function `permit_set_controller` is therefore defined in our implementation to always return `false`. (Indeed, the `Safe_controller` controls access to the `Company` action `set_controller` to ensure that it is not inappropriately removed as controller.) Note that changes to management of the company can still be made by use of the function `set_Safe_controller_controller` on the `Safe_controller`, as discussed below.

The above covers the functions A of `Company` that need to be placed under control using functions `permit_A`. The SAFE contract introduces the additional idea of an equity round and, as discussed above, we view this as a compound event that requires approval by the `Safe`. In particular, approval of the equity round is obtained by calling a function

```
function permit_equity_round(
    uint,                    // price of each share,
    uint,                    // pre-money valuation,
    Equity_round_investment[] memory,
                             // equity round investor investments
    uint[] memory,           // no. of shares to be issued
                                                to SAFE investors
    ShareType[] memory, // type of shares to be issued
                                                to SAFE investors
    address payable,    // controller after equity round succeeds
    uint,                    // deadline for completion of round
    uint)                    // index of this Safe in safes_issued
```

The parameters of the function include a detailed description of the structure of a proposed equity round: the price[5] at which shares are being sold to the new investors, the pre-money valuation on which that price is based, the number and type of shares to be issued to each of the new investors, as well as the number

---

[4] We are talking about "controller" in the sense of the management of the company and its smart contract realization here, not the effective control of the company resulting from a majority shareholding.

[5] The text of the Post-Money SAFE (but not the Pre-Money SAFE) allows that not all new investors in the equity round pay the same price. The SAFE is converted based on the minimum price in this event. We have made the simplifying assumption in the code that all investors pay the same price, as the alternative appears to occur only rarely. Both price and pre-money valuation are included as parameters because the Pre-Money SAFE with Cap Only does not specify a formula for computing price from pre-money valuation, and several alternatives are possible. See Chap. 4 for discussion of this.

and type of shares to be issued to each of the existing SAFE investors (as recorded in the variable `safes_issued` in the `Company` contract) in conversion of the SAFE. Type `Equity_round_investment` is a struct that captures an investor, the principal to be paid, and the class of shares to be issued to the investor.

To support automated enforcement of the SAFE holder's interest, the implementation in each subclass should return `true` just in case the proposal complies with the terms of the concrete SAFE. In particular, the number and type of shares that the inputs propose to issue to the holder of the SAFE represented by the `Safe` smart contract to which the call is being made, should be as described in the "Equity Financing" clause of the SAFE (or otherwise approved by the SAFE holder).

A difficulty we have already noted above is that the SAFE contract text uses open-textured concepts in describing the equity round. The terms of equity round (the pre-money valuation and share price) agreed by the founders and new investor directly determine the shareholding of the SAFE investor after the equity round, but the round is required to be 'bona fide'. What rights, if any, that the SAFE investor has to object to these terms are not clearly spelled out, except to the extent that they have an obligation to sign the documents associated with the equity round. If withholding their signature would delay completion of the round, they may be in a position to negotiate, but even this is uncertain.

As discussed in Sect. 6.4, this lack of precision in fact represents a risk to the SAFE investor, and that the founders and new investor have an opportunity to collude in ways that are detrimental to the interests of the SAFE investor. A smart contract implementation necessarily needs to address this lack of determinism, and implement a specific policy that embodies any actions that the SAFE holder may take with respect to the equity rounds. Multiple alternative policies might be implemented, each distributing the power to decide the round structure in different ways.

- *Full automation of the equity round clause:* this policy gives no rights to the SAFE investor, and simply encodes the formula for share issuance to the SAFE investor as a function of the pre-money valuation and price into the code for the `Safe` smart contract. The code for `permit_equity_round` returns true if and only if the number of shares issued to the SAFE investor satisfies this formula. The SAFE investor is guaranteed that their issuance will be in accordance with the formula, but not that the parameters of the equity round have been reasonably constructed.
- *No automation of the equity round clause:* In this approach, the SAFE investor first interacts with their `Safe` smart contract to approve the equity round proposal, and `permit_equity_round` returns `true` if and only if the proposal has been approved in this way. This fully protects the SAFE investor, but at the cost to the founders of enabling the SAFE investor to block even reasonably constructed equity rounds.
- *Partial automation of the equity round clause:* this policy is an intermediate case between the above two, that gives the SAFE investor rights to approve a variation to the number of shares determined by the formula, but no rights to block the round when the formula is satisfied. Approval is handled prior to the call to `permit_equity_round` by another function that takes as input the equity

round proposal and stores the fact that the investor has approved the proposal in the `Safe` contract. When this is the case, the call to `permit_equity_round` returns true, and when not, it returns true only when the equity round proposal satisfies the formula for issuance to the SAFE investor.

For the latter two approaches, the SAFE investor would make a call to the function `approve_equity_round` on the `Safe`, with appropriate parameters. Richer forms of policy, such as a blanket approval of proposals satisfying certain conditions within given time periods, can also be implemented in the same way. Whereas full automation would be an instance of increasing precision (see Sect. 14.1), the latter two approaches apply the strategy of reliance on a human oracle for dealing with open-textured terms (see Sect. 14.3).

In addition to the parameters of the function `permit_equity_round` discussed above, which are most relevant to the terms of the SAFE, some other parameters are included: the address of the controller of the `Company` smart contract once the equity round completes, a deadline for completion of the equity round, and (as a convenience for array lookups) the index of the `Safe` contract being called in the array `safes_issued` in `Company`.

The result of a call of `permit_equity_round` on a `Safe` contract relates only to that particular `Safe` contract—it is combined with results from all other issued `Safe` contracts to determine whether the round can commence. If so, execution of the round requires a number of additional steps before the round completes. Further details on this process are provided below, in Sect. 15.2.4.

If the equity round does eventually successfully complete, it is only then that shares are issued to the SAFE investors in conversion of their SAFEs. To signal to the `Safe` smart contract that the conversion has been performed, the `Company` calls a function `terminate_SAFE` with argument `Termination_reason.Equity_round` in the `Safe` contract, which sets variable `status` from `Active` to `Terminated` and records the termination reason as `Equity_round`.

There are some additional functions of `Safe` that relate to liquidity and dissolution events. We discuss these below in Sects. 15.2.5 and 15.2.6.

15.2.3 SAFE Controller

As discussed in Sect. 15.2.2, issuance of a SAFE constrains actions that the company may take, and the logic of these constraints is captured in the functions `permit_A` in the class `Safe`, which are used to indicate whether the action A of `Company` is permitted in a particular state of the company. Actual enforcement of these constraints is achieved by use of the Controller pattern. We introduce a new smart contract class `Safe_controller`, whose purpose is to act as the controller of a `Company`, in order to ensure that the company is compliant with all SAFEs it has issued. Before issuing a SAFE, a company should set its controller to be an instance of a `Safe_controller` smart contract and, to ensure that their interests will be protected by the code, a SAFE investor should not invest in a `Safe` unless a `Safe_controller` is in place.

The `Safe_controller` serializes the permitted company actions with the compound processes of issuing a SAFE and conducting an equity round. This is achieved by means of state variable with values in enumerated type

```
enum Safe_controller_status {No_event_active,
                             Safe_offered,
                             Equity_round_active,
                             Liquidity_event_pending,
                             Terminated}
```

and transitions as depicted in Fig. 15.4. Functions are enabled only at states from which they label an outgoing transition.

In state `No_event_active`, the `Safe_controller` contract functions as follows: it receives requests for actions A to be performed[6] on the `Company`. Before transmitting these to `Company`, it first calls the functions `permit_A` of all issued SAFEs in order to determine whether performing A is consistent with all these SAFEs. If not, then the request is denied, else the function A is called on `Company`.

The `Safe_controller` contract has some additional functionality that enables the company to issue a SAFE contract to an investor by means of an atomic swap process. To issue a SAFE, a `Safe` smart contract `safe` encoding its terms is first created, specifying a deadline. The function `offer_safe(safe)` is then called on the `Safe_controller`. The SAFE investor calls `accept_safe_offer` on the `Safe_controller`, attaching the required principal, in order to accept the offer. After the deadline, the offer can be withdrawn using function `withdraw_safe`. (These functions call similarly the named functions on the `Safe` to effect state transitions of the `Safe`.)

In addition to primitive actions that can be performed on the company, the `Safe_controller` also supports equity rounds through functions

```
start_equity_round,
finalize_equity_round,
abort_equity_round .
```

Function `start_equity_round` has the same parameters (except for the final convenience parameter, the index of the `Safe`) as function `permit_equity_round` discussed in Sect. 15.2.2. This function can be called to initiate an equity round. All the `Safe` contracts in the company's `safes_issued` data structure are called using `permit_equity_round`, to verify that the proposed equity round is consistent with the terms of all contracts (or has been otherwise approved by the

---

[6] One exception is action `add_safe`, which is masked and replaced by `offer_safe`, which calls `permit_add_safe`.

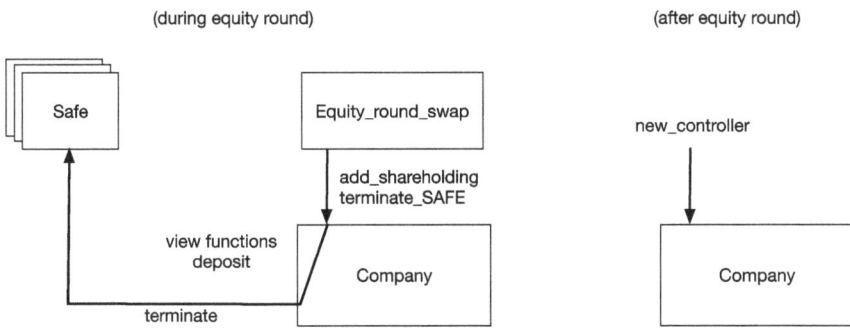

Fig. 15.3 Architecture configuration during and after an equity round

investor, as described above). If any of these calls returns false, the proposed equity round is rejected. Otherwise, an Equity_round_swap smart contract is created.

This Equity_round_swap smart contract is set to be the controller of the Company while it is active, but returns control to the Safe_controller if the equity round fails. Passing control to Equity_round_swap both enforces serialization by preventing company actions being performed via the Safe_controller and enables the Equity_round_swap smart contract to issue shares and convert the Safe contracts if the round succeeds. The changes of control of Company dynamically modify the architecture of Fig. 15.1, as depicted in Fig. 15.3.

We note that there is one limitation with this design, which might be understood as an instance of the strategy of "limitation of fact patterns". It assumes that all SAFE contracts define their Equity Financing events in the same way, so that if such an event occurs for one SAFE, it does so for all, and the SAFEs can all be terminated together. A similar point applies to liquidity events, discussed below. In fact, there are differences between the definitions of Liquidity Event used in Pre- and Post-Money SAFEs! (The latter treat direct listings as implying a Liquidity Event.) This might result in events in which only a subset of the set of SAFEs the company has issued terminate. Y Combinator already warns that combining SAFEs of different types is not recommended, and in earlier chapters we identify some specific problems that arise when this is done. As a pragmatic matter, therefore, our formalization is best applied for situations where all SAFEs issued by the company agree on the definition of what counts as an occurrence of the key events. (In principle, the Safe versions and the Safe_controller could be programmed to allow only a subset of the SAFEs to terminate when an Equity Financing or Liquidity Event occurs, and keep the Safe_controller in place after such an event. We have not pursued this idea in our implementation, and leave it as an issue for future work.)

15.2.4 Atomic Swap Implementation of the Equity Round

An atomicity issue similar to that solved by the atomic swap of assets between two parties (see Sect. 15.1.1) arises at the time of the equity round, but involving a larger number of parties (the Company, the SAFE investors, and the equity round investors).

This is particularly the case when the SAFE investors are given the opportunity to negotiate with respect to the conversion of their SAFE contracts to shares. Suppose that the shares are issued to the new investors using a process that is completely independent of the issuance of shares to the SAFE investors. Then, from each party's point of view, there is a risk that some other party will be granted a larger number of shares than expected. The rational new investor will have calculated the price of their investment based on expectations of the number of SAFE investor shares to be issued at the time of the equity round. If more shares than this are issued to the SAFE investors, the new investor will consider themselves to have been unfairly diluted.

A similar issue arises between different new investors, even in a situation where the SAFE investors are not given negotiation rights, and even when all new investors are sold shares at the same price. Simply varying the number of shares issued may result in changes to the power balance between these investors in company governance. An aggrieved party has recourse to the courts to attempt to rectify the situation if their dispute cannot be resolved by negotiation, but this entails litigation costs. Each would like to be assured that they will be issued the number of shares negotiated, or due from a SAFE, and that variations from the agreement will not alter their expected relative stake.

The methodology of atomic swaps can be applied to resolve this difficulty, bundling the individual share issuances into a single transaction, and ensuring that only the mutually agreed issuance of shares to each of the parties occurs in the equity round. Suppose that the parties with negotiation rights come to agreement on the number and type of shares to be issued to each of the SAFE and equity round investors. We can provide assurance to all parties that the equity round will be conducted in accordance with this agreement by codifying it using an atomic swap-like smart contract that first collects the new investment moneys, and then issues shares to all of the investors as per the agreement. In case some of the required consents or moneys are not obtained by a deadline, the contract aborts and the moneys can be recovered by the respective investors.

The `Equity_round_swap` smart contract in the implementation realizes this idea. It has local variables corresponding to the arguments of the function `start_equity_round` of the `Safe_controller`, already discussed in Sect. 15.2.3. These variables encode the proposed share issuance to each of the investors in the round, price and pre-money valuation. Additionally, the contract has variables `safe_controller` that records the address creating the `Equity_round_swap` smart contract (presumed to be a `Safe_controller` contract[7]), and `company`, recording the company smart contract for which the

---

[7] While a `Safe_controller` is the controller of the `Company`, an `Equity_round_swap` contract created from any address other than the `Safe_controller` will not be able to have

`Equity_round_swap` contract will issue shares. A `status` variable of enumerated type

```
enum SwapStatus {Awaiting_consent,
                 Finalized,
                 Aborted}
```

and initial value `Awaiting_consent` records the current state of equity round workflow.

The process whereby the `Equity_round_swap` is created, by invoking function `start_equity_round` in `Safe_controller`, has already been discussed in Sect. 15.2.3. The `Equity_round_swap` smart contract operates as follows:

- SAFE investors, to the extent that they have approval rights on the equity round, give these approvals prior to creation of the `Equity_round_swap` smart contract.
- While in state `Awaiting_consent`, the swap contract accepts `equity_investor_consents` function calls. An equity round investor uses this to consent to the proposed round and pay their principal. Only the correct investor (identified by their address) may make this call, and they must attach the correct principal when calling the function. Correct calls are recorded in the state of the `Equity_round_swap`, which also stores the principal payments.
- When all the equity round investor consents and principal payments have been received, anyone may call a function `finalize_equity_round` on the `Equity_round_swap` contract, which does the following:

  - Call `finalize_equity_round` on the `Safe_controller` contract. The effect of this call is to terminate the `Safe_controller`.
  - Next the function uses the `Company` function `add_shareholding` to issue shares, as per the proposed equity round, to each of the equity round investors and each of the holders of the SAFE contracts. (If new shares are to be added to the option pool, these would be created at this time.)
  - For each of the SAFE contracts, the company calls `Company` function `terminate_SAFE`. This in turn makes a call from the `Company` contract to the `terminate` function of the corresponding `Safe` contract, signaling that the SAFE has been converted.[8]

---

any direct effect on the `Company`. This is because it needs to first become the controller of the `Company` itself, and this requires that the `Safe_controller` transfer control. But the `Safe_controller` will only grant control to an `Equity_round_swap` that it has itself created.

[8] For security purposes, the `Safe` smart contract needs to verify that the conversion call is being made in the context of an equity round completion by an appropriate agent. There are various ways this could be realized. In our implementation, we use that fact that the sender of the call must be the `Company` smart contract, and that, once a `Safe_controller` has been installed as its controller,

- The function then transfers the total equity round investor money received to the `Company` contract.
- Next, the function calls `set_controller` on `Company` to set the controller of the `Company` to the party or smart contract (parameter `new_controller` in the `start_equity_round` call) agreed to be the controller after the equity round (capturing the post-round governance structure, see the discussion in Sect. 15.2.1).
- Finally, the `Equity_round_swap` contract is terminated by setting the variable `swap_status` to `Finalized`, so that it may perform no further actions.

- It may happen that, even though there was an agreement on the structure of the round, not all parties give the consent and payment required for the `finalize_equity_round` operation to be enabled. To cover this eventuality, the proposed equity round has a deadline. Once this has passed, a function `abort_equity_round` of `Equity_round_swap` may be called to do the following:

  - Call function `abort_equity_round` on the `Safe_controller`, to advise it that the equity round has aborted.
  - Restore control of the `Company` to the `Safe_controller`.
  - Set `swap_status` on `Equity_round_swap` to value `Aborted`. This enables the equity round investors to call a function `withdraw` on `Equity_round_swap` to obtain a refund of any money they had sent to this contract.

To see how these smart contracts interact with users, we outline a scenario.

Example 15.1 Acme Corp., a niche supplier of explosive tennis balls, seeks funding to expand its range of products. Acme's capitalization table is implemented as an instance of the `Company` smart contract. Acme founder and CEO, Mr. Q. Magoo, is the controller of that smart contract. He speaks with investor Mr. R. Runner, and they agree to funding via a SAFE. Acme formalizes this agreement as a `Safe` smart contract.[9]

Mr. Magoo installs a `Safe_controller` smart contract as controller of Acme (if one is not already installed) with Magoo as controller of the `Safe_controller`. He calls the function `offer_safe` with the SAFE smart contract as argument.

Mr. Runner views the SAFE, agrees with the formalization, verifies that Acme is now under the control of a `Safe_controller` contract, and calls `accept_safe_offer` in that contract with the agreed investment attached. That call results in the SAFE being added to Acme's extended capitalization table and the investment money is deposited in Acme's smart contract.

Company can only receive such a call from an `Equity_round_swap` later established as its controller.

[9] Acme and Runner may also sign a legal contract that expresses conditions agreed to by the parties to ensure that the `Safe` smart contract is a secure representation of the parties' intent and provides legal protections similar to those of a conventional SAFE. (See Sect. 17.4.).

A few years later, venture capitalist Mr. Wile E. Coyote—seeing a market for Acme's products, but also the need for improved product quality—and Acme agree to an equity financing. They negotiate a term sheet including pre-money valuation, a price per share, a post-investment controller for Acme, and the resulting shares after conversion for each SAFE.[10] Acme CEO, Mr. Magoo, calls `start_equity_round` with the term sheet information, which checks with all outstanding SAFEs that this funding is permitted (i.e., is in accord with the terms of the SAFE). Assuming the equity round is permitted, the `Safe_controller` contract creates an `Equity_round_swap` smart contract with the information about the equity round, and makes it controller of Acme.

Each equity investor (in this case, just Mr. Coyote) inspects this smart contract and verifies that it correctly represents the equity round term sheet before calling `equity_investor_consents` to consent to the equity round and contribute their investment. Once all equity investors have consented and invested, Mr. Magoo calls `finalize_equity_round`. This function issues shares to SAFE holders and equity investors, and deposits investor's money in Acme's account. It also sets Acme's controller to be the post-investment controller, and does some cleanup.

15.2.5 Liquidity Events

Liquidity events present a challenge to formal representation both because of indefinite language and their open structure. A liquidity event as defined in the Pre-Money and Post-Money SAFEs may occur in many different forms: IPO, direct listing, acquisition, corporate reorganization, disposal of assets, or change of control. Some of these types of liquidity events have a broad range of flexibility as to their structure.

For example, an acquirer may offer a complex package of cash and shares, with conditions relating to the rights of shareholders to accept the offer. (A compulsory acquisition of shares may apply after the acquirer has obtained a certain percentage of ownership in the company.) Similarly, corporate reorganizations may have many forms, including spin-outs, splits, mergers, acquisition of other entities, reincorporation, or change of domicile.

To address these difficulties in our smart contract representation of SAFEs, we have used two strategies. First, to address indefinite language, we defer to human input (see Sect. 14.3) to determine whether a liquidity event is occurring. Second, rather than attempt to develop an *ex ante* formalization for each of the many possible forms of liquidity event, we allow the precise terms of the liquidity event to be represented *ad hoc*, as a smart contract (applying programming of open structure,

---

[10] They may also sign a legal contract similar to a conventional equity round, but addressing the smart contract implementation of the financing. Note, however, that the `Safe_controller` and the `Equity_round_swap` code have already been locked in previously, to protect the interests of the SAFE investors, so the equity round investor is not able to negotiate on the use of this code.

see Sect. 14.5). We structure the sequence of events representing a liquidity event as follows:

1. Company management, represented as the controller of the Safe_controller, is trusted to provide the human input that a liquidity event is occurring. To do so, they call the function

```
function notify_liquidity_event(
                uint ndeadline,
                UD60x18 price,
                uint cash_to_be_distributed,
                address payable ncontroller
    ) public OnlyController
```

The effect of this function is to change the state of the Safe_controller to Liquidity_event_pending, indicating that a liquidity event is about to occur. The argument ndeadline gives a deadline by which SAFE investors are required to select whether they will Cashout or Convert their SAFE in this liquidity event.

To enable these investors to make their decision, a number of parameters of the liquidity event are provided. Parameter price gives the nominal share price at which the liquidity event is occurring (depending on the type of liquidity event, this may be the IPO price offered by the underwriter, the "fair market price" in case of direct listing, or the price offered by an acquirer).

The parameter cash_to_be_distributed represents the amount of cash available for distribution to SAFE investors who take the Cashout option in the liquidity event. (For security of these investors, it is required that this amount of cash be available in the Company contract, and in the state Liquidity_event_pending, spending of cash is prevented until the liquidity event is complete.)

Parameter ncontroller is the address of a new controller that will be installed as the controller of the company *after* the SAFEs have been either cashed out or converted, but *before* a distribution is made to the shareholders, including for the shares from converted SAFEs. This might be either the externally owned address of a human trusted to perform the liquidity event, or the address of a smart contract through which the distribution to shareholders, and the subsequent state of the company will be controlled.

To guarantee to the SAFE holders that the liquidity event will be performed as promised, these parameters are stored into the state of the Safe_controller. An event LiquidityEventNotified() is emitted to inform the SAFE holders that a liquidity event is pending. They can obtain details of the event from the parameter data stored into the state of the Safe_controller.

2. The SAFE investors have up to the indicated deadline to make their choice of option in the liquidity event. To do so, they call the function set_liquidity_choice on the Safe smart contract, providing the choice as an input.

As discussed in Chaps. 2 and 8, Post-Money SAFEs do not explicitly grant the SAFE investors a choice, but state that the choice is automatically determined by which delivers the maximum value of distribution. Our analysis in Chap. 8 shows that there are situations where this "optimal choice" is not uniquely determined, raising the question of how such an automatic choice should be implemented. To address this, our implementation does not attempt to make an automatic choice, but also grants to Post-Money SAFE holders the right to choose whether to Cashout or Convert. This represents a literal change to the terms of the legal SAFE, but we believe that it is one that both SAFE investors and founders will consider acceptable, given the impediments to automation. One can view this change as an application of the strategy of increasing precision, since the original contract does not state a resolution for this situation.

3. After the deadline, the function

```
function liquidity_event(
                uint[] calldata safes_payment,
                uint[] calldata safes_shares
                ) public OnlyController
```

can be called on the `Safe_controller`. The effect of this is to perform the cashout or conversion of the SAFEs, and to subsequently install the new controller from the `notify_liquidity_event` call as the controller of the `Company`, so that the remainder of the liquidity event, including the distribution to shareholders after the SAFE conversions, to be performed.

The argument `safes_payment` represents the amount of cash that will be distributed to SAFE investors who have chosen to Cashout. The value should be zero if the investor has chosen to convert instead. The argument `safes_shares` represents the number of shares that will be distributed to each of the SAFE holders in conversion of their SAFEs. In general, this will be zero for SAFE holders that have chosen to Cashout. However, SAFEs state that the distribution of cash may be reduced in the case of an intended "tax-free reorganization" of the company, and SAFE holders cashing out will receive shares for the balance of their Purchase Amount in that case. We have therefore implemented the `Safe` contracts so that SAFE investors may receive nonzero amounts of both cash and shares, totaling the Purchase Amount in value.

When the `liquidity_event` call is made on the `Safe_controller`, it first checks using calls to the functions

```
function permit_liquidity_event(
          uint[] calldata, // safes_payment
          uint[] calldata, // safes_shares
          UD60x18,         // price
          uint,            // cash_to_be_distributed,
          uint             // index of the Safe being called
          ) public virtual returns (bool);
```

on each of the `Safe` contracts, that the parameters of the liquidity event, from the `notify_liquidity_event` call, combined with those from the `liquidity_event` call, comply with the policies specified in each of the SAFE contracts being represented. If not, the `liquidity_event` call is reverted, otherwise the `Safe_controller` proceeds to paying out the specified cash amounts from `safes_payment`, and converting the SAFEs to shares as specified in `safes_shares`. For security reasons, the cash transfers are managed using a `Payout` contract, of the form described in Sect. 15.1.3. After this, it terminates the `Safe_controller`, and installs the new controller `ncontroller` from the `notify_liquidity_event` call as the new controller of `Company`. This new controller may then be used to perform the remainder of the effects of the liquidity event, in particular, the distribution to the shareholders and converted SAFE holders.

Example 15.2 In a slightly different universe than Example 15.1, Mr. Coyote wins the lottery and decides to buy Acme. After negotiating a price with Acme, Mr. Magoo calls `notify_liquidity_event` with details of the liquidity event.

After all SAFEs have called `set_liquidity_choice` to lodge their choice of Cashout or Convert, Mr. Magoo calls `liquidity_event`, providing the details of proposed payments and share issuances. This function gets permission from all SAFEs, issues shares and makes payments for all SAFEs (payments via a `Payout` smart contract), sets Acme's controller to be the post-liquidity-event controller, and does some cleanup.

The post-liquidity-event controller would then accept Coyote's payment and distribute it to stockholders (perhaps via a `Payout` smart contract), while transferring all shares to Mr. Coyote.[11] The controller would then install Mr. Coyote as controller of Acme.

15.2.6   Dissolution Events

The treatment of dissolution event is simpler than that of liquidity events, and handled in a single function on `Safe_controller`:

```
function dissolution(
            uint[] calldata safes_payment,
            uint[] calldata shares_payment
            )
```

---

[11] We have not developed a smart contract for such a post-liquidity-event controller, although the techniques we have described seem readily applicable in this case. As noted at the start of this subsection, in general liquidity events have too many possible variations to attempt to address them all in code.

Here, the parameter `safe_payment` contains the proposed payments to SAFE holders, and the parameter `shares_payment` contains the proposed payments to the shareholders. As with other calls, the `Safe_controller` first makes a call to the `Safe` smart contracts to check that the proposed distribution complies with the terms of those contracts. If so, the payouts are made, again via a `Payout` smart contract. After this, the state of the `Company` is set to `Dissolved`, and the `Safe_controller` state is also set to `Terminated`.

We have assumed that all assets of the company have been liquidated, that liquidation costs have been paid, and that the amount of cash to be distributed is available in the `Company` smart contract at the time the `dissolution` call is made.

15.3 Security Requirements

We now discuss how the architecture supports the reasoning required to justify a number of key correctness and security properties. We believe that the sketch below can form the basis for a formal proof of these properties, but we have not yet undertaken this. We ignore implementation-platform specific issues such as gas costs.

At the top level, each of the parties has their own interests, which the smart contracts are intended to enforce. To be assured that these properties will be enforced, each of the parties needs to needs to verify the components indicated.

- **SAFE investor:** The SAFE offer process can always be completed with either the SAFE active and registered by the company, and with the investor's principal transferred to the company, or with the SAFE withdrawn and no change to the investor's cryptocurrency holdings.
  Verification Requirements: `Safe_controller` (functions `offer_safe` and `accept_safe_offer` and `withdraw_safe`), `Company` and the investor's individual `Safe`.
- **SAFE investor:** Once the SAFE contract has been issued, in the first subsequent successfully completed Equity Financing, Liquidity or Dissolution event (if there is one), the company will issue shares (and/or cash) in a way that is consistent with the Equity Financing, Liquidity or Dissolution clause (respectively).
  Verification Requirements: The investor's individual `Safe`, `Company`, `Safe_controller`, `Equity_round_swap` and `Payout`.
- **All parties:** If an Equity Financing process is commenced, it can always be completed with either the process aborted and the company reverted to its state before the start of this process, or the process successfully completed.
  Verification Requirements: `Company`, `Safe_controller`, and `Equity_round_swap`.
- **Equity round investor:** All and only the shares promised in an equity round proposal will be issued if the round is successful, the agreed controller will be installed on `Company`, and the principal will be paid to the `Company`. If the

round is not successful, the money paid by the investor can be recovered.
Verification Requirements: `Equity_round_swap` and `Company`.

- **Company:** The company will grant no more shares than required by the contracts and Equity Financings that it offers, relinquish no more control than required by those agreements, and receive principal agreed.
  Verification Requirements: All `Safe` contracts, `Company`, `Safe_controller`, and `Equity_round_swap`.

- **All parties:** For security, no other parties should be able to cause a change of state that results in the SAFE investors, equity round investors or company losing cryptocurrency or a shareholding.
  Verification Requirements: `Company`, `Safe_controller`, `Equity_round_swap` and `Payout`.

These properties will require a number of background assumptions such as correct operation of the blockchain platform, and that parties do not share or have their cryptographic keys stolen.

Proof of these properties is facilitated by the architecture via its state-machine view of the contracts, given in Fig. 15.4. States correspond to the status variables in the contracts. Actions on the contracts are enabled only in the states indicated. An attempt to execute an action when not enabled is handled as a failed and reverted transaction on the blockchain. Some of the events triggering state transitions also trigger events in other machines. This has been indicated by naming the triggered event similarly, but in italic font. (For example, action `accept_safe_offer` is called by the SAFE investor on the `Safe_controller`, but triggers a call from there to the similarly named action *accept_safe_offer* on the `Safe`. Each of the actions `A`, `start_equity_round`, `notify_liquidity_event` and `dissolution` calls a similarly named `permit_*` function on the `Safe` contracts.) When an action calls an action on the `Company` smart contract, this has been indicated by including it in square brackets after the action name. For example, action `finalize_equity_round` on `Equity_round_swap` calls `set_controller` on `Company`. Correlated states in different contracts have been indicated by connecting them using a dashed line. (For example, `Safe_controller` will be in state `Equity_round_active` whenever there is an `Equity_round_swap` contract in state `Awaiting_consent`.)

Correctness and completeness of the state-machine model can be verified statically from the code (using its `require` statements and subroutine calls). Many invariants also follow from the state-machine model and a static analysis of write statements.

The verification that the parties need to conduct includes both source code correctness and checking that on-chain bytecode instances of the contracts have been correctly compiled from the source code and have the right parameter settings.[12] For

---

[12] Checking correspondence between on-chain bytecode and source code can be done using services such as Etherscan (http://etherscan.io) or Tenderly (http://tenderly.co), but requires trust in these services and the channels through which they are offered. On-chain factory contracts [5, 6] could be used to amortize, or delegate to a trusted third party, the effort that the parties need to put into verifying the contract source code and checking that it corresponds to on-chain bytecode.

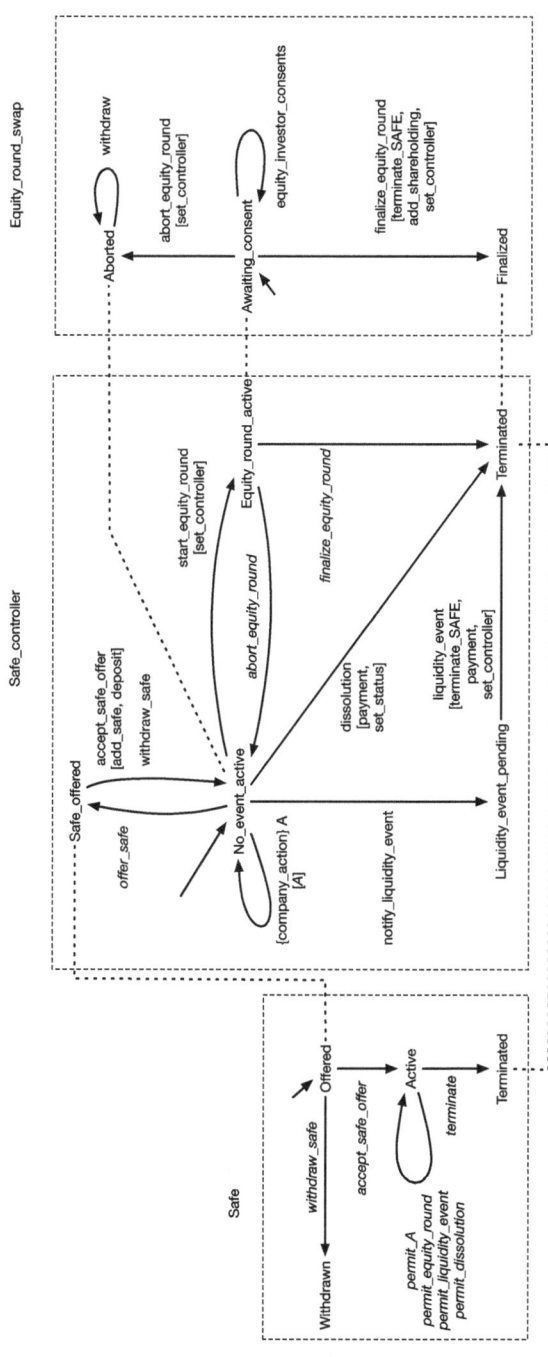

Fig. 15.4 Contract state transitions

example, the SAFE investor, before accepting the offer of a `Safe` *s* should verify the following:

- The `Safe` *s* has the expected setting of the parameters `controller` (this should be an address controlled by the investor), `company` and applicable terms such as `cap` and/or `discount`, and is in state `Offered`.
- A `Safe_controller` *sc* is the current controller of the `Company`. This controller is in state `Safe_offered`, as a result of a call of `offer_safe(s)`, and *s*.`safe_controller=` *sc*.
- The address *sc*.`controller` of the `Safe_controller` is under the control of management of the company.
- The bytecode of *s* has been compiled from source that correctly expresses the terms of the SAFE as agreed between the company and the investor. In particular, the `permit_equity_round` code returns `true` if and only if the proposed number of shares the SAFE investor will receive is as specified in the Equity Financing clause.
- The code in `Company` correctly represents and maintains shareholdings and issuance of `Safe` contracts. The code in `Safe_controller` correctly protects the rights of `Safe` holders. The code in `Equity_round_swap` ensures that an approved equity round proposal is eventually correctly acted upon if the consent and principal required for the round is obtained. The on-chain bytecode for these contracts is correctly compiled from the source code verified.

It is not to be expected that company management and investors will be competent to perform many of these types of verification, which require technical skills. Thus, as with all technology of any level of complexity, these users will typically place their trust in experts who are competent to perform such verification. Standardization of the components making up the architecture, and verification of the standard by well-known trustworthy parties such as standards bodies, regulators, registries and large organizations, would help to reduce the complexity of the trust decisions that users need to make when adopting the technology.

Our design of the smart contracts applies one of the principles of systems architecture, particularly in the context of secure systems development, which is to isolate critical functionality in components that are small enough to be verified (ideally, using formal methods), combined in the architecture in ways that ensure that the basis supporting key properties remains small and "local". Specific localization benefit is clearest for the `Safe` offer-acceptance process, the equity round process and the `Payout` contract.

Note that isolating money in `Equity_round_swap` ensures that it is not co-mingled with other company money and at risk of premature expenditure. Location in the `Equity_round_swap` contract and its receipt of control also enables the equity round investor to be assured of correct transfer of the money and issuance of shares without inspecting the code of the individual `Safe` smart contracts.

Handling of the SAFE investor's principal when investing could have been similarly managed using a swap contract, but this has not been done since it is attached to the `accept_safe` action on `Safe_controller`, which transfers it to the `Company` immediately on successful execution, and the blockchain reverts the transfer if the execution is unsuccessful.

The final security property is largely derivable from use of the Controller pattern and authorization checks in the form `require(msg.sender=..)` on each of the operations, with some exceptions that do not adversely affect the principal parties (e.g., anyone may `deposit` cryptocurrency into `Company`.)

15.4 Limitations of the Implementation

The representation of SAFEs as smart contracts we have presented above has made a number of assumptions and simplifications, and leaves various issues unaddressed. We list here some of the dimensions in which improvements could be made to develop a more satisfactory smart legal contract framework for SAFEs, and other types of instruments that a company might seek to use:

- We have assumed that the equity round can be fully described by pre-money valuation, price and share issuances to each investor class. Conceivably, other types of convertible instruments require additional parameters, but it is difficult to predict what these might be. A more abstract formulation of the equity round description, that enables additional parameters to be added by subclassing, might therefore be beneficial.
- The approach described assumes that all SAFEs will convert in the context of an equity round, so that the `Safe_controller` can be terminated when finalizing the equity round and control can be passed to the `Equity_round_swap`. Conceivably, different SAFEs could state different definitions of an Equity Financing event. A more subtle management of control is required if unconverted SAFEs or other types of convertible instrument could remain after the equity round, and the `Safe_controller` needs to be kept in place. A similar comment may apply to liquidity events.
- The criterion we have applied for existence of an equity round, that Preferred shares are being issued, could be open to manipulation, where a company colludes with equity round investors to have these accept Common shares instead, so that the SAFEs do not convert.[13] In the worst case, this stratagem could be used to forever prevent conversion of the SAFE contracts, depriving the SAFE investors of benefits (dividends and/or liquidity, depending on the SAFE type) from their investments. Since in any event, the SAFE contracts do not come with a guarantee that they will ever be converted, this risk seems inherent in the SAFE contracts, rather than being a fault of our smart contracts.

---

[13] Alternately, there have reportedly (personal communication, Vitaly Spassky and Romain de Spoelberch of Polymorphic Capital) been cases where a SAFE-like instrument converted in an equity round conducted as a crowdfunding that issued common rather than preferred stock, disadvantaging SAFE investors because they received shares with lesser rights than expected.

- We have treated the different types of shares that a company may issue using an enumerated type ShareType, without going into the details of the rights associated to different types of shares. It would be reasonable to also attempt to cover these using smart contracts that express these rights.
- We have represented stand-still provisions during an equity round overly simplistically as a block on payments while the Equity_round_swap is active. An alternate, arguably more accurate, way would be to represent the term sheet for the equity round as a smart contract, referred to by the Safe_controller and/or the Equity_round_swap.
- We have left a significant amount of power at the discretion of the company in liquidity events, by stating parameters of a liquidity event including the share price, cash available for distribution. The company's ability to set the controller to be used immediately after the Cashout amounts have been paid, which handles the distribution to shareholders, is also subject to manipulation. One way to have the smart contract prevent abuse of this power would be to implement standard controllers for some common types of liquidity event, such as an acquisition for a mix of cash and shares of the acquirer. Through use of on-chain factory contracts [5, 6] that keep a record of addresses to which they have spawned an instance, it could be made possible for Safe smart contract code to check that one of the pre-approved standard controllers is being used in the liquidity event. If not one of the approved ones, the SAFE investors could be asked to manually approve the controller before the Safe approves the liquidity event containing it.
- In principle, when issuing a SAFE contract, it may be necessary to check that the contract is consistent with other contracts signed. (For example, post-money SAFEs give out a proportion of the company, so the set of proportions needs to add up to less than 1.) Off-chain tools to assist with this task would be useful.
  However, if there is a diversity of instruments being used by the company, this check may be computationally complex, and may become undecidable with a rich enough representation format for SAFEs and other convertible instruments. An alternative is to not check, and leave failures of convertibility as a "run-time" issue, to be discovered at the time of the equity round. In case this problem is encountered, the company and investors are likely to want to make a compromise. The approach of verifying consent to the conversion, rather than a full automation, supports such compromises.
- In order to focus on the main issues, we have assumed away a number of aspects of cap tables that are commonly used by startups, such as the option pool, lockups and vesting schedules. A usable implementation would need to provide these.

15.5 Summary

Our architecture for a blockchain implementation of SAFE contracts consists of five smart contract classes:

- `Company`, which represents the company's cap table and financing commitments, such as SAFEs
- `Safe`, which represents a SAFE agreement between company and SAFE holder
- `Safe_controller`, which enforces obligations on the company to comply with SAFE agreements
- `Equity_round_swap`, which ensures a risk-free exchange of rights: SAFE holders exchange their SAFEs for shares, the new investor exchanges money for shares, and the company exchanges shares for money and release of obligations (SAFEs), and
- `Payout`, which provides a mechanism by which the company can meet its payment obligations arising from SAFEs, protecting both itself and the rights of the SAFE investors.

The architecture has been carefully designed to protect the rights and property of all parties, and to simplify reasoning about correctness.

The `Safe` contracts have a *declarative* form: they do not have operational effects, but only provide true/false answers as to whether an action to be performed on the `Company` complies with the SAFE contract they represent. The `Safe_controller` and `Equity_round_swap` perform the actual operations on the `Company`, and ensure compliance with the terms of all issued SAFEs.

We have presented only the minimal functionality of the `Company` class necessary to interact with SAFEs and perform equity financing. Other functions could be added which would provide an ability for sale of common shares, share buy-backs, SAFE replacement, etc.

Each smart contract, except `Payout`, has a state variable which is used to restrict the actions the smart contract can perform to those appropriate for that state. Figure 15.4 shows the transitions between states. Function calls cause these state variables to be updated (they are presented as labels on the arrows).

References

1. van der Meyden R, Maher MJ (2021) Architecture for smart SAFE contracts. In: 3rd conference on blockchain research & applications for innovative networks and services (BRAINS). IEEE
2. Herlihy M (2018) Atomic cross-chain swaps. In: Proceedings of ACM symposium on distributed computing. Version at https://arxiv.org/abs/1801.09515
3. OpenZeppelin (2024) Documentation: Ownership. https://docs.openzeppelin.com/contracts/2.x/api/ownership#Ownable. Accessed Dec 2024. Accessed Mar 2020
4. Cooley LLP (2024) 500 Startups KISS convertible debt & equity financing documents. https://www.cooleygo.com/documents/kiss-convertible-debt-equity-agreements/. Accessed Dec 2024

5. Xiwei X, Weber I, Staples M (2019) Architecture for blockchain applications, 1st edn. Springer Publishing Company, Incorporated
6. Factory Contracts (2024). https://research.csiro.au/blockchainpatterns/general-patterns/contract-structural-patterns/factory-contract/. Accessed Nov 2024
7. van der Meyden R (2019) On the specification and verification of atomic swap smart contracts. Abstract appears in IEEE international conference on blockchain and cryptocurrency, pp 176–179. An extended version is available at https://arxiv.org/abs/1811.06099
8. Popper N (2024) Hacker may have taken $50 million from cybercurrency project. The New York Times, 17 Jun 2016. See also http://en.wikipedia.org/wiki/The_DAO_(organization). Accessed Dec 2024
9. Deloitte (2024) Roadmap: Issuer's accounting for debt. https://dart.deloitte.com/USDART/home/publications/roadmap/debt. Accessed Dec 2024
10. DLA Piper (2024) Restrictions on transferability of shares. https://www.dlapiperintelligence.com/goingglobal/corporate/index.html?t=38-restrictions-transferability-of-shares. Accessed Dec 2024

Chapter 16
Evaluation

Having developed a general framework for the representation of SAFE contracts as smart contracts, and implemented a number of SAFEs in this framework, we now turn to an evaluation of this work. In particular, we consider the question of whether the smart contracts can replace the use of legal SAFE contracts. We begin with a discussion of applicable evaluation methodologies, and then focus on a comparative analysis of the way that the legal and smart contract SAFEs address a number of risks. We also consider whether our methodologies for dealing with indefiniteness adequately cover the content of the legal SAFE.

16.1 Evaluation Methodologies

In evaluating our SAFE smart contract implementation, a first question is: what specifically is the object of the evaluation? SAFE legal contracts were designed to address issues in equity financing, and one could evaluate to what extent these contracts, and our smart contract implementations of them, provide satisfactory solutions for this business purpose. As we do not aim to improve upon the business functionality of legal SAFE contracts, we will not attempt to cover this aspect. Instead, we take a comparative approach: assuming that the legal SAFEs are fit for purpose, to what extent do our smart contracts correctly express their functionality? We focus on coverage of the Equity Financing clause of SAFEs—the general issues for Liquidity and Dissolution events are similar.

Our smart contracts are structured into several components. A number of these, the `Company` contract representing the company, the `Safe_controller` the `Equity_round_swap`, and the `Payout` contract, are abstract and technical. Except for laying out the lifecycle and context of a SAFE, the details of the SAFE variants are not represented in these components. A consequence of this is that a formal specification of these components could be written, and formal methods, the

R. van der Meyden and M. J. Maher, *Simple Agreements for Future Equity (SAFE)*, Blockchain Technologies, https://doi.org/10.1007/978-981-96-3920-5_16

most rigorous verification methodology used in computer science, could be applied to give a very high integrity proof of the correctness of the design and implementation of these contracts with respect to this formal specification. (Formal methods are based in the use of mathematically precise notations and the use of automated and semi-automated tools to verify correctness of components with respect to their specifications.) A sketch of what would need to be proved in such a verification has been given in Sect. 15.3, but we leave the details to future work,[1] and will not address this aspect of evaluation in the remainder of this section.

A benefit of the way that we have decomposed the implementation is that it leaves the remaining components, the `Safe` contracts, as the sole components that directly related to the legal SAFE contracts, so that the evaluation of the extent to which the smart contracts correctly implement the legal SAFEs can be focused on this component. In particular, given the properties of the other components, we need to ask: does a SAFE smart contract respond `true` to the question of whether a particular operation is permitted exactly when the legal SAFE would imply that this operation is permitted?

An immediate issue is that we have applied the strategy of "limitation of fact patterns" (see Sect. 14.2): our smart contract implementation is able to generate a smaller set of fact patterns then the legal contract allows. This means that the two differ on the scope of their applicability. The appropriate comparison, therefore, is whether the two are equivalent *within the narrower scope allowed by the smart contract.*

One inherent problem with the question of equivalence is that the starting point, the legal SAFE, is an imprecise natural language document, whereas the smart contract code has a more precise operational semantics. Particularly where the indefinite terms have been formalized in code (rather than having their satisfaction determined by a process involving human agents), it is therefore not possible to set an exact standard for correctness of the smart contract with respect to the entire legal text.

However, where fragments of the natural language text are sufficiently precise, a correspondence to parts of the smart contract code can be established. The equations for the Equity Financing clause are indeed precise, so the correspondence can be checked by inspection, under the assumption that the event in question meets the definition of Equity Financing. Ideally, we would like to be able to establish the correspondence unconditionally, but we then must face the difficulty that the conditions under which this event applies involve open-textured concepts, as discussed in Sect. 13.2. The extent to which our approaches to handling open texture resolve this difficulty is discussed in Sect. 16.3.

At a broader level, adequacy of the formalization can be evaluated in a number of ways less demanding than a mathematical proof. For example, acceptance on inspection by a legal expert, or by users (company, investors) could be indicative of quality. A problem with this measure, however, is that it may be difficult to find appropriate individuals with sufficient knowledge both of the law and of coding (in

---

[1] The Solidity programming language, as presented in its documentation, falls short of having a full formal specification of its operational semantics, but work in this direction is in progress, e.g., [1].

the specific language used) to be able to make the judgments required. It would be beneficial, to this end, for the smart contract language to use a programming paradigm that structures programs in ways close to the natural language structures found in contracts. An imperative language such as Solidity does not meet this requirement.

Were these conditions and equations in a legal contract expressed in a *controlled natural language* [2], a fragment of natural language syntax that has been equipped with a formal semantics that has been justified to accurately capture the natural language semantics, then we could in fact establish a mathematical equivalence with smart contract code. A number of controlled natural languages targeting the specific domain of contract representation are under development [3, 4].

However, establishing adequacy of the controlled natural language with respect to the natural language, is an even more general and challenging problem than the problem of validating a particular text in the controlled natural language! Nevertheless, if the particular controlled natural language text expressing the contract has been written by an individual who understands both its formal semantics and the way that, qua natural language text, it will be interpreted by its intended audience of legal practitioners (lawyers and adjudicators) then there is more likely to be a strong correspondence between the formal and informal meanings.

Another difficulty with evaluation by inspection is that it can be difficult to envisage, or reason about, all the possible ways that a contract may be performed. This requires reasoning in the abstract, a cognitively more difficult task than reasoning about concrete facts. Even experts on the intended functionality of a system are likely to be better at answering questions about the "correct" behavior of the system in concrete examples than in the abstract. This motivates an evaluation methodology based on testing the behavior of the smart contracts in concrete test scenarios. One can determine whether the smart contract yields the same conclusions as the exemplar.

In principle, one potential source of test cases for smart contracts implementing legal SAFE contracts is case law. Obviously, the usefulness of case law for testing purposes requires that the contracts under consideration in the cases are not bespoke variants of a SAFE. We can hope to benefit here from the fact that SAFEs are intended as a standard contract (although variations have already proliferated). Unfortunately, SAFE contracts may be too new for case law to provide a satisfactory set of cases for this purpose: there appear to be very few relevant court rulings, and such as do exist are concerned mostly with legal issues that do not bear directly on how a SAFE should be interpreted.[2] Such contracts are "rarely litigated in court" ([5],

---

[2] Some examples are:

- JOHN LEVINSON AND ELLEN LEVINSON, Plaintiffs, v. TWIAGE, LLC, JOHN HUI, GREGORY P. SANTULLI, Defendants. Supreme Court, New York County, 2018. https://scholar.google.com/scholar_case?case=11246706013972417804. This case relates to claims an investor was fraudulently induced by a company to enter into a SAFE contract.
- CU*ANSWERS, INC., Plaintiff, v. G2LINK, LLC, Defendant. Case No. 2:18-CV-04525-JDW. United States District Court, E.D. Pennsylvania. December 31, 2019. https://scholar.google.com/scholar_case?case=7054561818531695690. This case concerns the interpretation not of the SAFE involved, but whether issuing the SAFE satisfies the "Qualified Financing" conditions for conversion of another contract, a Convertible Promissory Note.

footnote 186). A small number of worked examples of SAFE contract performance are available in "Safe Primer" documents provided by Y Combinator [6, 7].

The ultimate measure of acceptability is therefore likely to be the reactions of users to situations arising once the smart contract has been deployed. We can then gather metrics such as how often the smart contract results are contested by the parties, and how often an arbitrating authority overrules the on-chain result. As measures of adequacy of formalization, metrics are not without their problems, however. They may be measuring the litigiousness of the users or inclination of the arbitrator to overrule the on-chain result, rather than quality of the formalization. However, these metrics may be of some value for comparisons between two different smart contract formalizations of an informal contract, if one is less contested, or overruled, than the other. These measures also do not separate the issue of formalization quality from the fitness for business purpose, though the latter might rightly be considered to be the more important concern.

The inherent gap between a contract subject to indefiniteness and a smart contract implementation of that contract is familiar to software engineers in the form of the gap between informal requirements and more formal specifications and/or code. Software engineers address this problem using a methodology based on testing, and revision after deployment is the prevailing evaluation methodology. Development methodologies such as *agile* methods [8] aim to accelerate the feedback loop by which user reactions to a working system in concrete scenarios results in refinements of the code.

16.2   Risks Controlled

Both legal contracts and computer security mechanisms can be understood as attempts to control risk and make the possible futures more predictable. One approach to evaluating a proposal to implement a legal contract using smart contracts on a blockchain, is to consider the proposal's risk management properties, and compare them to an approach using a legal contract enforced only through the legal system.[3]

- INV Accelerator, LLC, Plaintiff, v. MX Technologies, Inc., et al., Defendants. No. 19-CV-2276 (AJN). United States District Court, S.D. New York. https://scholar.google.com/scholar_case? case=6312065034020918763. This case deals with the issue of whether a SAFE contract was entered into by the incubator INV Accelerator and a company Goldbean. Goldbean signed a participation agreement with INV Accelerator the terms of which stated that Goldbean would be issued with a SAFE for $ 0.5M, but that SAFE was never signed. More pertinent to the interpretation of the SAFE (but not the Equity Financing clause that is our main focus) is that Goldbean sought to escape the consequences of the Liquidity Event clause when it was subsequently acquired by MX Technologies, by changing the acquisition to a hiring of staff and a transfer of assets to MX Technologies, leaving the company as a shell.

[3] The practice of legal contracts may differ from the theory—business has a social dimension and often operates on trust. Contracting parties may prefer to use contracts for purposes of moral force,

Investment in early stage startups involves many risks, and SAFE contracts have been designed to provide protections against some specific financial risks. In particular, they provide downside protection for the investor in case the company is not able to sufficiently increase its valuation while allowing the investor to benefit from the upside of a high valuation. For the Equity Financing clause, the technical parts of the text of the legal SAFE are precise, and the smart contract code is obtained by a straightforward translation to code. Hence, subject to the necessary assumptions about the completeness and legal status of the on-chain representation of the company's cap table, discussed in Chap. 17, the SAFE smart contract provides similar protections. One proviso on this similarity relates to the open-textured concepts involved in the Equity Financing clause. We discuss these at greater length in Sect. 16.3.

Some risks to an investor are obviously not mitigated by a legal SAFE contract, and we should therefore not necessarily expect that smart contract implementations of a SAFE will be able to mitigate these risks. Risks outside the scope of a SAFE legal contract include behaviors by company management that are not illegal, but which disadvantage the SAFE holder. For example, founders may cause losses to SAFE investors by failure to develop the company, or by poor management of company funds. Founders may also disadvantage SAFE investors, decreasing the value of the shares to be received in advance of conversion of the SAFE, by paying out dividends,[4] or spinning out some of the assets of the company and using a stock distribution to shareholders. If the company is able to become profitable enough to fund its own operations and development, it may avoid ever conducting a priced equity round, preventing the SAFE investors from obtaining any benefit from their investment. These uncovered risks might well be priced in by sophisticated investors (via the values of the SAFE parameters, i.e., the cap and or discount) when purchasing a SAFE, so it can be questioned whether these are actually weaknesses of the instrument or deliberate omissions in its construction.[5] It is up to the SAFE investors (e.g., through their personal engagement with company management) to ensure that company governance effectively mitigates these risks. Where risks outside the scope of a legal contract can be mitigated by governance, these risks can be mitigated in exactly the same way when the legal contract is implemented as a smart contract.

Other risks outside the scope of a SAFE legal contract include illegal activity such as embezzlement of company funds by company staff, or provision of fraudulent data on the basis of which a valuation is established by an investor. Criminal law provides a measure of deterrent protection and recompense in this case. To the extent that there is a clear legal jurisdiction for the company's activities, or its interactions with

---

and avoid litigation in favor of negotiated settlements that differ from the enforceable terms of the contract [9].

[4] The Pre-Money SAFEs do not provide for SAFE holders to receive dividends. The Post-Money SAFEs do provide for dividends, but only in a diluted form (see Sect. 8.9).

[5] Attitudes of the courts to protecting convertible bond holder interests against those of shareholders, and the potential judicial bases for intervention are discussed by Bratton [10].

investors, the same protections are available when the SAFE is represented as a smart contract.

One risk outside of the scope of the legal SAFE that is addressed in the smart contract implementation is the risk of errors in the company's cap table due to company insider fraud, incompetence and/or outsider interference (e.g., malicious hackers of computer records). By placing shareholder records under the control of keys in the hands of investors or their proxies, a smart contract implementation distributes this risk away from the company or its registry, and replaces it by the residual risk of attacks on the blockchain validators and the mechanisms and processes by which cryptographic keys are protected.

It is in the spirit of the smart contract field to design novel mechanisms that mitigate risks by means of economic incentives and/or penalties. A common approach is "staking", or posting of cryptocurrency or token collateral by one or more of the parties to an interaction, with a smart contract enforcing that the collateral is lost in case evidence can be provided that the party has not adhered to required behaviors. Our aim in the present work has been to determine to what extent a legal SAFE can be implemented on its own terms as a smart contract, rather than to develop new mechanisms for the problem solved by legal SAFEs, so we have not attempted to develop such alternate mechanisms. However, it is questionable whether collateralization would be an appropriate mechanism, given that legal SAFE's are not in practice collateralized, and indeed to do so would be against the spirit of the high-risk tolerant investment posture of SAFE investors. The aim of a SAFE investment is to make funds available for company development, and locking funds up defeats the intent of the investment. Where the founders have non-monetary assets that they are prepared to put up as collateral for an investment, the resulting instrument would no longer be a SAFE. Our smart contracts could be extended to build in protections such as burn-rate constraints, but we do not pursue this, for similar reasons.

16.3   Risks Not Coverable by Smart Contracts Alone

As noted above, the Equity Financing clause in the legal SAFE contract contains "open texture" language which is difficult to express in code. Our proposed approaches to dealing with this difficulty for smart contract representation include limitation of fact patterns and either deferring decisions on the applicability of open texture concepts to (negotiated) human consideration, or replacing imprecise contract terms by precise ones in the interest of enabling automated enforcement. How satisfactory are these approaches for dealing with the Equity Financing clause? Do they enable the legal language to be entirely replaced by corresponding smart contract code? The following example (condensed from Example 6.1) suggests that our smart contracts leave the SAFE investor exposed to a risk against which the legal SAFE provides protection.

Example 16.1 A company has a single SAFE with SAFE holder Saffron and a new investor Neville wishes to purchase equity in the company. However, at Neville's valuation, Saffron would receive a majority of the shares. So Neville and the company founders collude to limit the shares that Saffron receives. They perform two equity rounds, the first one at an artifically high valuation. Saffron receives a small number of shares and her SAFE is terminated. Then, a second equity round is performed at a low valuation, giving Neville a significant proportion of the company shares. At the end, Neville and the founders are much better off than if there were a single equity round, and Saffron is much worse off. Example 6.1 has all the details.

Our general approach allows smart contract implementations of SAFE contracts to select a point on the spectrum between full automation of the equity financing in which the SAFE converts, and implementations of processes that give the SAFE investor bargaining rights in the construction of that equity round. Because the smart contract enforcement of the computations or processes is nuncospective (as discussed in Sect. 13.6), and the manipulation might not be detected until the second round, neither approach necessarily prevents the manipulation from occurring. However, it is reasonable to take the view that the legal SAFE contract protects against such collusion by requiring an Equity Financing to be a "bona fide transaction or series of transactions." Thus, the initial round of equity financing might not be considered *bona fide*. Alternatively, the two rounds of financing might be considered as a series of transactions, so that the number of shares due to Saffron would be based on a collective price. Such judgments can be formed only after the events have taken place. Enforcing such interpretations would require a legal judgment.

A related point applies to the way we have represented Liquidity Events in the smart contracts, giving discretion to company management to determine that such an event is occurring, and to set the parameters of the event to be performed on chain, such as the amount of cash available for distribution, and the nominal share price for the event. Again, there is a risk of manipulation by company management, against which investors would not be protected by the smart contract code alone.

There are therefore multiple reasons to retain the protections of the legal system even when working with smart contract implementations of the SAFE.

16.4 Summary

In the above, we have considered evaluation of our proposed approach to smart contract implementation of SAFE contracts from a number of angles. The conclusion of our discussion of evaluation methodologies is that the inherent gap between the indefiniteness of the legal SAFE contract and the definiteness of the smart contract implies that it is necessarily the case that, beyond the partial assurance provided by testing, we cannot obtain a complete assurance that the smart contract will behave in the same way as the legal contract in the context of the legal system. A consideration of risks leads to the conclusion that some risks must necessarily be addressed by the legal system. In particular, a legal SAFE provides protections against a two-round manipulation that the smart contract cannot fully address on its own. On multiple grounds, therefore, we conclude that in order to provide the full legal functionality of a legal SAFE, a smart contract implementation should be associated to a legal contract that provides protections that the smart contract omits. We therefore consider smart contract interactions with the legal system in the next section.

References

1. Jiao J, Kan S, Lin SW, Sanan D, Liu Y, Sun J (2020) Semantic understanding of smart contracts: Executable operational semantics of Solidity. In: IEEE symposium on security and privacy (SP), pp 1265–1282
2. Kuhn T (2014) A survey and classification of controlled natural languages. Comput Linguist 40(1):121–170
3. Lexon (2024) Plain text programming. Online, https://lexon.org. Accessed Nov 2024
4. Watt SJ, Goodenough O, Wong MW (2023) Deontics and time in contracts: An executable semantics for the L4 DSL. In: Sileno G, Spanakis J, van Dijck G (eds) Legal knowledge and information systems - JURIX 2023: the thirty-sixth annual conference, Maastricht, The Netherlands, 18-20 Dec 2023, Frontiers in Artificial Intelligence and Applications, vol 379. IOS Press, pp 119–124
5. Coyle JF, Green JM (2015) Contractual innovation in venture capital. Hastings Law J 66(1):133–183
6. Combinator Y (2016) SAFE Primer. https://web.archive.org/web/20180831020232/http://www.ycombinator.com/docs/SAFE_Primer.rtf
7. Combinator Y (2018) Post money safe user guide, https://www.ycombinator.com/docs/Post%20Money%20Safe%20User%20Guide.pdf. Accessed Dec 2024
8. Beck K et al (2001) Manifesto for agile software development. Online, http://agilemanifesto.org. Accessed Nov 2024
9. Levy KEC (2017) Book-smart, not street-smart: Blockchain-based smart contracts and the social workings of law. *Engaging Science, Technology, and Society*, 3:1–15
10. Bratton WW (1984) The economics and jurisprudence of convertible bonds. *Wisconsin Law Review*, pp 667–740

Chapter 17
Interaction with Law

In this chapter, we discuss a number of issues of law arising when representing SAFE contracts as smart contracts. A full treatment of this topic would be jurisdiction-dependent, and may need to delve into case law, legislation and regulation. We do not attempt to go to this level of detail, confining ourselves to the identification of some legal issues and options for addressing them.

17.1 Applicability of Law

A first question is whether or not law is applicable to smart contracts. Some have held that it is not. We might use the term "Smart Contract Absolutism," for the position that the state of the blockchain is definitive with respect to facts concerning ownership of on-chain assets and powers exercisable with respect to those assets, and not subject to any external legal jurisdiction.[1] On this view, the state of the blockchain cannot be incorrect, and an event on-chain is permissible if and only if it is enabled by the code of the underlying blockchain and its smart contracts.

In practice, this position has encountered a number of difficulties. One is that smart contract code may contain errors and security vulnerabilities, so that its behavior does not match the intentions of its designers and the expectations of its users. Errors sometimes have devastating consequences. In the attack on a security vulnerability of The DAO smart contract in June 2016 [1], around $US50M worth of value was stolen. Ironically, the adherent of Smart Contract Absolutism has deprived themselves of legal protections against such malfeasance. Interestingly, the Ethereum community

---

[1] The position is often expressed as "Code is Law." We prefer not to use this phrase to describe the position, since it appears to have been first used, by Lessig [4], to make a quite different point: that code implicitly imposes regulatory assumptions, and ought to be designed to build in Constitutional, i.e., legal values.

in general adopted the position that the attack was not part of the intent of the smart contract, and took action to defend it and retrieve stolen funds by making changes to what had been supposed to be an immutable blockchain. (Their interest in doing so may have been more to protect the still-nascent Ethereum platform, however. This precedent has not been followed for later attacks.)

A second difficulty for a Smart Contract Absolutist position taken with respect to a smart contract is that legal authorities may nevertheless assert that its operation falls within the scope of the law. An example of this is the SEC Report of Investigation on The DAO [5], in which the SEC asserted that The DAO tokens should be classified as securities subject to US securities laws (although no action was taken since The DAO had already been dissolved after being attacked).

In general, where a smart contract impinges on real-world matters in the scope of the law, one expects that the law will assert itself.[2] The legal system provides a social function of supporting trust in interactions between parties through the enforcement and remediation of contracts. Poncibò and DiMatteo [2] ask whether smart contracts offer the prospect that these roles of the legal system can be eliminated by use of smart contracts. They conclude that, although a reduced reliance on the legal system may be achievable, smart contracts cannot escape legal review, and the option of contracting parties to seek legal remediation cannot be eliminated.

This is likely to apply, in particular, to on-chain representations of company cap tables and equity instruments, which represent rights relating to persons and entities that are subject to a legal environment, and engage in real-world activities. We therefore expect that the law will assert itself with respect to smart contract implementations of cap tables and related instruments such as SAFE contracts.

To the extent that the parties avoid a legal system, and rely solely on the smart contract and blockchain for risk mitigation, some of the problems outlined in Chap. 13 will remain unaddressed.

17.2 Legal Status of Smart Contracts

If one does not take a Smart Contract Absolutist position, and accepts that matters that one is aiming to represent in a smart contract do fall within the scope of a particular legal jurisdiction, several further questions then arise.

From the point of view of the users of the smart contract, it becomes desirable to understand the precise legal standing of the matters that it purports to represent. In particular, for inherently legal facts such as those relating to ownership of shares in a company, it is desirable to understand how the cap table as represented on-chain is related to the cap table of the company "in law".

---

[2] Arguably, the Smart Contract Absolutist position is more sustainable for wholly virtual cryptocurrency holdings, which do not represent any fact in the real or legal world. Even here however, the fact that persons ascribe value to these holdings has led to the assertion of tax law over these holdings.

A first question is whether the law accepts that the blockchain has any relevance for questions relating to a company's cap table. In some cases, the law may in fact rule that a smart contract *cannot* be an official record of the cap table of a company. For example, regulators may require companies to use approved (non-blockchain) registries to represent their shareholder records, or insist that particular custodians be used to hold share certificates. If this is the case, then the blockchain can at best be a non-official copy of the official register. In general, such a regulatory stance is more likely to occur in the case of publicly traded companies, with private companies (including the startups that are most likely to use SAFEs) given significant freedom with respect to how they maintain their shareholder records.

Once it is allowed that the blockchain may record data relevant to the cap table in law, a second question is whether the cap table in law is determined solely by the state of the blockchain, or also by facts not represented on-chain. Shares may typically be issued by a company using paper certificates. If the company issuing shares on a blockchain retains the option to issue paper shares, from the point of view of the law, the official state of the cap table will be determined not just from the blockchain but also by the off-chain paper certificates. Other real-world facts concerning actions of the company's officers may also be pertinent.

This means that the on-chain representation of the cap table may differ from the legally accepted state. From the point of view of the holder of a SAFE contract represented on chain by a smart contract, such a divergence is problematic, since actions of the smart contract assume that the on-chain record is complete. For example, the Equity Financing clause makes use of the notion of Company Capitalization (the total number of shares issued). If the shares recorded on-chain are only a subset of the shares issued in law, then the calculations performed by the SAFE smart contract on-chain will not yield a correct share issuance to the SAFE contract holder at the time of the equity round.

To avoid this problem, some method for assuring the completeness of the on-chain representation is required. Possible approaches to such providing such assurance could include

1. public communications by the company promising to use the blockchain exclusively to represent its cap table, provided that the law accepts such statements as giving the blockchain representation legal force,
2. inclusion in legal contracts associated to an on-chain SAFE, and other share issuances, of terms to the effect that the on-chain representation of the cap table is the official representation, again assuming that the law considers the company to have the right to include such terms, or
3. the fact that the blockchain is operated by an entity recognized by the securities regulator as the official registry for the company.

None of these approaches *guarantees* that the on-chain representation is correct, but they do provide injured parties the option of seeking legal remediation when errors occur.

A similar question concerning legal status arises for the convertible instruments issued by a company: if these are represented using on-chain smart contracts, are these legally recognized? Particularly for Post-Money SAFEs, for which the conversion calculation makes reference to other SAFEs issued, it is again critical for correctness that the on-chain representation of SAFEs covers *all* SAFEs that the company has issued in law.

It would appear from the above considerations that a SAFE smart contract, both in order to obtain legal standing, and to be meaningful from the point of view of correctness with respect to the company's cap table in law, should be accompanied by a legal contract that contains terms relating to the legal standing of the on-chain representations of the cap table and the SAFE contract. This adds to the reasons that there should be a legal contract backing the smart contract that were already identified in Chap. 16. We discuss the possible form of such a legal contract in Sect. 17.4.

A further legal issue with respect to shares represented on-chain is that regulatory authorities may require that they be notified of certain changes to the cap table. This could be implemented either by the regulator directly reading the on-chain information, by using off-chain processes, or both, so does not impact either the legal contract or smart contract except inasmuch as it may be beneficial to include events in the smart contract that can serve as triggers for any off-chain processes.

17.3 Legal Activation

Contract law governs conditions under which parties cause a contract to be activated, and come to be committed to its terms. One requirement is mutual assent (also referred to as a "meeting of the minds"), that is, a mutual understanding and agreement to the contract. Often, this is demonstrated by a process of offer and acceptance, followed by signing the text of the contract.

The usual process for parties to become committed to a smart contract, is that one party places the smart contract on-chain, and one or more parties then participate in it by sending it transaction messages, which may have attached value. These messages are signed, but the content signed is not the "text of the contract": rather, the signature applies to the message being sent to the smart contract, which includes the function being called, the values of the parameters, and the amount of any attached cryptocurrency. The message serves to identify the sender (by public key) to the smart contract, and to authorize the blockchain miners to transfer the cryptocurrency amount from the control of the signing key to the control of the smart contract.

A process similar to this is used in the case of our implementation for the activation of a SAFE. The company first creates a `Safe` smart contract using the constructor of one of the specific subclasses representing particular types of SAFE. This smart contract is then "offered" to the SAFE investor by calling the operation `offer_safe` of `Safe_controller`, with parameters of the `Safe` contract and a deadline for acceptance. The SAFE investor "accepts" the offer by calling the operation `accept_safe_offer` of `Safe_controller`, and attaching their principal payment in cryptocurrency. The effect of this is that the payment is transferred to the `Company` smart contract, and the `Safe` is registered as one of the issued SAFE contracts.

A potential legal problem with this approach is that although it may establish certain *de facto* powers on the blockchain, the parties' assent to the terms of the contract is *implicit*, demonstrated by the fact that they participate in the contract's individual events. This approach is most consistent with the Smart Contract Absolutist perspective, in that the parties choose to participate in the smart contract and are, as a result, required to accept the behavior of the smart contract as authoritative. However, what is missing from a legal perspective is direct evidence of their assent to the contract as a whole, usually provided by having both parties sign the contract.

If the parties desire, or are required to, enter into a legal contract that either gives the smart contract legal force, or covers aspects of their agreement and dispute resolution needs that are not covered by the smart contract, then a more explicit process may be required in order to give effect to this contract. One option is to have the parties sign a paper contract off-chain, that includes enough detail about the smart contract to link the two.

Alternatively, if the legal jurisdiction in question accepts digital signatures,[3] we could handle explicit legal assent as follows. Fields are included in the smart contract that can store a signed copy of the legal contract (or a more space-efficient signed copy of the hash of the contract), one for each party. The smart contract can include operations by which the parties to the contract can save their signed copies into these fields. The remainder of the smart contract can be enabled once both parties have stored the correct signed content. This has the effect of activating the smart contract once both parties have signed.[4]

For some contracts the legal assent could be one-sided: the offering party creates on-chain a smart contract containing a signed copy of the offer document, and the accepting party interacts with the smart contract, possibly providing payment as consideration, but without providing a signature on the offer document. This may work for contracts in which the only obligations are on the side of the offeror: the accepting party is able to back claims for damages for non-performance using the signed statement by the offeror. However, the SAFE contract includes commitments from both parties. The SAFE investor is required to assent to the "Investor Representations," specifically with respect to legal capacity and to being accredited.[5] The Equity Financing clause, moreover, states an obligation of the SAFE Investor, that

---

[3] For digital signatures to have legal force, some means of binding the signature keys to legal identities is required, e.g., a legally recognized public key infrastructure.

[4] When legal activation is achieved by both parties creating (and somehow logging on chain) certain signed messages that represent their signature on the legal contract, each party will not wish to be legally committed until the other has also signed. Van der Meyden [3] discusses this issue, and develops formal characterizations of the offer and acceptance process and a process of "signature in counterparts."

[5] An alternative way for these to be established could be by including in the activation conditions of the smart contract that digital certificates attesting to these facts be provided to the smart contract.

they should execute (sign) the Equity Financing documents. It is less clear how the investor could be bound to this obligation if they have not explicitly signed the SAFE.

In most jurisdictions, it is not necessary for a contract to be signed, or even written, for a court to determine that the contract is binding. A signed, written contract simply eases the burden of establishing mutual assent, as well as delineating the details of the agreement. It is open to a court to infer that a contract exists based on the SAFE investor's interaction with the smart contract. However, for most investors, it is likely that a court will consider a written contract to be better evidence of mutual understanding, given the lack of familiarity of most investors with Solidity and its execution mechanism.

17.4   Towards a Smart Legal Contract

In the preceding sections, we have identified a number of reasons that a written legal contract should be used by the parties even when a smart contract has been used to implement a SAFE. These reasons include ensuring that the smart contract's representation of the cap table will be valid in law, ensuring that the smart contract correctly reflects the intended legal state of affairs, leaving interpretation of open-textured concepts to human processes, covering parts of the contract that are not implementable in the smart contract, and providing a fallback ensuring availability of remediation in cases of errors. We now consider the form and necessary content of such a contract. (Since some of the key provisions are likely to be jurisdiction-dependent, we do not attempt to develop detailed contract text, but focus on what needs to be expressed.)

To cover additional risks that are associated with the use of the smart contracts, ensure that these will be interpreted in law as they are intended, and enable the parties to rely upon the smart contract concerning the state of the company's cap table, we suggest that clauses with the following implications could be included in a contract between the SAFE investor and the Company.

- The Company warrants that the `Company` contract at (address), controlled by the `Safe_controller` at (address), will be operated in such ways as to ensure that it is maintained in a timely manner as a complete and correct record of the state of the Company's cap table and convertible instruments issued.
- The Company warrants that it will take all actions necessary (e.g., legal filings and notifications) to ensure that the record in the `Company` and `Safe_controller` smart contracts will be considered in law by (the applicable jurisdiction) as the official record of the Company's cap table.
- The Company will conduct its next equity round, liquidity event or dissolution event on-chain, by submitting a `start_equity_round`, `notify_liquidity_event` or `dissolution` transaction. (See Sects. 15.2.4, 15.2.5 and 15.2.6.)

- The Company will submit a `start_equity_round`, `notify_liquidity_ event` or `dissolution` transaction only when the smart contract is in a state completely and correctly representing the state of its cap table and convertible instruments issued.
- The Company warrants that in any submitted `start_equity_round`, `notify_liquidity_event` or `dissolution` transaction, the values of the parameters correctly and completely represent a valid Equity Financing, liquidity event or dissolution event, respectively, as defined in this contract. Values of parameters that are submitted at the Company's discretion (including pre-money valuation and price in an equity round, and price and cash available for distribution in a liquidity event) will be selected in good faith.
- The Investor warrants that they will take all actions necessary (e.g., monitor the blockchain or appoint a watchtower service) to ensure that they will be alerted in a timely manner of any states arising in the operation of the smart contracts that require submission of a transaction or transactions by the Investor. The Investor will react to such alerts by submitting a transaction, as required, by the deadline given.[6]
- (In a variant of the SAFE smart contract in which the investor has the opportunity to consent to the terms of the equity round, with the hash of the Equity Financing documents included in the parameters) the SAFE investor's submission of a `safe_investor_consents` transaction will imply that the investor assents to the terms of the Equity Financing documents. (See Sect. 19.1.)
- Disputes relating to the operation of the smart contract will be adjudicated in the first instance by an independent arbitrator appointed by (Arbitration Authority) according to (appointment process description).

We note that these provisions allow for some divergence between the on-chain record and the legal state, e.g., in case of bugs in the smart contract, malicious attacks, or legal rulings, but imply that the Company will correct the record in such events.

A number of options are available for the form of the legal contract, and the way that it relates to the smart contract code. Not all parts of the original SAFE will be represented in the smart contract—these parts should be retained as natural language text. For parts of the original SAFE that are represented in the smart contract, the question arises of whether to retain these parts in the legal contract, or to replace them either by code or references to code.

Grigg [6] has argued that there are a number of problems with the idea of a direct inclusion of code in contracts. First, code may not pass a "readability test" for contracts "in writing", since code is "opaque" even for experienced programmers, in the sense that the correctness of code in all circumstances can be difficult to ensure. Next, code will normally be accompanied by explanatory documentation in natural language, which would be more transparent and hence better suited for interpretation by courts than the code, for which they will require expert witnesses. Finally, where

---

[6] An alternative to this clause would state that the company has an obligation to inform the investor of the actions that need to be taken, and specify the acceptable means of communication.

both code and natural language text exists expressing the same content, one must dominate. The statement of dominance can only be expressed in natural language, so Occam's Razor suggests that natural language should dominate, although the natural language may explicitly defer to code.

We largely concur with these points, though we would note that the process of writing code and formally verifying its properties (the latter rarely undertaken in general, but increasingly being pursued for smart contracts) is likely to reveal ambiguities and unforeseen implications of the natural language text.[7] The improvements to code resulting from this process can be reflected back to yield improvements to the natural language text.

We would also argue that use of mathematical style, including equations, would clarify the natural language text of financial contracts. Being able to automatically extract a formal model of the contract from the contract itself (rather than having to manually construct one) would enable the parties to conduct formal analysis and simulation of the performance of the contract in order to decide whether to accept a contract offer. Controlled natural language represents a way to combine the benefits of natural language and automation, providing the possibility of a representation of the contract that is intelligible to humans and machines.

However, the current best compromise is a plain natural language contract with references to code. In the case of the SAFE, the legal contract that results from the above considerations then looks largely like the original SAFE contract plus the above provisions relating to operation of the smart contracts.

One might question, therefore, what benefit is obtained from the smart contract. But that a legal contract may be used to overrule a smart contract in some cases does not vitiate the benefits resulting from the smart contract, including improved efficiency, security and consensus between the company and its investors. In our implementation, in particular, we can point to the extra security provided by atomic transactions, and the ability to have a predictable, independent referee (the `Safe_controller`) to ensure that the company abides by many of its obligations in the SAFE contract.

However, the addition of the many technical clauses needed to tie the state of the on-chain cap table to the legal state of the cap table is detrimental to the design goals of the SAFE (brevity, simplicity, layperson-friendly). To address this, standard legal boilerplate could be developed to express these requirements, and as DLT becomes more common we can expect laypeople to become more familiar with the technology. In enterprise settings, for larger contracts, with different design goals, the addition of extra boilerplate might make little difference. It has long been commonplace for commercial arrangements between enterprises, automated through computer messaging systems, to include "Electronic Data Interchange Agreements" that describe the legal consequences arising from the use of the technology [7]. Nevertheless, for the SAFE, these additional clauses make the contract harder to understand.

---

[7] The flaws we found in the SAFEs during our analyses are an example of this. These points are summarized in Sect. 10.1.

Ricardian contracts [8] have been proposed as an integration of computable elements and natural language text that can be be accepted as contracts by the legal system. In their original form, these contracts applied to simple money-like instruments (e.g., deposits in a bank), and were structured as a digitally signed set of key-value records bundled with text for the instrument, so that a human recipient would receive a statement of its terms and conditions, while at the same time supporting automated processing. Richer forms have since been developed [9].

As a SAFE is not transferable, bundling of the SAFE smart contract and the legal text may not be necessary, but can nevertheless be achieved by including in the smart contract a (write-once) field for the signed (hash of the) legal text, smart contract source code and compiler information (so that on-chain bytecode can be verified to be a compilation of the source).

17.5 Adjudication

Once it is accepted that a smart contract will be accompanied by a legal contract, which may be used to overrule the smart contract, the issue arises of how disputes relating to the legal contract will be handled. The usual legal processes will of course apply, but these come at a high cost (in time as well as money), which counteracts the efficiency benefits that smart contracts are intended to provide.

If the parties are unable to negotiate a resolution on their own, a cheaper and more efficient process may be to make use of private courts and arbitration services, whose rulings are increasingly accepted by legal systems and their enforcement processes in a number of domains [10]. There have been proposals to develop an arbitration ecosystem specific to blockchain and smart contract environments, which would have the benefit that the arbitrators would have greater expertise in the technology than can be expected from existing court systems.

CodeLegit [11] has proposed an approach to arbitration of disputes relating to smart contracts using *Blockchain Arbitration Library* code that is combined with the smart contract. The parties to the smart contract are assumed to have signed a legal agreement containing the text

> Any dispute, controversy or claim arising out of or relating to this contract, or the breach, termination or invalidity thereof, shall be settled by arbitration in accordance with the Blockchain Arbitration Rules.

A party that wishes to dispute an event in the smart contract may make a function call that pauses the smart contract and starts the arbitration process. The process involves an *Appointing Authority*, who appoints an arbitrator (possibly with input from the parties concerning a list of candidates), and is responsible for reactivating, terminating or modifying the smart contract depending on the settlement or ruling of the arbitrator resulting from the arbitration process.

Kleros Courts [12] is a proposed crypto-economic protocol for adjudication. Arbitrators are required to stake cryptocurrency in order to participate in this ecosystem, and are incentivized to make "correct" decisions as a result of the protocol allowing decisions to be contested and referred to a larger group of arbitrators, with the original arbitrator penalized for failing to make the majority decision.

Both the technology and the ecosystem of blockchain arbitrators are nascent, and it will take time for them to reach an acceptable degree of trustworthiness (e.g., for reliability, correctness of decisions, and freedom from corruption). Their legal status, critical for enforceability of their decisions, also remains to be determined. As SAFE contracts ultimately derive their enforceability from a national or state legal jurisdiction, acceptability of the arbitration process to that jurisdiction is likely to be critical.

17.6 Enabling Response to Legal Orders and Variations

The legal system supports trust in multi-party interactions through enforcement of contracts and remediation of their breach. We now consider how the possibility of legal orders impacts the implementation of smart contracts, in general, and SAFE smart contracts, in particular. The implementations discussed above do not directly support variations required by legal orders, so accommodating the discussion of the present section in the code is left to future work.

Enforcement orders sometimes place no special demands on a smart contract. Where the history of an interaction is consistent with a legal contract, except that some party has not honored an obligation under that contract,[8] an enforcement ruling may assert the legal validity of the contract (if this is disputed by the offending party) and order that the obligation be complied with. When the agent of the obligated action is human, and the action occurs either in the real world or can be performed by that agent submitting a number of transactions to the blockchain, there are no special demands on the smart contract. Compliance with the order to perform the action suffices to transform the smart contract state so that it conforms to the legally required state.

Examples where this circumstance may arise in the case of smart contracts for SAFEs are the Company's obligation to perform an equity round on-chain by submitting an appropriate start_equity_round transaction (see Sect. 15.2.4), and the company's implicit obligation to notify the SAFE investor and provide equity round paperwork and the SAFE investor's obligation to sign this paperwork (see Sect. 19.1).

Similarly, obligations enforced through the smart contract are largely unproblematic. On event-driven platforms such as Ethereum, it is typically necessary for some

---

[8] The simplest case of this is obligations that are not subject to a deadline. It may also hold where a deadline for performance has passed, but late performance of the action would restore the smart contract to a state consistent with the remainder of the contract. Where non-performance by the deadline triggers contract conditions such as penalties that would remain in place even if the obligated action is performed late, the legal analysis of the situation becomes more complicated, and enforcement rulings may involve contract modifications, discussed later in this section.

party to submit a transaction to trigger an on-chain performance of the obligation. In effect, this transforms the obligation for an action to be performed into an obligation to submit a transaction. Commonly this is achieved by enabling the required transaction to be submitted by any party who might be disadvantaged by failure to perform the obligation, giving them the powers to ensure performance. (It is necessary that some party who has the power to submit the transaction is aware of the contract state and the obligation or opportunity to submit the transaction.)

An example of this is the obligation on the company to issue a certain number of shares to the SAFE investor in the event of an equity round. For the SAFE share issuance obligation, a call to the `finalize_equity_round` function guarantees conversion of the SAFE contract to shares, once consent of all required investors has been obtained and payment received. Any party may make this call. In particular, the SAFE investor has the power to enforce the obligation to issue shares once the equity round is ready to complete.

In automated conversion implementations, the *number* of shares issued is as prescribed by the SAFE contract. The sense in which the obligation is being enforced is more subtle in implementations in which the SAFE investor is required to consent to the equity round terms. Here, the SAFE investor may consent to a number of shares to be issued which differs from that described in the legal contract. In effect, the parties to the contract first document, on-chain, their agreement to a modification of the obligation, and it is the modified obligation that is enforced on-chain. (It would be beneficial for this process in the operation of the contract, and its legal interpretation, to be made explicit in the legal contract governing use of the smart contract.)

In general, however, rulings arising from contract disputes may either invalidate the contract, or require modifications to the contract. Rulings that the contract has been breached, and that the interests of one of the parties have been harmed, typically also involve some form of *remedy* or *compensation*.

There are competing views of how the appropriate compensation should be determined. One view is that the compensation should restore the injured party to a state before the breach occurred. Economists have taken the contrary view that compensation should be determined so as to ensure that the rational behavior of agents, given the compensation to be expected for breaches, will satisfy some criterion, such as "efficient breach," in which the contract is honored unless all parties (e.g., two parties with a sales contract and a third party that offers a higher price for the goods) benefit in the event of a breach by some party. A survey of economic analyses of compensation [13] has concluded that this perspective does not lead to a unique "correct" theory of compensation, since there are many criteria that such a theory might seek to optimize, leading to divergent conclusions about the appropriate compensation in a given case. In either case, analysis of concrete examples may involve considerations of *causality*, itself a subject of considerable philosophical dispute [14].

In computer science, there is a related area of work motivated by the concurrent processing of long running database transactions, where one transaction may cause another to fail. One response to this situation would be to "roll back" the interfering transaction, restoring the database to a prior state. However, this is problematic when some of the events have had effects in the real world, or in other systems that are not under the control of the database. It has been proposed to address this issue using "compensating transactions" to undo the undesired effects. Here also, there are many competing theories for how such compensating transactions should be determined [15], depending on what one chooses to optimize.

In the light of the unsettled nature of these theoretical approaches to compensation, and their inherent dependence on subjective choices about what should be optimized, automation of smart contract remediation and compensation grounded in a specific theoretical account of compensation would not appear to provide sufficient coverage of the requirement to be able to deal with court rulings. Adjudicators and courts will apply their own choice of approach, and may well make decisions that lack strong theoretical grounds. What is more critical is to enable compliance with the full range of potential legal revision and remediation orders.

The way in which legal revisions are dealt with at the level of smart contracts may depend on the underlying blockchain and its governance processes. Permissioned blockchains are the least problematic, since the validators are known entities and, provided that they accept orders from the contract's jurisdiction, can be ordered to make modifications to the blockchain. Matters are more complex on open blockchains, where the blockchain is immutable (except for very infrequent hard forks), validators are not necessarily known, and are internationally dispersed. Even on permissioned chains, the governance policy may prefer to minimize the extent to which validators are called upon to modify the blockchain and prefer that smart contracts be coded so as to accommodate legal modifications.

A classification of legal revisions has been discussed by Marino and Juels [16], who consider the capacity of smart contracts in Ethereum's Solidity language to be designed to accommodate various legal revisions. They identify two types of legal changes to a contract: *rescission* (rescinding the contract), and *modification*. Rescission annuls the contract, as if it never existed, whereas modification leaves the contract in place, but forces changes to its terms. Depending on the contract state, both types of orders may require compensation to reverse the effects of actions already performed with respect to the contract. Each type of change is caused by one of three types of powers:

- "by right", where one of the parties to the contract has rights in the contract to cause the change,
- "by agreement", where the parties agree to make the change to the contract, or
- "by Court", where an appeal to make the change is made to the court by one of the parties, and the court rules in favor of the request.

Thus, we have six notions, from "Rescission by Right" (also known as "Termination by Right") to "Modification by Court" (aka "Reformation"). The view in [16] is that the contract is embodied entirely in the smart contract, so contract modifications need to be applied directly to the smart contract. Features of Ethereum's Solidity language that enable realization of contract changes are identified, specifically:

- contract self-destruct as a means of termination,
- inclusion of a contract parameter that represents its termination state, and use of this parameter in function pre-conditions,
- updates of smart contract parameters, and
- variables of type function, which enable functions implementing a contract provision to be updated by assignment.

In general, these features need to be used very carefully since they potentially undermine the security guarantees that a smart contract was intended to provide. Some code patterns for each of the legal revision types is given in [16], but the arguments made for their correctness are sketchy and informal—we feel this direction deserves further study.

We now consider some of the specific issues raised by rescission and modification of SAFE contracts with respect to smart contract representations of the SAFE and the context of the state of the company as represented on-chain. One general observation relevant to all forms of change is that Miscellaneous provision (a) states that the contract may be amended only with approval of both the SAFE investor and the Company. This indicates that there is not a unilateral right of either party to Rescind or Modify, and all changes must be "by Agreement" or "by Court". We do not attempt here to resolve the issue just raised of how to manage access control so as to ensure that security properties of the smart contract are not undermined by changes to accommodate legal rulings.

When the SAFE contract has not yet converted to shares, its rescission can be implemented by simply removing the `Safe` smart contract from the list of issued SAFEs. In addition, the money paid by the SAFE investor must be returned (perhaps with interest). In case of modification when the SAFE has not yet converted, a simple substitution of a new `Safe` smart contract suffices.

Matters are more complex when the contract is to be changed after the SAFE has already converted, and has had effects on the cap table of the company. Consider the following scenario:

1. An elderly unaccredited investor A purchases a SAFE from startup company C for $200,000 at cap $1M. Company C failed to take any actions to verify that A was an accredited investor.
2. Company C performs well in its startup phase, and attracts an equity investor who contributes $1M at a pre-money valuation of $3M. In the equity round, investor A is issued shares valued at $600,000 at time of the round.
3. A large well-established and well-funded competitor enters the market for company C's product and rapidly steals its market share. The valuation of company C drops to $400,000.
4. Investor A dies and child B inherits the shares. B challenges the validity of the SAFE contract on the grounds that A was senile and unaccredited. The court agrees and rescinds the contract.

In this situation, more needs to happen than the mere deactivation of the `Safe` smart contract. Indeed, the contract has already terminated, and its effects are already reflected in the state of the blockchain. It seems likely that the court would order a monetary compensation to B, and that the shares issued to A be revoked.[9] The transaction could also be structured as a repurchase of shares by the company. Making a monetary payment is likely to be within the powers of the company. However, in general, investors would not want the company to have the power to arbitrarily revoke or repurchase a single investor's shares. This issue could be handled by requiring such changes to be justified by a signed order from the court. A Public Key Infrastructure and certificate reasoning process would be necessary if this is to be automated.

We remark that, in the specific situation of the present paper, a simpler expedient is available to resolve difficulties with amendment of smart contracts if the degree of flexibility required has not been programmed in, or the court ordered changes are blocked by the access control programmed in the smart contract. We have the advantage that the on-chain representations of the cap table are mere data, which are invested with value through an event in the legal domain: the company's declaration that the blockchain representation shall be its true record of the cap table. To effect a change, it suffices for the company to revoke this declaration and declare that henceforth, a smart contract newly created with the amended state serves as the company's cap table. Any unterminated SAFE contracts can similarly be represented as newly created smart contracts when setting up this new representation.[10] (In our implementation, transfer of any monetary value held in the `Company` smart contract is not constrained by the the `Safe_controller`, so is independent of the cap table representation. However, if transfers of money required approval of investors or SAFE holders, the interaction of such controls and legally required modifications to the cap table would require careful consideration.)

Obviously, the possibility of such a change weakens the sense in which the smart contract enforces security properties. In particular, it leaves the investors with only legal guarantees about the integrity of the cap table representation, rather than a strong technical guarantee. However, if there is any dispute about the correctness of the new cap table, with respect to what the Court has ordered, then that can be taken up in the legal domain. It is not within the power of the company to simply erase the old representation, so there is strong evidence to support any claims of fraud or error on the part of the company. At the cost of increased complexity and a change to the terms of the SAFEs, a governance process involving a vote of shareholders and SAFE holders could be implemented as part of the smart contract to protect against such radical powers.

---

[9] The former is to the detriment of other investors, the latter in their favor, since they consequently increase their percentage shareholding, but the net effect is unclear. It could be to the detriment of the other investors, for which they in turn may have grounds to claim breach of contract with the company, but such an action would likely be treated as a separate matter.

[10] This approach to dealing with legal rulings is similar to one that has been used to deal with security risks to smart contracts arising from adversarial attacks exploiting programming errors. It is common practice for the operators of a smart contract to control a "kill switch" that deactivates the smart contract and allows them to transfer the assets in the contract to a newly created contract. Here also, the other users of the smart contract are required to trust that the operator will not abuse this power.

17.7   Conclusion

In this chapter, we have discussed a number of the issues that arise when considering the legal status of smart contracts representing SAFE contracts: the very applicability of law, the legal status of the smart contracts, and their legal activation. Following our conclusion that, to have legal standing and to provide comparable risk coverage, the smart contracts need to be accompanied with a legal contract that describes requirements and interpretations of its operation, we have enumerated some of the key issues that the legal component of a smart legal contract for SAFEs should cover. We then discussed the question of how such a hybrid might be adjudicated, and how the need to respond to legal orders concerning the smart contract might be addressed.

We have not attempted to cover legal questions broader than the specific questions that arise for SAFE contracts, of which there are many, and a growing literature [17]. Even more so than technical issues in development of smart contract platforms, the understanding of such legal questions remains in a very unsettled state, with a large diversity of states of maturity of the answers, legislative responses and regulatory attitude in different jurisdictions.

References

1. Popper N (2016) Hacker may have taken $50 million from cybercurrency project. The New York Times, 17 Jun 2016. See also http://en.wikipedia.org/wiki/The_DAO_(organization). Accessed Dec 2024
2. Poncibò C, DiMatteo LA (2019) Smart contracts: contractual and noncontractual remedies. In: DiMatteo et al. [63], pp 118–140
3. van der Meyden R (2022) A formal treatment of contract signature. IEEE Trans Ser Comput 16
4. Lessig L (2000) Code is law, on liberty in cyberspace. Harvard magazine. https://www.harvardmagazine.com/2000/01/code-is-law-html. Accessed Dec 2024
5. US Securities and Exchange Commission (2017) SEC issues investigative report concluding DAO tokens, a digital asset, were securities. https://www.sec.gov/news/press-release/2017-131. Accessed Dec 2024
6. Grigg I (2020) Response to the UKJT's Public Consultation on crypotoassets, DLT and Smart contracts under English private law. Online, https://iang.org/papers/UKJT-response_on_SmartContracts-IanGrigg.pdf. Accessed Dec 2024
7. Baum MS, Perritt HH (1991) Electronic contracting. Publishing and EDI Law, Wiley Law Publications, New York
8. Grigg I (2004) The Ricardian contract. In: Proceedings of the first IEEE international workshop on electronic contracting. IEEE, pp 25–31
9. Clack CD (2018) Smart contract templates: legal semantics and code validation. J Digital Bank
10. Cannon AJ (2004) A pluralism of private courts. Civil Justice Quart 23:309–323
11. Codelegit (2024) https://datarella.com/codelegit-legal-libraries-for-smart-contracts/. Accessed Dec 2024
12. Kleros (2020) Dispute revolution: the Kleros handbook of decentralized justice. Kleros, augmented edition
13. Craswell R (2003) Instrumental theories of compensation: a survey. San Diego Law Rev 40(4). https://digital.sandiego.edu/sdlr/vol40/iss4/5. Accessed Dec 2024

14. Moore M (2003) For what must we pay? Causation and counterfactual baselines. San Diego Law Rev 40(4). https://digital.sandiego.edu/sdlr/vol40/iss4/6. Accessed Dec 2024
15. Colombo C, Pace GJ (2013) Recovery within long-running transactions. ACM Comput Surv 45(3):28:1–28:3
16. Marino B, Juels A (2016) Setting standards for altering and undoing smart contracts. In: Rule technologies. research, tools, and applications - 10th international symposium RuleML, proceedings, Springer LNCS, vol 9178, pp 151–166
17. Allen JG, Hunn P (eds) (2022) Smart legal contracts: computable law in theory and practice. Oxford University Press, Oxford

Chapter 18
Privacy and Platform Issues

Smart contracts can be implemented on a variety of platforms that differ in their approaches to privacy. In this chapter, we briefly consider the appropriateness of a number of platforms for deployment SAFE smart contracts. We begin with a discussion of policy appropriate to SAFE contracts and company cap tables, and then consider public and private blockchains as alternative platforms on which to deploy the smart contracts we have developed.

18.1 Policy

The degree of privacy of company cap tables is jurisdiction-dependent. Table 18.1 gives an example of a privacy policy for the cap table data held by a private company. (Many more or less fine-grained variants could be applied in practice.) The policy assumes that the cap table itself stores data concerning ownership of shares and SAFEs using investor pseudonyms, which can be linked to investor personal information (name and address) in a separate, private table. The policy considers the company constitution (which defines classes of shares and delineates their rights) as potentially private.

The company itself is permitted to access all information. For example, it needs shareholder identity for communications with shareholders, as well as regulatory compliance. Regulators may need detailed shareholder information (for some, or for all shareholders) in order to support policing of rules concerning, for example, foreign ownership and money laundering.[1] As the company is private, the general

---

[1] In Australia, proprietary companies are required to keep the Australian Securities and Investment Commission informed about their top 20 shareholders [1]. The shareholder register is only semi-private, since the public is entitled to inspect it, although the company may charge an inspection fee to non-shareholders.

© The Author(s), under exclusive license to Springer Nature Singapore Pte Ltd. 2025   253
R. van der Meyden and M. J. Maher, *Simple Agreements for Future Equity (SAFE)*,
Blockchain Technologies, https://doi.org/10.1007/978-981-96-3920-5_18

Table 18.1 Example of a privacy policy for cap table data (private company)

Visible to:	Company	Investor	Regulator	Public
Constitution (inc. share classes)	✓	✓	✓	×
Investor Name	✓	*	✓	×
Investor Address/email	✓	*	✓	×
Investor Pseudonym	✓	✓	✓	×
Shareholding by Pseudonym	✓	✓	✓	×
SAFEs Issued by Pseudonym	✓	✓	✓	×

public is not necessarily entitled to any information about the company. Individual investors (i.e., shareholders, SAFE investors, and potential new investors from whom funding is being sought) have an interest in understanding the broad structure of the cap table, but may not need the full details. For example, for purposes of determining a price in an equity round, they need only the total number of shares and SAFEs, and their detailed types. For this purpose, the cap table could also be released to them with shareholder identities masked using pseudonyms.

Even if only pseudonyms for investors are released, this could still allow anonymity of some shareholder data to be breached. For example, the founders might be easily identified as the largest shareholders. However, founder shareholding is information that the company might release in any event since it addresses investor questions concerning founder incentivization. There may also be circumstances in which it is necessary to release information to investors about other investors, such as when they need to negotiate with each other concerning the details of an equity round. The table acknowledges this by using '*' to indicate permission to access data in some circumstances.

In principle, the ability to link founders to their share numbers might be prevented through sybils: splitting a large shareholding into smaller pieces, each with a different pseudonym. On the other hand, this would impede the ability to determine information about concentration of control, which might be desirable (for example, to determine whether a liquidity event has occurred). Conceivably, therefore, the company might be required to release cap table data with each beneficial shareholder identified by a unique pseudonym.

18.2   Public Blockchains

Public blockchains such as Bitcoin and Ethereum, are open for use by any person, who may operate validator nodes or simply participate by contributing transactions. Users are represented using pseudonymous identities (public keys). Each user may have many pseudonyms. All data recorded on the blockchain is publicly visible—this includes bytecode of smart contracts.

Table 18.2 Visibility of cap table data (Ethereum implementation)

Visible to:	Company	Investor	Regulator	Public
Constitution (inc. share classes)	✓	✓	✓	✓
Investor Name	✓	*	✓	×
Investor Address/email	✓	*	✓	×
Investor Pseudonym	✓	✓	✓	✓
Shareholding by Pseudonym	✓	✓	✓	✓
SAFEs Issued by Pseudonym	✓	✓	✓	✓

The Solidity smart contracts we have developed similarly represent investors using pseudonyms: an Ethereum address (public key). This allows the cap table data to be split as a relation (Name, Address, Pseudonym), stored by the company off-chain, and a relation (Pseudonym, SAFE issued, Shares issued), stored on-chain in the `Company` smart contract. This would yield the visibility table for the implementation given in Table 18.2, in which '*' represents that the company may release this information to the investor, if required, by sharing the association between Ethereum addresses and shareholder and SAFE holder identity information. (A salted hash of personal information could be added to the smart contract to enable the recipient to verify the integrity of this information while keeping it private to other parties.)

This is close to satisfying an instance of the policy in Table 18.1. However, more information is visible to the public than is permitted by the policy, since the public is able to observe pseudonymous information about the company's investors. The company constitution itself, as well as the details of SAFE contracts, would also be visible to the extent that these are represented in smart contracts on the public blockchain. In some circumstances, this may be an acceptable implementation.

When there is a legal contract for a SAFE, and a digitally signed copy of this is placed in a field of the `Safe` smart contract, then the investor identity would be leaked, since it is recorded in the contract. Storing only the digital signature in the smart contract, and including random salt in the text, would avoid this, but this would also prevent the smart contract from verifying that the correct text has been signed. In place of the parties names, their pseudonyms could be used in the legal contract text, but this would require that legal contracts in this form to be considered valid in law. Certified data linking the pseudonyms with real identities would also need to be accessible to the legal system, but not to the general public, to make sense of this idea.

We remark that in the event there is a need for arbitration or litigation concerning any matter, the dispute is most likely to be between an investor and the company, since the investors do not have any direct legal relationships. The company will have identifying information of investors for compliance reasons, and the investor will also be able to identify the company and its key officers, assuming the investor has undertaken appropriate due diligence.[2] This implementation is therefore able to avoid the difficulty of applying legal enforcement against an anonymous party.

The cost of using a public blockchain as above is that more information is released to the public than may be desirable. In order to obtain an implementation in a public blockchain setting that more precisely implements the policy in Table 18.1, it may be possible to use zero-knowledge proof techniques [2, 3]. We leave an exploration of how this might be done while preserving the required integrity guarantees for the cap table and execution of the SAFE smart contracts, for elsewhere.

18.3   Permissioned Blockchains

Permissioned blockchains[3] differ from public chains in that they restrict the right to be a validator to a known set of parties. In this case, it is more efficient for the validators to use a classical byzantine consensus protocol, rather than a protocol based on "proof of work" or "proof of stake," as with open chains. Additionally, access control may also be applied to (non-validator) parties who submit transactions: on-boarding of such parties can be restricted, and they can be required to provide identifying information as part of the on-boarding process. Their interactions with the blockchain data can moreover be restricted to a particular view depending on their identity or role in the system. (Validators typically still observe the complete state of the blockchain, and are responsible for restricting the information sent to non-validator nodes to be limited to an appropriate view.)

In this setting, it is straightforward to implement the access control policy of Table 18.1. Heavyweight zero-knowledge techniques are not required. However, it is required that the participants trust the validators with private identity information and detailed knowledge of the cap table of the private company. This suggests that the most appropriate validators may be securities regulators and other related agencies such as registries or exchanges, who are already required to obtain this private data, rather than commercial entities which may be competitors of companies listed on-chain.

Collecting user identity during on-boarding also better supports the need for identity information when dealing with legal contracts and processes like adjudication and litigation. If the validators are known, and all subject to the same jurisdiction as applies to a legal contract between parties to a smart contract, the validators can moreover be forced to make changes to on-chain data as ordered by a legal authority.

---

[2] Thus, even though investors may be anonymous to each other in this implementation, the expected setting differs from that of many Initial Coin Offerings on open chains during the speculative ICO boom of 2016, in which all parties, including the company were anonymous to each other. This model, not surprisingly, was associated with high levels of fraud.

[3] There is a lack of consensus in the area on terminology. Related terms sometimes used synonymously are "private blockchain" and "consortium blockchain." For some, however, "private blockchain" implies ownership and control by a single entity, whereas "consortium blockchain" is used when the blockchain is operated by multiple, permissioned entities.

When validators are subject to multiple jurisdictions, this may be harder, but potentially, cross-jurisdictional treaties may apply to support this. Treaties dealing with securities law (e.g., the "Hague Convention on the law applicable to certain rights in respect of securities held with an intermediary" [4]) appear to have only limited adoption at present, however, so we expect that restricting validators to respecting a single jurisdiction may be necessary to ensure that international validators will comply with legal orders.[4]

A first explorations of the implementation of our approach to SAFE smart contracts on a permissioned blockchain has been conducted by Coulter [5]. This work uses Hyperledger Besu, an Ethereum client that supports development of enterprise blockchains with permissioning and privacy features. This enables implementation of privacy policies like the examples from the present chapter. In addition to these features, this project developed a graphical user interface for interacting with the smart contract implementation of SAFE contracts.

18.4   Conclusion

Further research in this direction is warranted, but in the near term, a permissioned blockchain is likely to be the most suitable platform on which to deploy our hybrid legal and smart contract approach to SAFE contracts, although they achieve this at the cost of decentralization and "trustlessness". However, it would also be worthwhile to explore implementations using zero-knowledge techniques on public blockchains to determine whether this trade-off is required.

References

1. Australian Securities & Investments Commission (2014) Information sheet 47: Company shareholders. https://asic.gov.au/for-business/running-a-company/company-shareholders/. Accessed Nov 24
2. Goldwasser S, Micali S, Rackoff C (1985) The knowledge complexity of interactive proof-systems (extended abstract). In: Sedgewick R (ed) Proceedings of the 17th annual ACM symposium on theory of computing, May 6–8, 1985, Providence, Rhode Island, USA. ACM, pp 291–304
3. Goldreich O (2001) Foundations of Cryptography, volume I: Basic Tools. Cambridge University Press
4. Anon (2006) Hague convention on the law applicable to certain rights in respect of securities held with an intermediary. https://www.hcch.net/en/instruments/conventions/full-text/?cid=72. Accessed Dec 2024
5. Coulter W (2021) Building smart SAFEs. Honours thesis, UNSW School of Computer Science and Engineering

---

[4] The international membership of the Libra Foundation accepting to be bound by Swiss law is an example of such an arrangement.

Chapter 19
Issues in Implementing Other SAFE Clauses

The primary focus of the previous chapters of Part IV of this book was to develop a smart contract implementation of the key Equity Financing, Liquidity, and Dissolution events of SAFE contracts. In this chapter, our aim is to determine the extent to which it would make sense to extend the implementation to cover other parts of the SAFE, and other documents that play a role in its performance. Where appropriate, we discuss the structure of an implementation, but we leave the development of code for any of these components to future work. Unless otherwise indicated, our discussion in this section is based on the Pre-Money SAFE with Cap Only, but we note some significant variances in the Post-Money versions of the SAFE where appropriate. Section 19.5 deals with Most Favored Nation SAFEs, which introduce some additional issues.

19.1 Execution of Equity Financing Documents

Section 1a(i) of the SAFE contract (see Sect. 2.1), requires that the SAFE investor "will execute and deliver to the Company all transaction documents related to the Equity Financing." (We have already noted in Sect. 13.3 that this potentially gives the SAFE investor some power to disrupt the equity round by withholding their signature, consequently leading to the SAFE investor obtaining an ability to negotiate on the construction of the round.) The documents related to the Equity Financing may be quite complex and may deal with matters such as reincorporation, revisions to be made to the company's governance structure on completion of the round, shareholder rights after the round, and indemnities and procedures relating to conduct of the round. Model documents for an equity round have been proposed by the (USA) National Venture Capital Association [1].

© The Author(s), under exclusive license to Springer Nature Singapore Pte Ltd. 2025   259
R. van der Meyden and M. J. Maher, *Simple Agreements for Future Equity (SAFE)*,
Blockchain Technologies, https://doi.org/10.1007/978-981-96-3920-5_19

Note that this clause states an *obligation* on the SAFE investor, to sign and return the Equity Financing documents. Implicitly, there is a related obligation on the company, to notify the SAFE investor of the Equity Financing and to provide the documents requiring signature. These actions may be essential to the legal status of the round and the obligations exist irrespective of whether the SAFE contract is interpreted as granting negotiation rights to the SAFE investor in the structure of the Equity Financing. In the case of the company's obligation to issue shares to the SAFE investor, it was feasible to enforce the obligation through a smart contract. By contrast, neither the obligation on the SAFE investor to sign the Equity Financing documents nor that on the company to notify the investor and provide these documents, can be enforced by a smart contract.

In the case of the investor, this is because the contents of the documents to be signed are open ended and not predictable at the time of the issuance of the SAFE contract, so it should be left to the investor to make the decision to sign. The investor may have sound reasons not to sign, for example, if there are reasons to believe that the round is not *bona fide*, but has been constructed to deprive the SAFE investors of rights (see Sect. 16.3).

Even if the signature is digital, there are technical difficulties in having a smart contract control the investor's signature key and sign on the investor's behalf. Although there exist *proxy signature* schemes whereby a user is able to delegate the ability to sign messages to another [2], implementing these on blockchain platforms where the contents of the smart contract are accessible to potentially untrustworthy validators (on permissioned blockchains), or even public (on open blockchains), would leak the delegate's signature key and enable parties other than the delegate to create a signature on behalf of the SAFE investor.[1]

In the case of the company, its obligation to notify the SAFE holder and provide the Equity Financing documents cannot be enforced by a smart contract since the contents of these documents will be negotiated with the new investors, so cannot be predicted at the time of creation of the smart contract for the SAFE. The SAFE holders would obtain notification of the equity round if they (or a delegate[2]) are constantly observing the blockchain, and the company calls the start_equity_round function, which posts the parameters of the equity round on-chain. However, this still does not mean that the company's implicit obligation to notify the SAFE holder of the Equity Financing is enforced, since the company may potentially conduct the equity round off-chain without notifying the SAFE holder. A legal agreement between the company and the SAFE holder is therefore required to ensure that the company does

---

[1] It may be possible to avoid this problem by means of a legal interpretation of such delegate signatures as being valid only if they can be proved to have been created by execution of the smart contract, together with a definition of the signature verification function that checks this property against the blockchain. However, in this case it is not clear that use of a sophisticated proxy signature is actually required, since simpler properties of the behavior of the smart contract may serve equally well. We leave exploration of this issue for future research.

[2] Service providers who maintain watch on the blockchain and inform users of particular on-chain events are called "watchtowers" in the blockchain community.

not conduct the equity round off-chain. This motivates some of the legal contract terms discussed in Sect. 17.4.

One approach for the company to provide the Equity Financing documents to the SAFE holder would be to include them in an additional parameter of the `start_equity_round` function. (Note that this does not imply that the obligation on the company to provide the documents is being enforced, since the company could still include an empty document in this field.) In an implementation in which approval of the equity round proposal by the `Safe` contract requires that the holder has consented to the round, the need for the holder's signature could then be eliminated if there is a legal agreement concerning operation of the smart contract that includes a clause stating that the SAFE holder granting their consent to the equity round will be interpreted as the holder assenting to the terms of the round expressed in these documents. On the automated conversion approach, the need for the SAFE holder's signature would remain. (On chains like Ethereum that charge for memory used by transactions, including large documents in the parameters of a transaction may be expensive or impossible because of transaction size limits, in which case it will be more efficient to include their cryptographic hash. There would then remain an obligation on the company to provide the Equity Financing documents to the SAFE holder in a form that would enable them to check the hash.)

19.2 Pro Rata Rights

Pro Rata Rights (PRR) are rights of an investor to purchase shares in a future equity round, intended to protect the investor from being diluted by future share issuance. PRR typically give the investor the option, but not an obligation, to purchase additional shares. In practice, only a fraction of the PRR shares need to be purchased.

In the Pre-Money SAFE, the Equity Financing clause obliges the company to issue PRR for the *next round* to the SAFE investor at the time of conversion, but this clause is replaced in the Post-Money SAFE by an optional side agreement (negotiable between the SAFE investor and the company) that grants the SAFE investor PRR for the round in which the SAFE converts.

The Pre-Money SAFE indicates the PRR are determined using *percentage basis*, which grants a right to maintain the percentage shareholding.[3] This is generally achieved by the company offering a fixed number of shares in the equity round, of which the PRR holders has rights to take up a percentage less than or equal to their percentage holding, with the remaining shares going to the new investors.

---

[3] Alternatives in use are the *dollar for dollar basis* (invest up to the original principal in the next round) and the *fixed sum* right (invest up to a fixed dollar amount). The latter are less common since they create tension by enabling the PRR holder to strictly increase their percentage holding in some circumstances.

PRR could be implemented in equity rounds conducted on-chain, using a smart contract for the PRR, together with an equity round atomic swap contract and controller smart contract similar to that in the existing code. When the equity round proposal is submitted by the company, the controller first checks that it contains options for each of the PRR holders, consistent with the rights encoded in their PRR smart contracts. The round proposal is rejected if this is not the case. Otherwise the equity round atomic swap smart contract is created. The PRR holding investors first interact with this swap contract by paying for the fraction of their PRR that they will take up, followed by consent and payment by the new investors. (A new deadline needs to be added for the PRR stage, to ensure the new investors get adequate time to meet the ultimate deadline.)

A technical issue arises out of the fact that pro rata rights are exercised *after* the SAFE has terminated. More importantly, for the implementation, they are exercised after the SAFE controller, which enforces the rights of SAFE holders, has been terminated. While PRR could be represented as smart contracts, and implemented as outlined above, the new controller could ignore these elements unless they are embedded in code it must run. For this reason, the equity round swap smart contract must manage the PRR. It can keep a record of the PRR that become live after the conversion of a SAFE and, in the next equity round, verify that the PRR have been respected; if not, it aborts the equity round. PRR that expire are deleted. To handle PRR endowed in a side-letter, this implementation would need to be extended, allowing the company to add PRR for a SAFE holder to the swap contract.

However, this approach does not take into account that PRR may create tensions between new investors and those holding PRR. Sometimes, the new investors want existing shareholders to exercise their PRR rights to show commitment to the company and validate the new investors' valuation. In other cases, the new investor prefer that these rights are not taken up so that they more cheaply attain their desired stake in the company.

Some richer schemes than the above could also be implemented as a way to deal with these tensions between PRR holders and new investors. For example, making PRR transferable would enable new investors to compensate PRR holders for giving up their rights. (However, it should be noted that this may conflict with share transfer limitations intended to limit changes of control.)

19.3 Company and Investor Representations

The sections "Company Representations" and "Investor Representations" of the SAFE relate to the factual background of the contract, and concern matters that underpin its legal validity. Generally, the representations relate to matters that are not likely to be represented in detail on-chain, and are probably best left as natural language text in a legal contract to be signed by the parties. (Section 17.3 discusses interactions between signatures on legal documents and smart contract transactions.)

Should any of the representations made by either the company or the investor be false, legal grounds may exist for the other party to have the contract declared void and claim damages, although they may also choose to renegotiate its terms. Section 17.6 discusses issues of amendment of smart contracts and handling of legal rulings in a broader context.

More specifically, the company represents the following (we summarize text of the SAFE and comment on the possibility of automation):

- That the company is properly registered, in good standing and has the power to operate its properties. Some of this information may (in future) become available on a variety of blockchain systems in some jurisdictions, but oracles and/or cross-chain information transfer may be necessary for any automation.
- That the company has the power to execute and perform the SAFE contract, the SAFE has been duly authorized, and the SAFE will be "a legal, valid and binding obligation of the Company." Apart from authorization, these are general, broad-ranging claims that might be referencing statutes, regulations, legal judgments, or existing contractual obligations. Even if all such references are available on the chain, evaluating their applicability will require handling the open texture of the references.
- The company is not in violation of specific kinds of regulatory, legal, or contractual obligations that "could reasonably be expected to have a material adverse effect on the Company." This is an open-textured expression. The obligations themselves are likely to be broadly open-textured, and so offer only limited opportunity for automation. In addition, some of these obligations may be confidential, and not independently verifiable for this reason alone.
- The company's performance of the contract will not violate any statutes or regulation or result in adverse consequences for the company. (Impediments to automation are similar to the previous point.)
- The only consents required for performance of the contract are those internal to company or by securities law. Again, this requires knowledge of the external legislation and regulatory environment, and is not easily amenable to automation.
- The company holds IP rights (trademarks, copyright, patents, etc.) necessary to conduct its business. (This depends on the declaration by the company being not just correct but also complete, so is best handled by a legal agreement rather than online verification.)

The SAFE investor representations, in brief, are:

- That the investor has the legal capacity to enter into the contract, accepts its obligations, with exceptions for bankruptcy or other credit and equity laws.
- That the investor is an accredited investor (as defined by the applicable jurisdiction) and understands the risks involved in the SAFE and that the SAFE and the shares to be issued have not been registered and are not transferable until registration or exempted from registration.

The conditions for these representations are likely to be broadly open-textured, and are best handled in a legal contract. Equities law and regulation may limit the extent to which such representations may be relied upon by the company and place obligations on companies to verify investor accreditation. For example, in the USA, companies are required to verify investor accreditation status, with Safe Harbor provisions stating sufficient conditions for a company to have complied with its verification obligations. There are some moves in some jurisdictions to standardize investor accreditation information and make it available on blockchain (e.g., in Australia there has been discussion of the use of Tax Office identifiers as a basis for this, motivated by the Australian Stock Exchange blockchain project among others [3], and in the USA, VerifyInvestor, a subsidiary of the tZero blockchain-based security trading platform, offers securities law compliant attestations of investor accreditation [4]), so the representations concerning accreditation could be replaced by on-chain evidence.

In general, the representations provide statements that assure the other party that the basis for the contract is sound, without providing any specifics. This is in keeping with the lightweight nature of SAFEs in comparison with equity rounds, where more due diligence is performed (see Chap. 1). The role of the representations is to provide a basis for future legal action/defense if the claims turn out to be questionable. Since smart contracts seem fixed to nuncospective actions, we have discussed what can be done in verifying these claims, rather than waiting for evidence that one is untrue. But this is a much more demanding aim than the legal contract requires. It seems better, from complexity, cost, and computational effort standpoints, to perform verification where it is convenient, but to continue to require the representations in a legal contract.

In any case, matters such as the investor's acceptance of obligations and limitations of the SAFE and their understanding and acceptance of its risks and limitations requires the investor's signature, which is best placed on natural language text in a legal contract.

19.4 Miscellaneous Provisos

The section "Miscellaneous" of the SAFE contract contains provisos dealing with a diverse range of matters concerning operation and interpretation of the contract.

Proviso (a) states that "Any provision of this instrument may be amended, waived or modified only upon the written consent of the Company and the Investor." From the point of view of smart contract implementations of any part of the contract, this implies that the implementation needs to admit amendments that may occur. We discussed this issue in a broader context in Sect. 17.6.

Proviso (b) concerns sufficient conditions for notice required by the SAFE, stating that methods such as courier, email, and registered mail are adequate. When implementing the SAFE using a smart contract, this clause could be revised to indicate the conditions for adequate communication via on-chain events. See Sect. 19.1 for related discussion.

Proviso (c) states that certain rights are *not* implied by the (Pre-Money) SAFE, specifically, the rights to vote or receive dividends,[4] or any other rights of a shareholder (including the right to receive notice of meetings) until shares have been issued. Since none of these rights are explicitly mentioned in the remainder of the SAFE, this clause could be considered redundant, but it acts to preempt legal interpretations of such rights as implicit in the SAFE. Beyond excluding from the code implementations for such interpretations, this proviso does not have any effect on the smart contract.

Proviso (d) is similarly negative and states that the SAFE, and any associated rights of either party, are not transferable without approval of both parties, with some exceptions for transfers to entities controlled by the SAFE investor, and rights of the company to make transfers in relation to reincorporation or changes of domicile. The set of entities related to the SAFE investor and the structure of ownership and control is open ended and could be complex, so it is best to handle the exceptions in a coarse grained way, leaving verification of the conditions to the company. Transfers in relation to reincorporation are discussed in Sect. 17.6.

Proviso (e) states that in case any part of the SAFE is deemed invalid, the remainder of the SAFE will remain operative. With respect to a code representation of the SAFE, this suggests that there should be a close correspondence between the clauses of the SAFE and the smart contract representation. Given the large divergence of structure between the legal text and its representation in imperative code, it is inherently difficult to achieve this. Thus, the most reasonable means of dealing with this proviso is to create a new smart contract and substitute it for the old. This could be achieved by extending the `SAFE_controller` to allow for such substitutions. However, as other investors may have made their investment decisions based on the contract being revised, careful consideration should be given to governance of this process, to prevent it being used maliciously to make changes that adversely affect the other investors.

Proviso (f) states the jurisdiction governing the SAFE contract "without regard to the conflicts of law provisions" (meaning that the parties are prevented from arguing that the jurisdiction may identify a conflict of laws and apply law from another jurisdiction). This is largely a matter for the legal contract with few interactions with the smart contracts, except possibly where the jurisdiction is an on-chain arbitration service (see Sect. 17.5 for discussion of a number of approaches to this).

Proviso (g) only appears in the Post-Money SAFEs. It commits the parties to the position that the SAFE is characterized as stock, for all federal and state income tax purposes (including their tax returns). Such commitments concern actions taken outside the transaction, and predominantly by the SAFE holder. Enforcing this commitment would require control of the SAFE holder's income tax return, an unreasonable and—given the complexity of the tax code—impractical proposition.

---

[4] The Post-Money SAFE alters these terms in granting rights to receive dividends.

19.5   Most Favored Nation SAFEs

The main feature of the MFN SAFE is the ability to amend the SAFE to the same terms as a convertible security issued by the company after the purchase of the MFN SAFE. Key to this feature is the following provision.[5]

> **"MFN" Amendment Provision.** If the Company issues any Subsequent Convertible Securities with terms more favorable than those of this Safe (including, without limitation, a valuation cap and/or discount) prior to termination of this Safe, the Company will promptly provide the Investor with written notice thereof, together with a copy of such Subsequent Convertible Securities (the "MFN Notice") and, upon written request of the Investor, any additional information related to such Subsequent Convertible Securities as may be reasonably requested by the Investor. In the event the Investor determines that the terms of the Subsequent Convertible Securities are preferable to the terms of this instrument, the Investor will notify the Company in writing within 10 days of the receipt of the MFN Notice. Promptly after receipt of such written notice from the Investor, the Company agrees to amend and restate this instrument to be identical to the instrument(s) evidencing the Subsequent Convertible Securities.

"Subsequent Convertible Securities" (henceforth abbreviated to SCS) is defined by

> **"Subsequent Convertible Securities"** means convertible securities that the Company may issue after the issuance of this instrument with the principal purpose of raising capital, including but not limited to, other Safes, convertible debt instruments and other convertible securities. Subsequent Convertible Securities excludes: (i) side letters or ancillary agreements that do not amend or modify the terms of such convertible securities; and (ii) the following types of securities: (A) options issued pursuant to any equity incentive or similar plan of the Company; (B) convertible securities issued or issuable to (1) banks, equipment lessors, financial institutions or other persons engaged in the business of making loans pursuant to a debt financing or commercial leasing or (2) suppliers or third party service providers in connection with the provision of goods or services pursuant to transactions; and (C) convertible securities issued or issuable in connection with sponsored research, collaboration, technology license, development, OEM, marketing or other similar agreements or strategic partnerships.

A number of issues are raised when attempting to formalize these clauses. The notification requirement raises issues similar to those already discussed in Sect. 19.1. Open-textured concept are again an issue: the definition of SCS contains many expressions that do not have precise definitions: e.g., "convertible securities," "principal purpose of raising capital," and "other Safes." Similarly, the MFN Amendment Provision uses imprecise expressions such as "promptly" and "may reasonably be requested."

A further difficulty is the interpretation of the phrase "amend and restate this instrument to be identical to the instrument(s) evidencing the SCS." It is not completely clear what is meant by this, and "identical" is unlikely to be precisely what is meant, since at least the principal and beneficiary of the SCS are likely to differ in SCS and the new instrument to be granted to the SAFE investor. Substituting these

---

[5] In this section, we refer to the text of the Post-Money MFN SAFE, version 1.3.

parameters of the MFN SAFE for those in the SCS is likely to be involved in the amendment. However, it may not be clear what the corresponding parameter of the SCS is. For example, what corresponds to the principal parameter if the new instrument involves a (contingent) sequence of principal payments delivered over time? Other parameters of the SCS may not correspond to any parameter of the SAFE. For example, what if the SCS specifies a proxy of the SCS investor for some actions, or a designated party for arbitration of disputes? Further, what if the jurisdiction differs? Should this be modified to the jurisdiction of the SAFE? This might not make sense, if some of the terms of the SCS were constructed with a specific alternate jurisdiction in mind.

If the modification is to be automated, it would seem that the smart contracts for the Subsequent Convertible Securities that the company is permitted to issue need to be constrained to smart contracts with a known interface, that enables the modification to be constructed by substitution of a limited set of parameters. A further issue is that the source code of the smart contract needs to be available, so that it can be checked that it will behave as expected by the investor. On the Ethereum chain, only the bytecode will be available, which may make it difficult to determine automatically that the code is for the required interface, and to make the required substitution.

If the smart contract is paired with a legal agreement, the latter needs in any case to be provided to the SAFE investor off-chain, so source code for the smart contract can be provided to the SAFE investor as part of this communication. The SAFE investor will want to examine both to determine whether or not to notify the Company that they are taking up their option to switch to the new instrument. Some negotiation may be necessary to come to an agreement concerning the appropriate modification. This makes it reasonable to implement the on-chain activity using an on-chain proposal and acceptance for the modified contract (typically only the final agreement would be put on-chain). Either party could be the proposer, since both would need to agree.

19.6 Summary

Sections 3, 4, and 5 of the SAFE contract, involving company and SAFE investor representation and miscellaneous other provisions, are difficult to enforce as a smart contract. They involve open-textured concepts as well as a breadth of considerations that are either infeasible or impractical to implement. Implementing the execution of equity financing documents is also difficult.

In contrast, the pro rata rights of shareholders and SAFE holders seem feasible to enforce using the techniques we have demonstrated. MFN SAFEs offer additional challenges, but it seems an approximation of the MFN SAFE could be implemented in circumstances where the company is constrained to offer only a narrow range of convertible securities, which are minor variations of existing SAFE contracts.

References

1. National Venture Capital Organisation NVCA (2020) Model legal documents. https://nvca.org/model-legal-documents/. Accessed Apr 2020
2. Boldyreva A, Palacio A, Warinschi B (2012) Secure proxy signature schemes for delegation of signing rights. J Cryptol 25(1):57–115
3. Chanticleer (2020) Australia's new investor passport. Australian Financial Review. https://www.afr.com/chanticleer/australia-s-new-investor-passport-20201008-p5635p. Accessed Oct 2020
4. Verify Investor (2020). https://www.verifyinvestor.com. Accessed Oct 2020

Chapter 20
Conclusion

20.1 Summary

The material presented in this book is the outcome of a project aimed at understanding, through a case study, the process of developing a smart contract from a financial legal contract. SAFE contracts were selected for this purpose because they appeared sufficiently short and simple. In the final analysis, matters proved to be significantly more complex than anticipated.

A first step in the project was to develop a sufficiently precise understanding of the operation of the financial terms of SAFE contracts in order to be able to write smart contract code correctly implementing these terms. To determine how best to structure, the code and which parts of the operation of the legal contract should be represented in the smart contract itself, and which parts should be represented "off-chain", we also had to understand the way that these terms are interpreted in practice. Work to this end led to the material presented in Parts II–III. The actual development of the smart contract code—including problems faced, techniques used, and its evaluation—is presented in Part IV.

One of the principal conclusions from the effort is that, unless we modify the powers of the SAFE investor or encode a right to third party arbitration, this particular contract type is not fully amenable to implementation solely as a smart contract, but requires a "smart legal contract": a combination of smart contract code with a background legal contract. This combination is necessary in order to protect against certain risks, and to govern aspects of the operation of the smart contract that require human judgment and action in good faith.

We do not view this conclusion as implying that the effort was not worthwhile. Computer code, due to its mechanistic nature, is much less forgiving of errors and misunderstandings than the legal system, so the attempt to represent a contract as code forces a much sharper focus on details than is usually applied to contracts. Our

© The Author(s), under exclusive license to Springer Nature Singapore Pte Ltd. 2025   269
R. van der Meyden and M. J. Maher, *Simple Agreements for Future Equity (SAFE)*,
Blockchain Technologies, https://doi.org/10.1007/978-981-96-3920-5_20

effort to understand SAFE contracts in order to formalize them has led to the identification of weaknesses of these contracts, and allowed us to develop an improved
understanding of their workings.

Since their publication by Y Combinator, SAFE contracts have become somewhat
of a venture finance industry standard. While the substantial effort to develop a
formalization and conduct detailed formal analyses may not be warranted for bespoke
or low value contracts, we think that it can be worthwhile for frequently used contract
forms, since the cost will be amortized over the many applications. It is not uncommon
for computer industry standards bodies to solicit work on the formal analysis of
the artifacts they standardize, and we expect that such efforts could become more
common also in the legal domain, particularly as contracts and legislation come to
be more frequently implemented using code. Already for the finance industry, the
International Swaps and Derivatives Association has long sought to develop standards
for derivatives contracts and the interfaces to be used by financial institutions to
automate their processing [1, 2].

Our analysis to develop a detailed understanding of SAFEs has uncovered a number of subtleties. A first issue that needed to be faced in formalizing SAFEs was
the operation of the Equity Financing clause. We addressed this first in Part II for
scenarios in which the company has issued just a single SAFE, which converts in
an Equity Financing event. Since conversion of a SAFE dilutes the new investor, a
naive understanding of the conversion process (indeed, the one that Y Combinator
itself appears to have first used) raises questions about the rationality of the SAFE
as a financial instrument. On this understanding, contrary to expectations, it is not
conservative for either the SAFE investor or the new investor, potentially returning
to them shares in the company that are worth less than their investment amounts.

We argued that understanding this discrepancy is aided by considering alternative
ways to do accounting for convertible instruments such as SAFEs, as Liabilities or on
the Cap Table. Each alternative leads to a different set of equations that apply to the
round. Mixing equations from different sets results in some of the misunderstandings
underpinning the discrepancy. Related to the accounting viewpoints is an ambiguity
in the term "pre-money valuation" and its relationship to the share price at which the
equity round should be conducted.

We then applied the general analysis to the specific case of Pre-Money and Post-
Money forms of the SAFE note with Cap and no Discount. The ambiguities in the
understanding of pre-money valuation leave open a variety of conversion methods
at the time of the equity round, discussed in Sects. 4.2–4.5 for the Pre-Money SAFE
and in Chap. 5 for the Post-Money SAFE. We examined how these methods perform,
under the two accounting viewpoints, with respect to the desideratum that an equity
round should be conservative for both the new investor and the SAFE investor.

The understanding that there are multiple ways applied in practice to determine
the relationship between pre-money valuation and price of the equity round had an
impact on the smart contract we developed: rather than compute a price, our smart
contract for the SAFE verifies a constraint on how these parameters relate to the
number of shares issued to the SAFE investor. The formal analysis of Post-Money

SAFEs, in particular, further justifies this approach, since, rather than describe a computation, it explicitly states a constraint.

We also considered game-theoretic aspects of the choice of investment instrument with regard to conversion in an equity round. Our analysis in Sect. 7.6 compares the SAFE conversion methods of Sects. 4.2–4.5 and Chap. 5 both for a scenario where the players are "fully rational", and for a scenario where they defer resolution of the conversion method until after a pre-money valuation has been fixed.

Among the conversion methods that might be applied to ensure that the equity round is conservative for both SAFE and equity round investors is to spread the new investor's money across two equity rounds, with the SAFE converting in the first round. This idea is potentially applicable in a malicious "attack" in which the company and the equity round investors collude against the interests of the SAFE investor. In Sect. 6.4 we derived the optimal (but legally questionable) way for the founders and new investor to "game" the SAFE by structuring an investment into two rounds so as to minimize the share going to the SAFE investor.

For the case of fully rational, honest players, we reached the conclusion that many of the conversion methods are essentially equivalent in their outcomes with respect to proportional shareholdings of the players after the equity round. The "Discounted Valuation" conversion method gives a clean understanding of this canonical outcome. We therefore applied this method in Part III to also analyze equity rounds in situations in which the company has issued multiple SAFEs. Here we found that while, for single SAFE scenarios, there is an equivalence between Pre- and Post-Money SAFEs, a similar relationship holds for multiple SAFEs only under certain limited conditions (having a common cap) and in a weaker sense (at the level of the scenario, rather than for each SAFE independently). Moreover, we also identified issues with the convertibility of Post-Money SAFEs with Cap using the Discounted Valuation method—there are unexpected cases where the equations that a Post-Money SAFE requires to be solved in fact have no solutions.

We also analyzed the effects of liquidity events (IPO, acquisition, etc.) on outcomes for SAFE holders. In such events, SAFE investors each have a choice to make, but one investor's choice can affect another investor's payout. We approached this issue with a game-theoretic analysis, which shows that so long as all SAFEs in the scenario have the same type, there is a stable set of optimal choices that can be made by the investors that yields them a "maximal payout". However, this is not the case when the set of SAFEs issued have mixed types, where there is not always a stable "maximal solution". This analysis also impacted our design choices in the smart contract code. One option could have been to have the smart contract code automatically compute and apply the maximal payout in a liquidity event. As this cannot be guaranteed to exist, our implementation instead leaves it to investors to make the choice, even for Post-Money SAFEs, in which the legal text states that the investor should get the maximal payout from their two options.

The subtleties around the lack of solutions of equations necessary for the operation of multiple SAFEs, in both Equity Financing and Liquidity events, also lead us to propose, in Chap. 10, alternate SAFE terms, that are designed to ensure that a unique payout always exists. We take this to be a benefit from the effort of formalization and detailed analysis. Guaranteed performability in all possible situations is of particular significance in settings where actors secure their agreements using immutable smart contracts.

In Part IV, we turned to applying the detailed understanding from Parts II–III to the investigation of smart contracts for SAFE contracts and the environment in which they operate. Our answer to the question of whether SAFE contracts can be smart is a *qualified* "yes".

In the course of the investigation, we identified, in Chap. 13, a number of challenges to this objective: the indefiniteness and incompleteness of legal text, the openness of the situations in which SAFE contracts apply, and the inability of smart contracts, unlike the legal system, to apply retrospective reasoning after events have unfolded. We discussed some strategies for dealing with these difficulties in Chap. 14: increasing precision, limitation of fact patterns, use of human oracles, propose-and-verify implementations, and representation of open scenarios at "run-time" as code objects using higher-order programming. Using these strategies, we developed a smart contract architecture for representing a company using SAFEs, which we have implemented in the Solidity smart contract programming language.

In evaluating our smart contract representation of SAFEs, in Chap. 16, we noted that the indefiniteness of legal text also limits the extent to which one can claim that a smart contract is equivalent to an original legal contract. We can, however, do a comparison with respect to risks covered. While there is a significant overlap, the reason we need to qualify our answer that smart contracts can be used to represent SAFEs relates to the inability of smart contracts to act retrospectively.

As already noted, in Sect. 6.4, we showed that there is an "attack" in which the company and the new investors collude to structure the investment into two equity rounds to disadvantage SAFE investors. The original legal SAFEs have text that provides protection against this risk, but a smart contract, on its own, cannot defend against it, unless the investor is given rights to block a round that they do not have in the original contract. For this, and other reasons, we recommend that the smart contract be used in conjunction with a legal contract that constrains the way that it is used. We have sketched some of the required content for such a contract in Chap. 17, and discussed some of the associated legal issues, but have left open the development of the (necessarily jurisdiction-dependent) text for such a contract.

20.2 Future Work

Both our work on the analysis of SAFE contracts, and on their implementation as smart contracts, leaves open a number of questions for future work.

We have focused, for much of the book, on Pre- and Post-Money SAFEs with Cap Only. We believe the methodology we have applied to the analysis and implementation of these contracts can be applied to the other SAFE contract types, but we have not attempted to carry this through. Chapter 19 discusses some of the ways that our work in Part IV is incomplete, such as lacking a treatment of options, pro rata rights, and a richer model of share classes, and coverage of Most Favored Nation SAFEs. Our game-theoretic analysis of liquidity events is based on discrete Nash equilibria as a solution concept: it might also be of interest to consider mixed equilibria, particularly in settings such as mixed SAFE types where no discrete equilibrium exists. Equilibria notions from cooperative game theory might also be worth exploring, both for Equity Financing and Liquidity events. We have also not attempted to apply probabilistic methods to analyze SAFEs from the point of view of valuation under assumptions about the future performance of company.

Our formal reasoning in this book has been done using pen and paper. Formal representation of an artifact as code opens up the possibility of applying formal verification methods and tools. One class of such tools (model checkers [3]) can fully automate simple reasoning about code, another (theorem provers [4]) provide semi-automated support for the development of proofs, in a formal logic, of more expressive properties of code. Both types of tools can help to obtain much higher levels of assurance of the security and correctness of code. For the Solidity programming language, formal verification tools are still at early stages of development. While we sketched out some security properties and the way that the architecture supports reasoning about them in Sect. 15.3, we have not attempted to carry out a proof of either the security properties or the results from Parts II–III using any such tool. This would be an interesting area for future work, as a case study in the use of such tools.

Arguably, the ultimate measure by which we should evaluate a smart (legal) contract should be whether it proves its value to its users in practice: does it yield efficiencies, eliminate risks, or reduce disputes to an extent that it comes to be adopted in the marketplace. We have not attempted to deploy our smart contracts in practice, or explore this empirical question, so we also leave this for future work.

We hope, in any case, that what we have presented in this book is of value both to readers interested in a better understanding of SAFE contracts, and to readers interested in understanding, through a case study, the nature of smart legal contracts for financial applications.

While formalization of law, envisioned by Leibniz [5], has long been an object of study, it has been a niche area of research, and progress has been slow. With the rise of blockchain systems as smart contract platforms, there is now significant impetus for more rapid advancement.

We are still early in the development of a discipline of "smart legal contract engineering", the object of which will be the knowledge and tools needed to create computational legal artifacts to secure multi-party agreements, protecting legal rights of the parties while supporting efficient and correct performance, delivering the benefits mooted in the Preface. Our view, based on the present case study, is that such a discipline will draw on a skillset from what are currently independent areas, drawing on areas including mathematical finance, accounting, game theory, law, and computer science. We hope that the present work provides a useful example to help spur the development of such a discipline.

References

1. International Swaps and Derivatives Association. https://www.isda.org. Accessed Nov 24
2. Clack CD, McGonagle C (2019) Smart derivatives contracts: the ISDA master agreement and the automation of payments and deliveries. *CoRR*, abs/1904.01461
3. Clarke EM, Grumberg O, Peled DA (1999) Model checking. MIT Press, London, Cambridge
4. Harrison J, Urban J, Wiedijk F (2014) History of interactive theorem proving. In: Siekmann JH (ed) Computational logic, Handbook of the history of logic, vol 9. North-Holland, pp 135–214
5. Artosi A, Pieri B, Sartor G (eds) (2013) Leibniz: logico-philosophical puzzles in the law. Springer

Appendix
Working for Comparisons of Proportional Shareholding

In this appendix, we justify the rankings of outcomes of proportional shareholdings for the three parties given in Sect. 7.6. We calculate the relationship between the outcomes for each party in each of the three cases. In each case, there are just two possible outcomes for the new investor, and it is easy to see that the standard SAFE calculation never gives them the best outcome. We use the following abbreviations: Two-Round = $2R$, Standard = Std, Discounted Valuation = DV, Dollars Invested = DI, Zero-Round (Standard) = $ZR(Std)$, and Zero-Round (Discounted) = $ZR(DV)$.

- **Assumption:** $v_{pre} < c$. Here, we have the following comparisons for the Founders:

  $2R$ versus Std

  $$\equiv \frac{c}{c+m} \cdot \frac{v_{pre}}{v_{pre}+m_{new}} \text{ versus } \frac{v_{pre}}{v_{pre}+m+m_{new}}$$

  $$\equiv c \cdot (v_{pre}+m+m_{new}) \text{ versus } (c+m) \cdot (v_{pre}+m_{new})$$

  $$\equiv cv_{pre}+cm+cm_{new} \text{ versus } cv_{pre}+cm_{new}+mv_{pre}+mm_{new}$$

  $$\equiv cm \text{ versus } +mv_{pre}+mm_{new}$$

  $$\equiv c \text{ versus } v_{pre}+m_{new}$$

This may be in either order, depending on m_{new}.

  Std versus $ZR(Std)$

  $$= \frac{v_{pre}}{v_{pre}+m+m_{new}} \text{ versus } \frac{v_{pre}}{v_{pre}+m} \cdot \frac{v_{pre}}{v_{pre}+m_{new}}$$

  $$= (v_{pre}+m_{new})(v_{pre}+m) \text{ versus } v_{pre}(v_{pre}+m+m_{new})$$

  $$= m_{new}m \text{ versus } 0$$

so $Std > ZR(Std)$.

© The Editor(s) (if applicable) and The Author(s), under exclusive license to Springer Nature Singapore Pte Ltd. 2025
R. van der Meyden and M. J. Maher, *Simple Agreements for Future Equity (SAFE)*, Blockchain Technologies, https://doi.org/10.1007/978-981-96-3920-5

$ZR(Std)$ versus DV

$$= \frac{v_{pre}}{v_{pre} + m} \cdot \frac{v_{pre}}{v_{pre} + m_{new}} \text{ versus } \frac{v_{pre} - m}{v_{pre} + m_{new}}$$

$$= v_{pre}^2 \text{ versus } (v_{pre} + m)(v_{pre} - m)$$

$$= v_{pre}^2 \text{ versus } v_{pre}^2 - m^2$$

so $ZR(Std) > DV$.

$2R$ versus DV

$$= \frac{c}{c + m} \cdot \frac{v_{pre}}{v_{pre} + m_{new}} \text{ versus } \frac{v_{pre} - m}{v_{pre} + m_{new}}$$

$$= cv_{pre} \text{ versus } (v_{pre} - m)(c + m)$$

$$= 0 \text{ versus } v_{pre}m - mc - m^2$$

$$= c + m \text{ versus } v_{pre}$$

So $2R > DV$ since $v_{pre} < c$.
For $2R$ versus $ZR(Std)$, the rightmost terms are identical, so

$2R$ versus $ZR(Std)$

$$= \frac{c}{c + m} \text{ versus } \frac{v_{pre}}{v_{pre} + m}$$

$$= cv_{pre} + cm \text{ versus } cv_{pre} + v_{pre}m$$

$$= c \text{ versus } v_{pre}$$

So $2R > ZR(Std)$ in this case.
In this case, $DV = ZR(DV)$ and $DI = Std$ by inspection. So, for the founders, we have $DV = ZR(DV) < ZR(Std) < DI = Std$, and $DV < 2R$ but the relation of $2R$ to $ZR(Std)$ and Std depends on m_{new} and m.
In case of the SAFE investor, we have $2R < ZR(Std) < DV$ and $DI = Std < DV$ by inspection.

Std versus $ZR(Std)$

$$= \frac{m}{v_{pre} + m + m_{new}} \text{ versus } \frac{m}{v_{pre} + m} \cdot \frac{v_{pre}}{v_{pre} + m_{new}}$$

$$= (v_{pre} + m)(v_{pre} + m_{new}) \text{ versus } (v_{pre} + m + m_{new})v_{pre}$$

$$= v_{pre}^2 + v_{pre}m_{new} + v_{pre}m + mm_{new} \text{ versus } v_{pre}^2 + v_{pre}m + v_{pre}m_{new}$$

$$= mm_{new} \text{ versus } 0$$

so $Std > ZR(Std)$.

So the SAFE investor order on outcome values is $2R < ZR(Std) < Std = DI < DV$.

- **Assumption:** $c \leq v_{\text{pre}} < c + m$.

For the founders: we have the following comparisons. By inspection $2R = ZR(Std)$ and $DV = ZR(DV)$.

DV versus $2R$

$$= \frac{v_{\text{pre}} - m}{v_{\text{pre}} + m_{\text{new}}} \text{ versus } \frac{c}{c + m} \cdot \frac{v_{\text{pre}}}{v_{\text{pre}} + m_{\text{new}}}$$

$$= (v_{\text{pre}} - m)(c + m) \text{ versus } c v_{\text{pre}}$$

$$= v_{\text{pre}} c + v_{\text{pre}} m - mc - m^2 \text{ versus } c v_{\text{pre}}$$

$$= v_{\text{pre}} m \text{ versus } mc + m^2$$

$$= v_{\text{pre}} \text{ versus } c + m$$

so we have $DV < 2R$ by the assumption.

$2R$ versus DI

$$= \frac{c}{c + m} \cdot \frac{v_{\text{pre}}}{v_{\text{pre}} + m_{\text{new}}} \text{ versus } \frac{c}{c + m} \cdot \frac{v_{\text{pre}} + m}{v_{\text{pre}} + m + m_{\text{new}}}$$

$$= \frac{v_{\text{pre}}}{v_{\text{pre}} + m_{\text{new}}} \text{ versus } \frac{v_{\text{pre}} + m}{v_{\text{pre}} + m + m_{\text{new}}}$$

$$= v_{\text{pre}}(v_{\text{pre}} + m + m_{\text{new}}) \text{ versus } (v_{\text{pre}} + m)(v_{\text{pre}} + m_{\text{new}})$$

$$= v_{\text{pre}} m_{\text{new}} \text{ versus } (v_{\text{pre}} + m) m_{\text{new}}$$

$$= 0 \text{ versus } m m_{\text{new}}$$

so we have $2R < DI$.

DI versus Std

$$= \frac{c}{c + m} \cdot \frac{v_{\text{pre}} + m}{v_{\text{pre}} + m + m_{\text{new}}} \text{ versus } \frac{v_{\text{pre}}}{v_{\text{pre}} + \frac{v_{\text{pre}} m}{c} + m_{\text{new}}}$$

$$= c(v_{\text{pre}} + m)(v_{\text{pre}} + \frac{v_{\text{pre}} m}{c} + m_{\text{new}}) \text{ versus } v_{\text{pre}}(c + m)(v_{\text{pre}} + m + m_{\text{new}})$$

$$= c v_{\text{pre}}^2 + v_{\text{pre}}^2 m + c v_{\text{pre}} m_{\text{new}} + c m v_{\text{pre}} + m^2 v_{\text{pre}} + c m m_{\text{new}}$$
$$\text{versus } v_{\text{pre}}^2 c + v_{\text{pre}} c m + v_{\text{pre}} c m_{\text{new}} + v_{\text{pre}}^2 m + v_{\text{pre}} m^2 + v_{\text{pre}} m m_{\text{new}}$$

$$= c m m_{\text{new}} \text{ versus } v_{\text{pre}} m m_{\text{new}}$$

$$= c \text{ versus } v_{\text{pre}}$$

Hence, we have $DI < Std$ using assumption $c \leq v_{\text{pre}}$.

So, for the founders, we have $DV = ZR(DV) < 2R = ZR(Std) < DI < Std$ in this case.

For the SAFE investor, we have $2R = ZR(Std)$ and $DV = ZR(DV)$.
The outcome for $2R$ can be written as

$$\frac{m}{v_{\text{pre}} + m_{\text{new}}} \cdot \frac{v_{\text{pre}}}{c + m}$$

which is the outcome for DV times a number not larger than 1, by assumption.
Hence, $2R < DV$.
The outcome for Std can be written as

$$\frac{m\, v_{\text{pre}}}{v_{\text{pre}}c + v_{\text{pre}}m + m_{\text{new}}c}$$

which has the same numerator as $2R$. The denominator of $2R$ multiplies out to the
larger number $v_{\text{pre}}c + v_{\text{pre}}m + m_{\text{new}}c + m_{\text{new}}m$, so we have $2R < Std$.
To compare DV with Std, we multiply the denominator and numerator of DV by
v_{pre}, so that we get

$$
\begin{aligned}
DV \text{ versus } Std &= v_{\text{pre}}c + v_{\text{pre}}m + m_{\text{new}}c \text{ versus } v_{\text{pre}}^2 + v_{\text{pre}}m_{\text{new}} \\
&= v_{\text{pre}}(c + m) \text{ versus } v_{\text{pre}}^2 + m_{\text{new}}(v_{\text{pre}} - c)
\end{aligned}
$$

Since, by assumption, we have $c \le v_{\text{pre}} < c + m.$, we have $v_{\text{pre}}^2 < v_{\text{pre}}(c + m)$ and
$0 \le v_{\text{pre}} - c$. Hence, the order between these expressions depends on m_{new}.

$ZR(Std)$ versus DI

$$
\begin{aligned}
&= \frac{m}{c + m} \cdot \frac{v_{\text{pre}}}{v_{\text{pre}} + m_{\text{new}}} \text{ versus } \frac{m}{c + m} \cdot \frac{v_{\text{pre}} + m}{v_{\text{pre}} + m + m_{\text{new}}} \\
&= \frac{v_{\text{pre}}}{v_{\text{pre}} + m_{\text{new}}} \text{ versus } \frac{v_{\text{pre}} + m}{v_{\text{pre}} + m + m_{\text{new}}} \\
&= v_{\text{pre}}(v_{\text{pre}} + m + m_{\text{new}}) \text{ versus } (v_{\text{pre}} + m_{\text{new}})(v_{\text{pre}} + m) \\
&= v_{\text{pre}}^2 + v_{\text{pre}}m + v_{\text{pre}}m_{\text{new}} \text{ versus } v_{\text{pre}}^2 + v_{\text{pre}}m + m_{\text{new}}v_{\text{pre}} + m_{\text{new}}m \\
&= 0 \text{ versus } m_{\text{new}}m
\end{aligned}
$$

So $ZR(Std) < DI$.

DI versus DV

$$= \frac{m}{c+m} \cdot \frac{v_{\text{pre}} + m}{v_{\text{pre}} + m + m_{\text{new}}} \text{ versus } \frac{m}{v_{\text{pre}} + m_{\text{new}}}$$

$$= (v_{\text{pre}} + m_{\text{new}})(v_{\text{pre}} + m) \text{ versus } (c + m)(v_{\text{pre}} + m + m_{\text{new}})$$

$$= v_{\text{pre}}^2 + v_{\text{pre}}m + m_{\text{new}}v_{\text{pre}} + m_{\text{new}}m$$

$$\text{versus } c(v_{\text{pre}} + m + m_{\text{new}}) + mv_{\text{pre}} + m^2 + mm_{\text{new}}$$

$$= v_{\text{pre}}^2 + m_{\text{new}}v_{\text{pre}} \text{ versus } c(v_{\text{pre}} + m + m_{\text{new}}) + m^2$$

$$= (v_{\text{pre}} - c)(v_{\text{pre}} + m_{\text{new}}) \text{ versus } (c + m)m$$

All terms on the last line are positive since we are in a case where $c \le v_{\text{pre}}$. Choice of m can result in either order, so these terms are not directly comparable without further information.

DI versus Std

$$= \frac{m}{c+m} \cdot \frac{v_{\text{pre}} + m}{v_{\text{pre}} + m + m_{\text{new}}} \text{ versus } \frac{\frac{m v_{\text{pre}}}{c}}{v_{\text{pre}} + \frac{v_{\text{pre}}m}{c} + m_{\text{new}}}$$

$$= c(v_{\text{pre}} + m)(v_{\text{pre}} + \frac{v_{\text{pre}}m}{c} + m_{\text{new}}) \text{ versus } v_{\text{pre}}(c + m)(v_{\text{pre}} + m + m_{\text{new}})$$

$$= cv_{\text{pre}}^2 + mv_{\text{pre}}^2 + cv_{\text{pre}}m_{\text{new}} + cv_{\text{pre}}m + m^2 v_{\text{pre}} + cmm_{\text{new}}$$

$$\text{versus } cv_{\text{pre}}^2 + cv_{\text{pre}}m + cv_{\text{pre}}m_{\text{new}} + v_{\text{pre}}^2 m + v_{\text{pre}}m^2 + v_{\text{pre}}mm_{\text{new}}$$

$$= cmm_{\text{new}} \text{ versus } v_{\text{pre}}mm_{\text{new}}$$

$$= c \text{ versus } v_{\text{pre}}$$

Since we are in a case where $c \le v_{\text{pre}}$, we have $DI < Std$.

Thus, for the SAFE investor, we have $2R = ZR(Std) < DI < Std$ and $2R = ZR(Std) < DV = ZR(DV)$, with the order between $DV = ZR(DV)$ and each of DI and Std possibly either way.

Since we are in a case where $c < v_{\text{pre}}$, we have $v_{\text{pre}} + m_{\text{new}} \le v_{\text{pre}} + m + m_{\text{new}} \le v_{\text{pre}} + \frac{v_{\text{pre}}m}{c} + m_{\text{new}}$ so the order for the new investor is straightforwardly $2R = DV = ZR(DV) = ZR(Std) < DI < Std$.

- **Assumption $c + m \le v_{\text{pre}}$:** The table in this case differs from that for the case $c \le v_{\text{pre}} < c + m$ only in the rows for DV and $ZR(DV)$, and the assumption $c \le v_{\text{pre}}$ used in a number of the comparisons for that case continues to hold. Hence, there is no change from the previous case in the ordering of any of the rows other than $DV, ZR(DV)$. By inspection, we have $2R = DV = ZR(DV) = ZR(Std)$ for all agents in this case. Thus, for the founders, the outcomes are ordered $2R = DV = ZR(DV) = ZR(Std) < DI < Std$ and for the SAFE investor the outcomes are ordered $2R = DV = ZR(DV) = ZR(Std) < DI < Std$.

Index

A

Accounting, 8, 39–40, 201
Accounting view, 35–39, *see also* cap table view; liability view
Ambiguity, 146, 174–176, 183, 186
Angel investors, 4, 16, 75
Atomic swap pattern, 195–196, 200, 212, 214
Automated decision-making, 191

B

Bitcoin, 160–164, 169
Blockchain, 159–163
Bridging loan, 5, 7

C

Capital stock, 19, 62, 110, 116, 137, 138, 141
Cap table view, 36–39, 60
Carolynn Levy, 8, 9
Change of Control, 20, 179, 262
Common stock, 4, 5
Company, 200, 202–204, 207–209, 211, 213, 216, 218, 221, 222, 224, 227, 229, 241, 242, 250, 255
Company Capitalization, 19, 24, 25, 42
Conservative, 33, 37, 44–46, 49, 57, 59, 64, 69–75, 90, 93, 96, 100, 152
minimally, 94
Controller pattern, 196–199, 202, 204, 207, 211, 225

Convertible note, 5, 7–9, 75, 135, 136, 141
Crowdfunding, 4, 9, 11

D

Debenture, 5, 6
Deferred interpretation, 75–80
DI, *see* Dollars Invested method
Dilution, 8, 10, 35, 43, 45, 47, 50, 55, 57, 58, 139, 153
Direct listing, 25, 139, 146
Discount, 8, 9
Discounted Valuation method, 50–53, 55, 60, 74, 77, 84, 90–103, 147
Discount Price, 19
Dissolution event, 17, 22–23, 27, 136, 139–142, 150, 176, 220–221
Dividends, 6, 10, 18, 23, 28, 63, 73, 136–139, 146
Dollars Invested method, 53–55, 60, 77
DV, *see* Discounted Valuation method

E

Equity financing, 4–9, 16, 147
Equity Financing event, 16–20, 24–25, 149
Equity round, *see* equity financing
Equity round equations, 31–33, 39, 152
Equity_round_swap, 200, 204, 213, 213, 213–217, 221, 222, 224, 225, 227, 229
ERC-20, 166, 169
ERC-721, 170

© The Editor(s) (if applicable) and The Author(s), under exclusive license to Springer Nature Singapore Pte Ltd. 2025
R. van der Meyden and M. J. Maher, *Simple Agreements for Future Equity (SAFE)*, Blockchain Technologies, https://doi.org/10.1007/978-981-96-3920-5

Ethereum, 162, 164–169

F
Fact pattern, 184–186, 204, 213, 230, 234

I
Incompleteness, 176–178, 183, 186, 187
Inconsistency, 34–35, 47, 174, 177
Initial Public Offering, 16, 20, 139

K
Keep It Simple Security (KISS), 8, 12

L
Liability view, 35–36, 38–39, 60
Liquidity Capitalization, 137
Liquidity event, 17, 20–22, 25–27, 109–110, 136, 149, 217–220
    game, 110–118
Liquidity Price, 28, 137

M
Mixed SAFE types, 106–107, 132–135, 142, 145, 156

N
Nash equilibrium, 118–135
    algorithms, 126, 129, 132
    existence, 119, 128, 129, 131, 135
    optimum, 118, 123, 126–131, 135
    unique, 119–121, 132
New Safe, *see* post-money SAFE
Nuncospectivity, 180–181, 235, 264

O
Open texture, 173–174, 176, 191–192
Option pool, 10, 21, 24, 26, 43, 139

P
Paul Graham, 8
Payout, 200, 204, 220–222, 224, 227, 229

Payout pattern, 198–199
Percent-Ownership method, 47–50, 52, 77–95
Post-Money SAFE, 10, 23–73, 141, 147
Preferred stock, 4–6, 18–19

Pre-Money SAFE, 10, 11, 17–23, 72–73, 141
Pre-money valuation, 9
Privacy, 253–257
Pro rata rights, 10, 11, 80, 84, 261–262, 276, 278
PRR, *see* pro rata rights
PSAFE, 148–150

S
Safe, 200, 203–213, 215, 216, 218–222, 224, 226, 227, 230, 240, 249, 255, 261

Safe_controller, 200, 204, 209, 211–218, 220–222, 224–227, 229, 240, 242, 244, 250, 265
Seed funding, 5, 16
Simple Agreement for Future Equity (SAFE), 5
    cap and discount, 11, 19, 22, 25, 114, 115
    cap only, 113–118
    discount, 19, 22, 25, 27, 114–115
    MFN, 11, 19–20, 22–25, 27, 113, 115, 137, 139, 265–267
Simple Agreement for Future Tokens (SAFT), 12
Smart contract, 163–164, 166–171
Smart legal contract, 242–245
Solidity, 166–169, 197, 230
Standard equity round, 31–35
Standard method, 43–46, 71, 74, 77
Std, *see* Standard method

T
Taxation, 7, 9, 10, 12, 39, 238, 265
Termination, 27–28
Term sheet, 43, 47, 53, 54, 179, 204
Two stage method, 77

V
Valuation cap, 8, 9
VC, 4, 16, 75

Y
Y Combinator, 6, 8–12, 43, 61, 107

Z
Zero-Money Round method, 55–60, 74, 77
    discounted Valuation, 56–57, 60, 66
    standard, 57–60, 71
ZMR, *see* Zero-Money Round method